Quantum Dots Handbook

Quantum Dots Handbook

Edited by **Eva Murphy**

NY RESEARCH PRESS
New York

Published by NY Research Press,
23 West, 55th Street, Suite 816,
New York, NY 10019, USA
www.nyresearchpress.com

Quantum Dots Handbook
Edited by Eva Murphy

International Standard Book Number: 978-1-63238-381-5 (Hardback)

Printed in the United States of America.

Contents

Preface

Every book is initially just a concept; it takes months of research and hard work to give it the final shape in which the readers receive it. In its early stages, this book also went through rigorous reviewing. The notable contributions made by experts from across the globe were first molded into patterned chapters and then arranged in a sensibly sequential manner to bring out the best results.

This book gives innovative and resourceful techniques for calculating the optical and transport characteristics of quantum dot structures. The book deals with the importance of transport and electronic characteristics of quantum dot structures. This is a collaborative initiative, providing primary research such as the ones conducted in physics, chemistry, and material science. This book serves as a source of reference for this field.

It has been my immense pleasure to be a part of this project and to contribute my years of learning in such a meaningful form. I would like to take this opportunity to thank all the people who have been associated with the completion of this book at any step.

Editor

Transport and Electronics Properties of Quantum Dot Systems

1

Quantum Injection Dots

Eliade Stefanescu
Center of Advanced Studies in Physics of the Romanian Academy
Romania

1. Introduction

In optoelectronics, quantum dots are essential elements for coupling a device to an electromagnetic field in the infrared domain of the frequency spectrum. Such a dot is a small semiconductor region, with a forbidden band specific to a given application, embedded in the active i-region of the p-i-n junction of a laser structure (1), or in the sensitive i-region of the p-i-n junction of a photovoltaic structure (2). These quantum dots have the advantage of

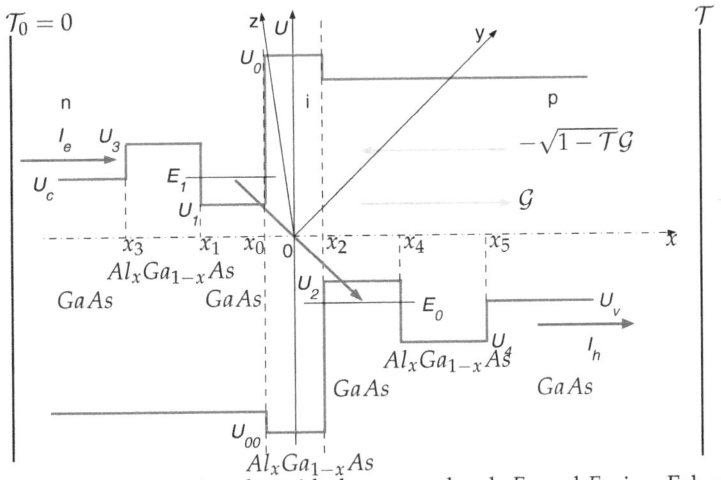

Fig. 1. Quantum injection dot with the energy levels E_1 and E_0, in a Fabry-Perot resonator with transmission coefficients of the mirrors $\mathcal{T}_0 = 0$ and $\mathcal{T} > 0$. By quantum transitions between these levels, a super radiant electromagnetic field with two counter-propagating waves of amplitudes $\mathcal{G}$ and $\sqrt{1-\mathcal{T}}\mathcal{G}$ is generated.

being feasible for a large class of applications (transition frequencies). However, they have the disadvantage that, being embedded in a bulk region, any injection or photovoltaic process includes transport processes through the active (sensitive) region, which are very dissipative. In such a system of quantum dots, an injected electron in the bulk active region of a laser structure has a large probability to recombine with an electron hole before reaching a quantum dot, where this electron becomes an active one, contributing to the laser field generation. Similarly, an optically excited electron in a quantum dot of a photo-voltaic bulk active region, has a large probability to meet another quantum dot, and to decay in the ground state of this quantum dot before reaching the quasi-neutral zone of the n-region, where this electron brings its contribution to the generated current.

In this chapter, we deal with a different kind of quantum dots (figure 1), as basic elements of a new class of power optoelectronic devices, devoted to the conversion of the environmental heat into usable energy (3–6). Such a quantum dot is substantially different from a bulk quantum dot (1; 2). These dots are deployed in double arrays of donor-acceptor pairs of impurity atoms placed in quantum wells at the two sides of the i-region of an n-i-p junction. The quantum dot density has an appropriate value to include the whole internal field of the semiconductor junction in the quantum dot region, i.e. between the two impurity arrays forming this region. Otherwise, spatial or mobile charge layers arise at the external sides of the two quantum wells, altering the distance $E_1 - E_0$ between the two ground energy levels of these wells. Thus, in a process of current injection, or of an optical current generation, the electrons traverse the internal field region by quantum transitions between the two levels of interest (see figure 1).

In section 2, we derive explicit expressions for electric potentials, energy levels in these potentials, wave-functions, the quantum dot density, and transition dipole moments, which determine the strength of the coupling of a quantum dot to a quasi-resonant electromagnetic field, and to the dissipative environment. We obtain operation conditions for the characteristics of the separation barriers.

In section 3, we consider the dissipative couplings of a quantum injection dot. We describe the dissipative dynamics of such a quantum dot by a master equation for a system of Fermions, coupled to a complex environment of other Fermions, Bosons, and the free electromagnetic field (7). This equation depends on analytic dissipative coefficients, describing correlated transitions of the system particles with the environmental particles, transitions stimulated by thermal fluctuations of the self-consistent field of the environmental Fermions, and the non-Markovian dynamics induced by these fluctuations. We derive explicit expressions of the dissipation coefficients as functions of universal constants and physical properties of the semiconductor structure: effective masses of electrons and electron holes, concentrations of donors and acceptors in the conduction regions, transition frequency, dipole moments, crystal density, elasticity coefficient, geometrical characteristics of the semiconductor structure, and temperature.

In section 4, we consider the dissipative dynamics of an electromagnetic field interacting with the quasi-free electrons of a semiconductor structure. By a method previously used in (7), we obtain a master equation with coefficients depending on frequency and the effective masses, transition dipole moments, and densities of states of these electrons. We derive field equations coupled to the polarizations of the system of active Fermions, with explicit expressions of the coupling and absorption coefficients, as functions of the physical properties of the active quantum dot system, and characteristics of the dissipative environment.

In section 5, we derive equations for the density matrix elements of a quantum injection dot interacting with a quasi-resonant electromagnetic field. We obtain equations for the amplitude mean-values of the forward and backward electromagnetic waves, propagating in a Fabry-Perot cavity which includes a system of such quantum dots. We derive optical equations for a system of quantum injection dots in a resonant Fabry-Perot cavity, with an additional term in the population equation, for describing a current injection in the semiconductor structure (3–5).

In section 6, we present the concept of quantum heat converter, as a device based on systems of quantum injection dots. This device is conceived in two versions: (1) longitudinal quantum heat converter, where the electromagnetic field propagates in the direction of the injected current, i.e. perpendicularly to the semiconductor chip (3; 4), and (2) transversal quantum heat converter, where the electromagnetic field propagates perpendicularly to the injected current, i.e. in the plane of the semiconductor chip (3; 5). Any of these versions could be preferred in specific applications. We derive analytical expressions of the super radiant power, as a function of the injected current, dissipative coefficients, coupling coefficient, number of super radiant transistors, transmission coefficient of the output mirror, and the geometrical characteristics of the device. We get operation conditions for these parameters. We describe the super radiant dissipative dynamics of a quantum injection dot, when a step current is injected in the device.

In this chapter, we present an analytical model depending only on material and geometrical characteristics, temperature, and universal constants.

2. Physical model

The essential problem of any optoelectronic device is the coupling of an electric current to an electromagnetic field. As a function of the roles played by these physical quantities into the input-output characteristic, one can conceive two kinds of devices. In the conventional optoelectronics, a device with the electric current as an input and the electromagnetic field as an output, if this field is not coherent, is called LED. If the output field is coherent, it is called laser. Conversely, if the electric current is the output, depending on an input field, the device is called photodiode, or photovoltaic cell if it is a power device, devoted to the electric energy production.

The new field of the heat conversion optoelectronics, includes two similar kinds of device. In this case, the radiant device is called quantum heat converter (3–5), while the photovoltaic device is called quantum injection system (6). Both kinds of such devices are based on an electron-field interaction with a potential V in the total Hamiltonian

$$H = H_0^S + H^F + V, \tag{1}$$

including a term for an electron with the electric charge $-e$ in an electromagnetic field with the vector potential $\vec{A}$,

$$H_0^S + V = \frac{(\vec{p} + e\vec{A})^2}{2M} + U(\vec{r}), \tag{2}$$

and the Hamiltonian of this field which, for the system represented in figure 1, is of the form

$$H^F = \hbar\omega(a_+^+ a_+ + a_-^+ a_- + 1), \tag{3}$$

where $a_+ - a_+^+$, $a_- - a_-^+$ are the creation-annihilation operators of the two counter-propagating waves. In equation (2), we distinguish the Hamiltonian of the electron in the potential $U(\vec{r})$ of an active quantum dot

$$H_0^S = \frac{\vec{p}^2}{2M} + U(\vec{r}) = \sum_i \varepsilon_i c_i^+ c_i, \tag{4}$$

and the interaction potential

$$V = \frac{e}{M}\vec{p}\vec{A}, \tag{5}$$

while the term in $\vec{A}^2$ is negligible in the non-relativistic approximation. This potential depends on the electron momentum

$$\vec{p} = \mathrm{i}M \sum_{ij} \omega_{ij}\vec{r}_{ij}c_i^+ c_j, \tag{6}$$

and the vector potential

$$\vec{A} = \frac{\hbar}{e}\vec{K}\left(a_+ e^{\mathrm{i}kx} + a_+^+ e^{-\mathrm{i}kx} + a_- e^{-\mathrm{i}kx} + a_-^+ e^{\mathrm{i}kx}\right), \tag{7}$$

of the electric field

$$\vec{E} = \mathrm{i}\frac{\hbar\omega}{e}\vec{K}\left(a_+ e^{\mathrm{i}kx} - a_+^+ e^{-\mathrm{i}kx} + a_- e^{-\mathrm{i}kx} - a_-^+ e^{\mathrm{i}kx}\right). \tag{8}$$

In these expressions, M is the electron mass, $c_i^+ - c_j$ are Fermion operators, ω_{ij} are transition frequencies of the active electron, $\vec{r}_{ij}$ are dipole moments, ω is the frequency of the field, and

$$\vec{K} = \vec{1}_y\sqrt{\alpha\frac{\lambda}{\mathcal{V}}} \tag{9}$$

is a vector in the y-direction of this field, depending on the wavelength λ, the fine structure constant $\alpha = \frac{e^2}{4\pi\varepsilon_0\hbar c} \approx \frac{1}{137}$, and the quantization volume $\mathcal{V}$.

From (5) and (6), we notice that the electron-field interaction of the system depends on the transition dipole moment

$$\vec{r}_{01} = \vec{r}_{10} = \int_{\mathcal{V}_s} \Psi_0\vec{r}\Psi_1 d^3\vec{r}, \tag{10}$$

where

$$\Psi_1(x,y,z) = \psi_1(x)\phi_1(y)\chi_1(z) \tag{11a}$$
$$\Psi_0(x,y,z) = \psi_0(x)\phi_0(y)\chi_0(z) \tag{11b}$$

are eigenfunctions of the Hamiltonian (4), for the ground states of the two quantum wells. For the potential $U(\vec{r})$, we distinguish seven regions, of four $GaAs$ quantum wells, and three $Al_xGa_{1-x}As$ potential barriers, determined by the impurity concentrations of these regions (see figure 1). For the two thick conduction regions with the potentials U_c and U_v, we use a three-dimensional model, with a quantization volume $\mathcal{V}_n = 1/N_D$ for the n-region of donor concentration N_D, and $\mathcal{V}_p = 1/N_A$ for the p-region of acceptor concentration N_A. In these quantization volumes, for an electron with the effective mass M_n, and an electron hole with the effective mass M_p, we consider the densities of states (8)

$$g^{(n)}(E_\alpha) = \mathcal{V}_n\frac{\sqrt{2}M_n^{3/2}}{\pi^2\hbar^3}\sqrt{E_\alpha} \tag{12a}$$

$$g^{(p)}(E_\alpha) = \mathcal{V}_p\frac{\sqrt{2}M_p^{3/2}}{\pi^2\hbar^3}\sqrt{E_\alpha}, \tag{12b}$$

and the occupation probabilities of these states with the kinetic energies E_α:

$$f^{(n)}(E_\alpha) = \frac{1}{e^{(U_c+E_\alpha)/T}+1} \approx e^{-(U_c+E_\alpha)/T} \quad (13a)$$

$$f^{(p)}(E_\alpha) = \frac{1}{e^{(-U_v+E_\alpha)/T}+1} \approx e^{-(-U_v+E_\alpha)/T}, \quad (13b)$$

where approximate expressions are taken into account for the usual case of a non-degenerate semiconductor. Considering the integral of the number of particles occupying the states of a quantization volume, one gets the two potentials

$$U_c(T) = T\ln\frac{N_c(T)}{N_D}, \qquad N_c(T) = 2\left(\frac{\sqrt{M_nT/2\pi}}{\hbar}\right)^3 \quad (14a)$$

$$U_v(T) = -T\ln\frac{N_v(T)}{N_A}, \qquad N_v(T) = 2\left(\frac{\sqrt{M_pT/2\pi}}{\hbar}\right)^3. \quad (14b)$$

For the very thin layers of the quantum wells and potential barriers, we use a two-dimensional model. For a quantization area A_e, in n and p regions, one gets the densities of states

$$g^{(1)} = A_e\frac{M_n}{\pi\hbar^2} \quad (15a)$$

$$g^{(2)} = A_e\frac{M_p}{\pi\hbar^2}. \quad (15b)$$

By using the Fermi-Dirac distribution in the particle number integral, we obtain expressions similar to (14) for the potentials of the two $GaAs$ quantum wells, as a function of the surface quantum dot density N_e

$$U_1(T) = -T\ln\left(e^{\frac{\pi\hbar^2 N_e}{M_nT}} - 1\right) \quad (16a)$$

$$U_2(T) = T\ln\left(e^{\frac{\pi\hbar^2 N_e}{M_pT}} - 1\right). \quad (16b)$$

Similar expressions are obtained for the separation barriers, as functions of the donor and acceptor arrays with concentrations N_3, N_4 embedded in the very thin $Al_xGa_{1-x}As$-layers of these barriers,

$$U_3(T) = -T\ln\left(e^{\frac{\pi\hbar^2 N_3}{M_nT}} - 1\right) \quad (17a)$$

$$U_4(T) = T\ln\left(e^{\frac{\pi\hbar^2 N_4}{M_pT}} - 1\right), \quad (17b)$$

and for the potential barrier U_0 between the two quantum wells,

$$U_0(T) = -T\ln\left(e^{\frac{\pi\hbar^2 N_0}{M_nT}} - 1\right), \quad (18)$$

with a slight donor concentration N_0, controlling this potential. For these wells, we consider harmonic wave-functions with exponential tails in the neighboring barriers

$$\psi_1(x) = A_1 \cos\left[k_1(x_0 - x) - \arctan\frac{\alpha_1}{k_1}\right], \quad x_1 \le x \le x_0 \tag{19a}$$

$$\psi_1(x) = A_1 \sqrt{\frac{E_1 - U_1}{U_0 - U_1}} e^{-\alpha_1(x - x_0)}, \quad x_0 \le x \le x_2 \tag{19b}$$

$$\psi_1(x) = A_1 \sqrt{\frac{E_1 - U_1}{U_3 - U_1}} e^{-\alpha_3(x_1 - x)}, \quad x_3 \le x \le x_1 \tag{19c}$$

and

$$\psi_0(x) = A_0 \cos\left[k_0(x - x_2) - \arctan\frac{\alpha_0}{k_0}\right], \quad x_2 \le x \le x_4 \tag{20a}$$

$$\psi_0(x) = A_0 \sqrt{\frac{U_2 - E_0}{U_2 - U_{00}}} e^{-\alpha_0(x_2 - x)}, \quad x_0 \le x \le x_2 \tag{20b}$$

$$\psi_0(x) = A_0 \sqrt{\frac{U_2 - E_0}{U_2 - U_4}} e^{-\alpha_4(x - x_4)}, \quad x_4 \le x \le x_5, \tag{20c}$$

while the tails beyond these barriers are neglected. These wave-functions depend on the wave-numbers

$$k_1 = \frac{1}{\hbar}\sqrt{2M_n(E_1 - U_1)} \tag{21a}$$

$$k_0 = \frac{1}{\hbar}\sqrt{2M_p(U_2 - E_0)}, \tag{21b}$$

attenuation coefficients

$$\alpha_1 = \frac{1}{\hbar}\sqrt{2M_n(U_0 - E_1)} \tag{22a}$$

$$\alpha_0 = \frac{1}{\hbar}\sqrt{2M_p(E_0 - U_{00})} \tag{22b}$$

$$\alpha_3 = \frac{1}{\hbar}\sqrt{2M_n(U_3 - E_1)} \tag{22c}$$

$$\alpha_4 = \frac{1}{\hbar}\sqrt{2M_p(E_0 - U_4)}, \tag{22d}$$

and normalization coefficients

$$A_1 = \sqrt{2}\left[x_0 - x_1 + \frac{\hbar}{\sqrt{2M_n}}\left(\frac{1}{\sqrt{U_0 - E_1}} + \frac{1}{\sqrt{U_3 - E_1}}\right)\right]^{-1/2} \tag{23a}$$

$$A_0 = \sqrt{2}\left[x_4 - x_2 + \frac{\hbar}{\sqrt{2M_p}}\left(\frac{1}{\sqrt{E_0 - U_{00}}} + \frac{1}{\sqrt{E_0 - U_4}}\right)\right]^{-1/2}, \tag{23b}$$

while the energy eigenvalues are given by the equations:

$$E_1 - U_1 = \frac{\hbar^2}{2M_n(x_0 - x_1)^2}\left(\arctan\sqrt{\frac{U_0 - E_1}{E_1 - U_1}} + \arctan\sqrt{\frac{U_3 - E_1}{E_1 - U_1}}\right)^2 \tag{24a}$$

$$U_2 - E_0 = \frac{\hbar^2}{2M_p(x_4 - x_2)^2}\left(\arctan\sqrt{\frac{E_0 - U_{00}}{U_2 - E_0}} + \arctan\sqrt{\frac{E_0 - U_4}{U_2 - E_0}}\right)^2 . \tag{24b}$$

We take the two energy eigenvalues as $E_1 = U_c$, $E_0 = U_v$. In this case, the whole internal potential of the n-i-p junction is included on the distance d between the two charge layers of the quantum dot region, which means a quantum dot surface density

$$N_e = \varepsilon\frac{E_1 - E_0}{e^2 d}, \tag{25}$$

while this distance can be approximated as

$$d = \frac{1}{2}(x_2 - x_0 + x_4 - x_1). \tag{26}$$

The energy levels $E_1 = U_c$ and $E_0 = U_v$ can be obtained from (14), as functions of the donor and acceptor concentrations N_D and N_A of the conduction regions. By choosing appropriate values for the separation and quantum dot barriers, from (17) and (18) one gets the surface concentrations N_3, N_4 and N_0 of these barriers. With the widths $x_1 - x_3$ and $x_5 - x_4$, the separation barrier must have a higher penetrability P than the necessary value to provide the injected current I through the device area A_D, which means that the density of this current must be smaller than the thermal current $\frac{1}{6}eN_D v_T P$ emergent from a unit volume with the thermal velocity $v_T = \sqrt{T/M_n}$, and crossing the barrier. We get the conditions

$$\alpha_3(x_1 - x_3) < \frac{1}{2}\ln\left(\frac{eN_D A_D}{6I}\sqrt{\frac{T}{M_n}}\right) \tag{27a}$$

$$\alpha_4(x_5 - x_4) < \frac{1}{2}\ln\left(\frac{eN_A A_D}{6I}\sqrt{\frac{T}{M_p}}\right). \tag{27b}$$

Thus, a quantum injection dot is a two-level system, with the energy levels E_0 and E_1, in a quantization volume shaped as a parallelepiped with the basis area $\mathcal{A} = \frac{1}{N_e}$ and the height $x_5 - x_3$. In this volume, we consider the two wave-functions (19)-(20) for the coordinate x, while, for the coordinates y and z in the plane of the quantum dot array, we consider the wave-functions $\phi_1(y)$, $\phi_0(y)$ and $\chi_1(y)$, $\chi_0(y)$, describing a thermal motion with an energy mean-value T. For a longitudinal device, when the electromagnetic field propagates in the direction x of the injected current, the electric component E_y of this field is coupled with the component y_{01} of the transition dipole moment of the system between thermal states. For a transversal device, when the electromagnetic field propagates in a direction y perpendicular to the direction x of the injected current, i.e. in the plane of the quantum dot array, the electric component E_x of this field is coupled with the transition dipole moment x_{01} between the two states (19)-(20) of the system. These dipole moments also determine the dissipative couplings of the quantum dot. They essentially depend on the width $x_2 - x_0$ of the quantum dot barrier of height U_0, which determines the overlap of the two wave-functions $\psi_1(x)$ and $\psi_0(x)$. When this width is chosen for reasonable values of the electron-field and dissipative couplings, and N_3, N_4 for reasonable values of the separation barriers satisfying the conditions (27), from equations (16) and (24)-(26) the geometrical characteristics x_0x_1, x_2, x_3, x_4 are obtained.

As an example, for a concentration $N_D = N_A = 3.16 \times 10^{16}\ cm^{-3}$ of a super radiant junction, working at temperature $T = 10\ ^0C$, we get a transition frequency $E_1 - E_0 = 0.1866\ eV$. A quantum dot $Al_{0.37}Ga_{0.63}As$-barrier of $U_0 = 0.5\ eV$ is obtained for a surface concentration of $N_0 = 6.4243 \times 10^6\ m^{-2}$ donors, and separation barriers of $U_3 - U_c = U_v - U_4 = 0.05\ eV$ are obtained for the surface concentrations of these barriers $N_3 = 8.01 \times 10^{13}\ m^{-2}$ and $N_4 = 2.552 \times 10^{13}\ m^{-2}$. For a width $x_2 - x_0 = 5.5\ nm$ of the quantum dot barrier, we get a quantum dot surface concentration $N_e = 1.476 \times 10^{16}\ m^{-2}$ and the widths of the two quantum dot wells $x_0 - x_1 = 4.189\ nm$ and $x_4 - x_2 = 1.576\ nm$, while separation barriers with widths $x_1 - x_3 = 10\ nm$ and $x_5 - x_4 = 3\ nm$ satisfy the conditions (27) for an injected current $I = 45\ A$ in a device with an area $A_D = 4\ cm^2$.

3. Dissipative dynamics of quantum injection dots

We consider a quantum injection dot with two energy levels E_1, E_0, coupled to a super radiant field and a complex dissipative environment of a semiconductor structure as it is represented in figure 1: (1) the quasi-free electrons of the n-conduction region $x < x_3$, (2) the quasi-free holes of the p-conduction region $x > x_5$, (3) the phonons of the crystal at temperature T, and (4) the photons of the free electromagnetic field existing at temperature T (see figure 2). In (3)

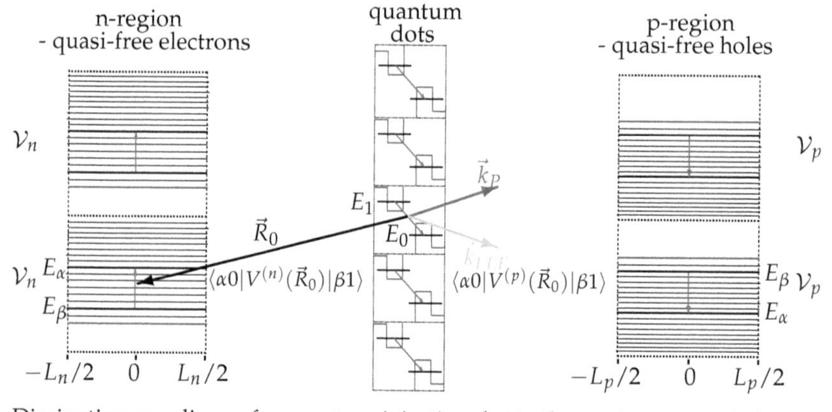

Fig. 2. Dissipative couplings of a quantum injection dot to the environment. A decay $|1\rangle \rightarrow |0\rangle$ of the active electron is correlated with: (1) a transition $|\beta\rangle \rightarrow |\alpha\rangle$ of a quasi-free electron in a quantization volume $\mathcal{V}_n$, (2) a transition $|\beta\rangle \rightarrow |\alpha\rangle$ of a quasi-free hole in a quantization volume $\mathcal{V}_p$, (3) a phonon creation with a wave vector $\vec{k}_P$, and (4) a photon creation with a wave vector $\vec{k}_{FEF}$.

we showed that, for a semiconductor structure with the characteristics mentioned at the end of the preceding section, the decay rate corresponding to the phonon environment is dominant, the decay rate due to the conduction electrons and holes is smaller, while the decay rate given by the free electromagnetic field is negligible.

We consider a system of interest, including a system of Fermions S with the Hamiltonian H_0^S and an electromagnetic field F with the Hamiltonian H^F, in a dissipative environment with the Hamiltonian H^E. Taking into account a potential of interaction V between the Fermion system S and the field F, and a system-environment potential V^E, the dynamics of the total system is described by an equation of motion of the form:

$$\frac{d}{dt}\tilde{\chi}(t) = -\frac{i}{\hbar}\left[\varepsilon\tilde{V}(t) + \varepsilon\tilde{V}^E(t), \tilde{\chi}(t)\right]. \tag{28}$$

In this equation, tilde denotes operators in the interaction picture, e.g.

$$\tilde{\chi}(t) = e^{\frac{i}{\hbar}(H^E + H_0^S + H^F)t}\chi(t)e^{-\frac{i}{\hbar}(H^F + H_0^S + H^E)t}. \tag{29}$$

According to a general procedure disclosed in (9), we take a total density of the form

$$\tilde{\chi}(t) = R \otimes \tilde{\rho}(t) + \varepsilon\tilde{\chi}^{(1)}(t) + \varepsilon^2\tilde{\chi}^{(2)}(t) + \ldots \ , \tag{30}$$

where $\rho(t)$ is the reduced density matrix of the system of interest, while R is the density matrix of the dissipative environment at the initial moment of time, $t = 0$, the time-evolution of the environment being described by the higher-order terms $\tilde{\chi}^{(1)}(t), \tilde{\chi}^{(2)}(t), \ldots$. The parameter ε is introduced to handle the orders of the terms in this series expansion, and is set to 1 in the final results. The reduced density of the system is

$$\tilde{\rho}(t) = Tr_E\{\tilde{\chi}(t)\}, \tag{31}$$

while the higher-order terms of the total density have the property:

$$Tr_E\{\tilde{\chi}^{(1)}\} = Tr_E\{\tilde{\chi}^{(2)}\} = \ldots \ = 0. \tag{32}$$

If initially the environment is in the equilibrium state R, the density matrix of the total system is of the form $\chi(0) = R\rho(0)$. We suppose that at time $t = 0$, due to the interaction V of the system of Fermions with the electromagnetic field, or due to a non-equilibrium initial state $\rho(0) \neq \rho_T$, a time-evolution begins, while the reduced density satisfies a quantum dynamical equation of the form

$$\frac{d}{dt}\tilde{\rho}(t) = \varepsilon\tilde{B}^{(1)}(\tilde{\rho}(t), t) + \varepsilon^2\tilde{B}^{(2)}(\tilde{\rho}(t), t) + \ldots \ . \tag{33}$$

From the dynamic equation (28), with expressions (30)-(33), we obtain the quantum master equation

$$\begin{aligned}\frac{d}{dt}\rho(t) = &-\frac{\mathrm{i}}{\hbar}[H, \rho(t)] - \mathrm{i}\sum_{ij}\zeta_{ij}[c_i^+ c_j, \rho(t)] \\ &+ \sum_{ij}\lambda_{ij}([c_i^+ c_j\rho(t), c_j^+ c_i] + [c_i^+ c_j, \rho(t)c_j^+ c_i]) \\ &+ \sum_{ijkl}\zeta_{ij}\zeta_{kl}\int_{t-\tau}^{t}[c_i^+ c_j, e^{-\mathrm{i}[\phi(t') + \frac{1}{\hbar}H_0^S(t-t')]}[c_k^+ c_l, \rho(t')]e^{\mathrm{i}[\phi(t') + \frac{1}{\hbar}H_0^S(t-t')]}]dt',\end{aligned} \tag{34}$$

where the coefficients

$$\zeta_{ij} = \frac{1}{\hbar}\sqrt{\frac{1}{Y^F}\int_{(\alpha)}\langle\alpha i|(V^F)^2|\alpha j\rangle f_\alpha^F(\varepsilon_\alpha)g_\alpha^F(\varepsilon_\alpha)d\varepsilon_\alpha} \tag{35}$$

describe transitions stimulated by the thermal fluctuations of the self-consistent field of the Y^F environmental Fermions in a certain quantization volume - hopping potential (10), $\phi(t')$ is a phase fluctuation operator, while the coefficients

$$\lambda_{ij} = \lambda_{ij}^F + \lambda_{ij}^B + \gamma_{ij} \tag{36}$$

describe correlated transitions of the system Fermions with environment particles, including explicit terms for Fermions, Bosons, and photons of the free electromagnetic field. These terms depend on the dissipative two-body potentials V^F, V^B, the densities of the environment states $g^F(\varepsilon_\alpha), g^B(\varepsilon_\alpha)$, the occupation probabilities of these states $f^F(\varepsilon_\alpha), f^B(\varepsilon_\alpha)$, and temperature T. For a rather low temperature, $T \ll \varepsilon_{ji},\ j > i$, these terms become

$$\lambda_{ij}^F = \frac{\pi}{\hbar}|\langle\alpha i|V^F|\beta j\rangle|^2[1 - f^F(\varepsilon_{ji})]g^F(\varepsilon_{ji}) \tag{37a}$$

$$\lambda_{ji}^F = \frac{\pi}{\hbar}|\langle\alpha i|V^F|\beta j\rangle|^2 f^F(\varepsilon_{ji})g^F(\varepsilon_{ji}), \tag{37b}$$

for the Fermion environment,

$$\lambda_{ij}^B = \frac{\pi}{\hbar}|\langle\alpha i|V^B|\beta j\rangle|^2[1 + f^B(\varepsilon_{ji})]g_\alpha^B(\varepsilon_{ji}) \tag{38a}$$

$$\lambda_{ji}^B = \frac{\pi}{\hbar}|\langle\alpha i|V^B|\beta j\rangle|^2 f^B(\varepsilon_{ji})g_\alpha^B(\varepsilon_{ji}) \tag{38b}$$

for the Boson environment, and

$$\gamma_{ij} = \frac{2\alpha}{c^2\hbar^3}\left|\vec{r}_{ij}\right|^2 \varepsilon_{ji}^3(1 + \frac{1}{e^{\varepsilon_{ji}/T} - 1}) \tag{39}$$

for the Boson environment of the free electromagnetic field, where $\vec{r}_{ij}$ are the transition dipole moments. The dissipative terms of the master equation (34) with coefficients (36)-(39) describe correlated transitions of the system and the environment particles, with energy conservation, $\varepsilon_{ji} = \varepsilon_{\alpha\beta}$, in agreement with the quantum-mechanical principles and the detailed balance principle (11). The non-Markovian part of this equation takes into account the fluctuations of the self-consistent field of the environment Fermions, with the coefficients (35).

A significant component of the dissipative dynamics comes from the Coulomb interaction of the active electrons, mainly located in the interval (x_3, x_5), with the conduction electrons and holes in the conduction regions $(-\infty, x_3)$ and $(x_5, +\infty)$, respectively (figure 1). We use the notations $\vec{r}$ for the position vector of an active electron, and $\vec{R}_0 + \vec{R}$ for the position vector of a dissipative electron (hole), where $\vec{R}_0$ is the position vector of an arbitrary n (p) cluster, and $\vec{R} = \vec{1}_x X + \vec{1}_y Y + \vec{1}_z Z$ is the position of an electron (hole) in this cluster (figure 3). In this case, the Coulomb potential in a first-order approximation of the two-body term $\vec{R}\vec{r} = Xx + Yy + Zz$ is

$$V^C(\vec{R},\vec{r}) = \frac{\alpha\hbar c}{|\vec{R}_0 + \vec{R} - \vec{r}|} \approx \frac{\alpha\hbar c}{|\vec{R}_0|}\left(1 + \frac{Xx + Yy + Zz}{R_0^2}\right). \tag{40}$$

From this expression, only the second term, bilinear in the coordinates of an active electron and of an electron (hole) of the environment, yields contributions in the two-body transition matrix elements of the decay (excitation) rates (37):

$$V^F(\vec{R}_0,\vec{R},\vec{r}) \doteq V^{(n)}(\vec{R}_0,\vec{R},\vec{r}) = -V^{(p)}(\vec{R}_0,\vec{R},\vec{r}) = \frac{\alpha\hbar c}{|\vec{R}_0|^3}(Xx + Yy + Zz). \tag{41}$$

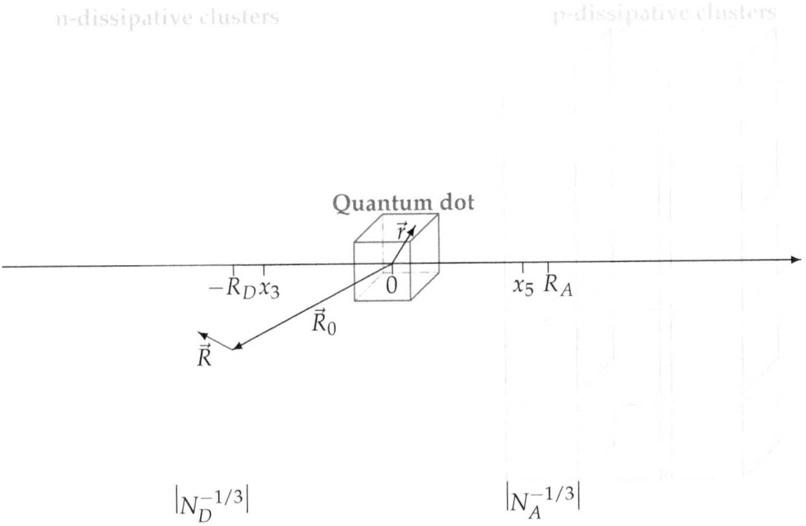

Fig. 3. The electron of a quantum injection dot is coupled to the quasi-free electrons of the n-dissipative clusters (n-region) and quasi-free holes of the p-dissipative clusters (p-region) by an electric dipole-dipole interaction: $V^F(\vec{R}_0, \vec{R}, \vec{r}) = \frac{\alpha\hbar c}{|\vec{R}_0|^3}\vec{R}\vec{r}$.

From the wave-functions derived in the preceding section, we obtain the dipole moment of a quantum dot:

$$x_{01}^{(\Psi)} = c_{01}^{(x)}\left(\frac{x_2 - x_0}{2} - \frac{1}{\alpha_0 - \alpha_1}\right) \tag{42a}$$

$$y_{01}^{(\Psi)} = z_{01}^{(\Psi)} = c_{01}^{(x)}\frac{\hbar}{2\sqrt{M_n T}} \tag{42b}$$

$$y_{10}^{(\Psi)} = z_{10}^{(\Psi)} = c_{01}^{(x)}\frac{\hbar}{2\sqrt{M_p T}}, \tag{42c}$$

as a product of the overlap function

$$c_{01}^{(x)} = \frac{A_1 A_0}{\alpha_0 - \alpha_1}\sqrt{\frac{(E_1 - U_1)(U_2 - E_0)}{(U_0 - U_1)(U_2 - U_{00})}}\left(e^{-\alpha_1(x_2 - x_0)} - e^{-\alpha_0(x_2 - x_0)}\right). \tag{43}$$

and a quantity that we call the state separation distance. At the same time, with the initial and the final energies $E_\beta = T/2$, $E_\alpha = E_\beta + \varepsilon_{10}$, we obtain the dipole moments for the n-zone

$$X_{\alpha\beta}^{(n)} = Y_{\alpha\beta}^{(n)} = Z_{\alpha\beta}^{(n)} = \frac{\hbar}{\varepsilon_{10}}\sqrt{\frac{2\varepsilon_{10} + T}{M_n}} \approx \sqrt{\frac{2\hbar}{M_n\omega_0}}, \tag{44}$$

and for the p-zone,

$$X_{\alpha\beta}^{(p)} = Y_{\alpha\beta}^{(p)} = Z_{\alpha\beta}^{(p)} = \frac{\hbar}{\varepsilon_{10}}\sqrt{\frac{2\varepsilon_{10} + T}{M_p}} \approx \sqrt{\frac{2\hbar}{M_p\omega_0}}, \tag{45}$$

where we used the notations $\varepsilon_{10} \doteq \hbar\omega_0 \doteq E_1 - E_0$. With the quantum dot dipole moments (42), and the environment dipole moments (44)-(45), we calculate the matrix elements of the two-body potential (41). With these matrix elements, from (37) with the densities of states (12) and the occupation probabilities of these states (13), one obtains the dissipative coefficients for the coupling of a quantum dot with a dissipative cluster. By integration over all clusters of both hemispheres of the n and p conduction regions, with the quantization volumes $\mathcal{V}_n = \frac{1}{N_D}$ and $\mathcal{V}_p = \frac{1}{N_A}$ as differential volumes in these integrals, we obtain the dissipation coefficients (3):

$$\lambda_{01}^{(n)} = \frac{4\alpha^2 c^2 \sqrt{2M_n}(\varepsilon_{10} + \frac{T}{2})|c_{01}^{(x)}|^2 \mu_{01}^2}{3\left(\frac{N_D^{-1/3}}{2} - x_3\right)^3 \varepsilon_{10}^{3/2}(e^{-(U_c+\varepsilon_{10})/T} + 1)} \tag{46a}$$

$$\lambda_{10}^{(n)} = \frac{4\alpha^2 c^2 \sqrt{2M_n}(\varepsilon_{10} + \frac{T}{2})|c_{01}^{(x)}|^2 \mu_{01}^2}{3\left(\frac{N_D^{-1/3}}{2} - x_3\right)^3 \varepsilon_{10}^{3/2}(e^{(U_c+\varepsilon_{10})/T} + 1)} \tag{46b}$$

$$\lambda_{01}^{(p)} = \frac{4\alpha^2 c^2 \sqrt{2M_p}(\varepsilon_{10} + \frac{T}{2})|c_{01}^{(x)}|^2 \mu_{01}^2}{3\left(\frac{N_A^{-1/3}}{2} + x_5\right)^3 \varepsilon_{10}^{3/2}(e^{-(-U_v+\varepsilon_{10})/T} + 1)} \tag{46c}$$

$$\lambda_{10}^{(p)} = \frac{4\alpha^2 c^2 \sqrt{2M_p}(\varepsilon_{10} + \frac{T}{2})|c_{01}^{(x)}|^2 \mu_{01}^2}{3\left(\frac{N_A^{-1/3}}{2} + x_5\right)^3 \varepsilon_{10}^{3/2}(e^{(-U_v+\varepsilon_{10})/T} + 1)}, \tag{46d}$$

where

$$\mu_{01}^2 = \left(\frac{x_2 - x_0}{2} - \frac{1}{\alpha_0 - \alpha_1} + \frac{\hbar}{\sqrt{M_n T}}\right)\left(\frac{x_2 - x_0}{2} - \frac{1}{\alpha_0 - \alpha_1} + \frac{\hbar}{\sqrt{M_p T}}\right). \tag{47}$$

is the square of the separation distance of the two states $\Psi_0(\vec{r})$ and $\Psi_1(\vec{r})$. From (35), by similar calculations we obtain the fluctuation coefficients of a quantum dot in the self-consistent field of dissipative clusters:

$$\left[\zeta_{11}^{(n)}\right]^2 = \frac{\alpha^2 c^2 M_n^{3/2} T^{3/2}}{360\pi\sqrt{2\pi}\hbar^3} \cdot \frac{N_D^{1/3}[A_1^2(x_0^3 - x_1^3) + \frac{1}{Ne}]}{N_e N_c\left(\frac{N_D^{-1/3}}{2} - x_3 + \frac{x_0 + x_1}{2}\right)^5} \tag{48a}$$

$$\left[\zeta_{11}^{(p)}\right]^2 = \frac{\alpha^2 c^2 M_p^{3/2} T^{3/2}}{360\pi\sqrt{2\pi}\hbar^3} \cdot \frac{N_A^{1/3}[A_1^2(x_0^3 - x_1^3) + \frac{1}{Ne}]}{N_e N_v\left(\frac{N_A^{-1/3}}{2} + x_5 - \frac{x_0 + x_1}{2}\right)^5} \tag{48b}$$

$$\left[\zeta_{00}^{(n)}\right]^2 = \frac{\alpha^2 c^2 M_n^{3/2} T^{3/2}}{360\pi\sqrt{2\pi}\hbar^3} \cdot \frac{N_D^{1/3}[A_0^2(x_4^3 - x_2^3) + \frac{1}{Ne}]}{N_e N_c\left(\frac{N_D^{-1/3}}{2} - x_3 + \frac{x_4 + x_2}{2}\right)^5} \tag{48c}$$

$$\left[\zeta_{00}^{(p)}\right]^2 = \frac{\alpha^2 c^2 M_p^{3/2} T^{3/2}}{360\pi\sqrt{2\pi}\hbar^3} \cdot \frac{N_A^{1/3}[A_0^2(x_4^3 - x_2^3) + \frac{1}{Ne}]}{N_e N_v\left(\frac{N_A^{-1/3}}{2} + x_5 - \frac{x_4 + x_2}{2}\right)^5}. \tag{48d}$$

The Markovian dissipative coefficients (46) describe a very strong, exponential decrease of the decay rate with the width $x_2 - x_0$ of the quantum dot barrier, given by the square of the overlap function (43), and a strong decrease with the distances x_3 and x_5 of the quantum dot separation from the two conduction regions, which enter at the denominators with power 3. The non-Markovian coefficients (48) describe a very strong decrease of the fluctuation rate with the separation distances x_3 and x_5, which enter at the denominators with power 5.

The coupling of a quantum dot electron with a vibrational mode α is described by the potential matrix element (3):

$$V_{01\alpha}^{EP} = V_{10\alpha}^{EP} = -\hbar\omega_0^{3/2} \frac{M\vec{r}_{01}\vec{1}_\alpha}{\sqrt{\mathcal{M}\hbar/2}}, \tag{49}$$

where $\mathcal{M}$ is the mass of the phonon quantization volume $\mathcal{V}_\mathcal{P}$. For the density of phonon states of energy $\hbar\omega_0$, we obtain an expression similar to (12):

$$g_P(\hbar\omega_0) = \mathcal{V}_\mathcal{P} \frac{\sqrt{2}\mathcal{M}^{3/2}}{\pi^2\hbar^3} \sqrt{\hbar\omega_0}. \tag{50}$$

We consider the sound velocity v from the phonon wavelength expressions

$$\lambda_P = \frac{v}{\nu} = \frac{2\pi v}{\omega} = \frac{2\pi\hbar v}{\varepsilon_{10}}, \qquad \lambda_P \equiv \frac{2\pi}{k_P} = \frac{2\pi\hbar}{\sqrt{2\mathcal{M}\varepsilon_{10}}}, \tag{51}$$

and the crystal density

$$D \equiv \frac{\mathcal{M}}{\mathcal{V}_\mathcal{P}} = \frac{2\pi^2\hbar^2}{\mathcal{V}_\mathcal{P}\lambda_P^2\varepsilon_{10}}. \tag{52}$$

With (49)-(52), from (38) we obtain the decay (excitation) rates

$$\lambda_{01}^P = \frac{E_e^2\varepsilon_{10}^5}{\pi\hbar^6c^4v^3D} \cdot \frac{|c_{01}^{(x)}|^2\mu_{01}^2}{1 - e^{-\varepsilon_{10}/T}} \tag{53a}$$

$$\lambda_{10}^P = \frac{E_e^2\varepsilon_{10}^5}{\pi\hbar^6c^4v^3D} \cdot \frac{|c_{01}^{(x)}|^2\mu_{01}^2}{e^{\varepsilon_{10}/T} - 1}, \tag{53b}$$

where $E_e = Mc^2$ is the rest energy of the electron, and v is the sound velocity, which can be calculated from the Young elasticity coefficient E and the crystal density D:

$$v \approx \sqrt{\frac{E}{D}}. \tag{54}$$

We notice that both systems of dissipation coefficients (46) and (53) are proportional to the squares of the state separation distance and overlap function. Expressions (53) describe a very strong dependence of the decay rates on the transition energy ε_{10}, being proportional to this energy with power 5. However, they are valid for phonon wavelengths much larger than the distance between the atoms of the crystal lattice. Otherwise, the number of the density modes can no more be treated as a quasi-continuous function of frequency, and the probability of any non-resonant interaction vanishes (Mösbauer effect). We also found that for the rather low transition energies specific to the quantum injection dots, the decay rate due to the phonon coupling is rather low, e.g. for the structure presented at the end of the preceding section with $\varepsilon_{10} = 0.1866\ eV$, we got $\lambda_{01}^P = 2 \times 10^7\ s^{-1}$. As we found by direct calculations (3), for the quantum injection dots, which are separated by potential barriers from the conduction electrons, the decay rate due to these electrons is much lower than the decay rate due to the coupling to the phonons of the crystal lattice vibrations.

4. Dissipative dynamics of electromagnetic field

The operation of a semiconductor device with quantum dots interacting with a quasi-resonant electromagnetic field is based on the transparency of the host semiconductor structure, with a band to band transition frequency much higher than the quantum dot transition frequency. That means that this electromagnetic field is absorbed by the host semiconductor structure mainly by intra-band transitions, essentially depending on overlap functions of thermal states with excited states populated (depopulated) by these transitions.

The dynamics of this electromagnetic field, with a potential $\tilde{V}(t)$ of interaction with a quantum dot and a potential $\tilde{V}^E(t)$ of interaction with a conduction electron (hole), is described by a system of equations of the form (28), (30), (31) (33), which, in the second-order approximation, provides

$$\frac{d}{dt}\tilde{\rho}(t) = \tilde{B}^{(1)}[\tilde{\rho}(t),t] + \tilde{B}^{(2)}[\tilde{\rho}(t),t] \tag{55a}$$

$$\tilde{B}^{(1)}[\tilde{\rho}(t),t] = -\frac{\mathrm{i}}{\hbar}Tr_E[\tilde{V}(t)+\tilde{V}^E(t), R\otimes\tilde{\rho}(t)] \tag{55b}$$

$$\tilde{\chi}^{(1)}(t) = \int_0^t \left\{-\frac{\mathrm{i}}{\hbar}[\tilde{V}(t')+\tilde{V}^E(t'), R\otimes\tilde{\rho}(t')] - R\otimes\tilde{B}^{(1)}[\tilde{\rho}(t'),t']\right\} dt' \tag{55c}$$

$$\tilde{B}^{(2)}[\tilde{\rho}(t),t] = -\frac{\mathrm{i}}{\hbar}Tr_E[\tilde{V}(t)+\tilde{V}^E(t), \tilde{\chi}^{(1)}(t)]. \tag{55d}$$

The interaction potential $\tilde{V}(t)$ is obtained from (5)-(6), while the dissipative potential $\tilde{V}^E(t)$ is given by the similar expressions

$$V^E = \frac{e}{M}\vec{P}\vec{A} \tag{56}$$

and

$$\vec{P} = \mathrm{i}M\sum_{\alpha\beta}\omega_{\alpha\beta}\vec{R}_{\alpha\beta}c_\alpha^+ c_\beta. \tag{57}$$

From (5)-(7) and (56)-(57), we derive expressions of these potentials depending only on the positive transition frequencies $\omega_{ji}(j>i)$ and $\omega_{\alpha\beta}(\alpha>\beta)$, and take into account the so called "rotating-wave approximation", which includes only conservative processes, when an electron excitation is correlated only with a photon annihilation, while an electron decay is correlated only with a photon creation:

$$V = \mathrm{i}\sum_{j>i}\hbar\omega_{ji}\vec{K}\vec{r}_{ij}\left[c_j^+c_i\left(a_+e^{\mathrm{i}kx}+a_-e^{-\mathrm{i}kx}\right) - c_i^+c_j\left(a_+^+e^{-\mathrm{i}kx}+a_-^+e^{\mathrm{i}kx}\right)\right] \tag{58}$$

$$V^E = \mathrm{i}\sum_{\alpha>\beta}\hbar\omega_{\alpha\beta}\vec{K}\vec{R}_{\alpha\beta}\left[c_\alpha^+c_\beta\left(a_+e^{\mathrm{i}kx}+a_-e^{-\mathrm{i}kx}\right) - c_\beta^+c_\alpha\left(a_+^+e^{-\mathrm{i}kx}+a_-^+e^{\mathrm{i}kx}\right)\right]. \tag{59}$$

We consider the time-dependent expressions of these operators in the interaction picture,

$$\tilde{a}(t) = ae^{-\mathrm{i}\omega t}, \qquad \tilde{a}^+(t) = a^+e^{\mathrm{i}\omega t} \tag{60}$$

$$\tilde{c}_i^+(t)\tilde{c}_j(t) = c_i^+c_je^{-\mathrm{i}\omega_{ji}t}, \qquad \tilde{c}_j^+(t)\tilde{c}_i(t) = c_j^+c_ie^{\mathrm{i}\omega_{ji}t} \tag{61}$$

$$\tilde{c}_\beta^+(t)\tilde{c}_\alpha(t) = c_\beta^+c_\alpha e^{-\mathrm{i}\omega_{\alpha\beta}t}, \qquad \tilde{c}_\alpha^+(t)\tilde{c}_\beta(t) = c_\alpha^+c_\beta e^{\mathrm{i}\omega_{\alpha\beta}t}, \tag{62}$$

and take equations (55) for the mean-values of the electron operators. We retain only the slowly time varying terms, obtained from the resonance condition $\omega_{\alpha\beta}=\omega$, while the rapidly varying terms are neglected. We consider the summations over the environmental states as integrals over a quasi-continuum of states, with the densities $g(\varepsilon_\alpha)$ and $g(\varepsilon_\beta)$ and occupation probabilities $f(\varepsilon_\alpha)$ and $f(\varepsilon_\beta)$, and neglect the thermal energies $\varepsilon_\beta \sim T$ in comparison with the transition energy $\hbar\omega$. We obtain the quantum master equation

$$\begin{aligned}\frac{d}{dt}\rho(t) = & - i\omega[a_+^+a_+ + a_-^+a_-, \rho(t)] \\ & + \sum_{j>i} \omega_{ji}\vec{K}\vec{r}_{ij}\left[\langle c_j^+ c_i\rangle (a_+e^{ikx} + a_-e^{-ikx}) - \langle c_i^+ c_j\rangle (a_+^+e^{-ikx} + a_-^+e^{ikx}), \rho(t)\right] \\ & + \Lambda\left\{[a_+\rho(t), a_+^+] + [a_+, \rho(t)a_+^+] + [a_-\rho(t), a_-^+] + [a_-, \rho(t)a_-^+]\right\},\end{aligned} \tag{63}$$

with the dissipation coefficient

$$\Lambda = \pi\hbar\omega^2 g(\hbar\omega)(\vec{K}\vec{R}_{\alpha\beta})^2, \tag{64}$$

depending on quantities which, according to (9), (44), and (12), are of the form

$$K = \sqrt{\alpha\frac{\lambda}{\mathcal{V}}} \tag{65}$$

$$R_{\alpha\beta} = \sqrt{\frac{2\hbar}{M\omega}} \tag{66}$$

$$g(\hbar\omega) = \mathcal{V}^S\frac{\sqrt{2}M^{3/2}}{\pi^2\hbar^3}\sqrt{\hbar\omega}. \tag{67}$$

Unlike the master equation for an electromagnetic field mode derived in (12), we considered explicit expressions of the electron-field potential of interaction, and neglected the thermal energy in comparison with the transition energy. With (65)-(67), the dissipation coefficient (64) takes a form

$$\Lambda = \Omega\frac{\mathcal{V}^S}{\mathcal{V}} \tag{68}$$

depending on the quantity

$$\Omega = 4\alpha\sqrt{\frac{2Mc^2\omega}{\hbar}}, \tag{69}$$

and the two quantization volumes, $\mathcal{V}$ of the electromagnetic field and $\mathcal{V}^S$ of the dissipative electron system. We consider a quantization volume $\mathcal{V}^S = V_n = \frac{1}{N_D}$ for an n-type region, or $\mathcal{V}^S = V_p = \frac{1}{N_A}$ for a p-type region. From physical point of view, a quantization volume of the electromagnetic field $\mathcal{V}$ means a measuring process corresponding to a confinement in this volume. The electromagnetic field can not be quantized in a volume $\mathcal{V}^S$, but in a much larger one, with much larger dimensions than the field wavelength λ. For an electromagnetic field, we consider a unit quantization volume $\mathcal{V} = 1_V = 1_L^3 = 1m^3$, because, in this case, the radiation density is equal to the electromagnetic field density times the light velocity, $S = w_E c$ $[W/m^2]$, where $w_E = \frac{\varepsilon_0 E^2}{2}$ $[J/m^3]$ is calculated with this quantization volume in the expression (8)-(9) of the field.

We notice that the master equation (63) describes an electromagnetic field quantized in a unit volume $\mathcal{V} = 1m^3$ in interaction with an electron system occupying a much smaller volume $\mathcal{V}^S = V^{(n)}, V^{(p)}$. A system of N dissipative electrons can be taken into account by multiplying the dissipative coefficient Λ with N. However, in such a description, the electromagnetic field is considered of a constant amplitude inside the quantization volume $\mathcal{V} = 1_V$, while, in fact, this amplitude undertakes a spatial variation due to the interaction with the system quantized in a volume $\mathcal{V}^S \ll \mathcal{V}$, i.e. propagation characteristics as the absorption coefficient and the refractive index inside the field quantization volume are not taken into account. We take into account the spatial variation of the electromagnetic field by considering this field as being given by an x dependent density matrix, as product of density matrices for the two counter-propagating waves, $\rho(x,t) = \rho_+(x,t)\rho_-(x,t)$, and taking the dissipative terms as integrals over the paths traveled by these waves. Considering a distribution of N_D dissipative clusters over the the thickness L_D of the device, from the master equation (63), we get

$$\frac{d}{dt}\rho_+(x,t) = -i\omega[a_+^+ a_+, \rho_+(x,t)] + \sum_{j>i} \omega_{ji}\vec{K}\vec{r}_{ij}\left[\langle c_j^+ c_i\rangle a_+ e^{ikx} - \langle c_i^+ c_j\rangle a_+^+ e^{-ikx}, \rho_+(x,t)\right] \tag{70a}$$

$$+\frac{\Omega_D}{1_L}\int_0^x \left\{[a_+\rho_+(x',t'), a_+^+] + [a_+, \rho_+(x',t')a_+^+]\right\} e^{-ik(x-x')}dx'$$

$$\frac{d}{dt}\rho_-(x,t) = -i\omega[a_-^+ a_-, \rho_-(x,t)] + \sum_{j>i} \omega_{ji}\vec{K}\vec{r}_{ij}\left[\langle c_j^+ c_i\rangle a_- e^{-ikx} - \langle c_i^+ c_j\rangle (a_-^+ e^{ikx}, \rho_-(x,t)\right] \tag{70b}$$

$$+\frac{\Omega_D}{1_L}\int_x^{L_D} \left\{[a_-\rho_-(x',t'), a_-^+] + [a_-, \rho_-(x',t')a_-^+]\right\} e^{-ik(x'-x)}dx',$$

depending on the dissipative coefficient

$$\Omega_D = \Omega\frac{1_L^2 L_D}{1_L^3} = \Omega\frac{L_D}{1_L}, \tag{71}$$

obtained by summation over the dissipative clusters with the volume $1_L^2 L_D$ in the quantization volume 1_L^3. The exponential factors in the integrals describe the delay of the field propagating from the coordinate x' of a dissipative element to the coordinate x of the density matrix of this field at this coordinate. These equations describe the dissipative dynamics of a forward electromagnetic wave, propagating from $x' = 0$ to $x' = x$, and of a backward electromagnetic wave, propagating from $x' = L_D$ to $x' = x < L_D$. With these equations, we calculate the mean-values of the field operators

$$a_+(x,t) = \langle a_+\rangle = Tr\{a_+\rho_+(x,t)\} = \mathcal{A}_+(x,t)e^{-i\omega t} \tag{72a}$$

$$a_-(x,t) = \langle a_-\rangle = Tr\{a_-\rho_-(x,t)\} = \mathcal{A}_-(x,t)e^{-i\omega t}. \tag{72b}$$

For an array of two-level systems with the coordinate x, interacting with the electromagnetic field with the frequency $\omega \approx \omega_0$, we define the time slowly-varying amplitude of the polarization $\mathcal{S}(x,t)$ by the relations

$$\langle c_i^+ c_j\rangle = \rho_{ji}(x,t) = \frac{1}{2}\mathcal{S}(x,t)e^{-i\omega t}, \tag{73}$$

$$\mathcal{S}(x,t) = \mathcal{S}_+(x,t)e^{ikx} + \mathcal{S}_-(x,t)e^{-ikx}, \tag{74}$$

where $\mathcal{S}_+(x,t), \mathcal{S}_-(x,t)$ are slowly-varying in space and time amplitudes of the polarization induced by the two counter-propagating waves of the field. Having in view Heisenberg's uncertainty principle

$$\Delta k\Delta x \geq \frac{1}{2}, \qquad \Delta\omega\Delta t \geq \frac{1}{2}, \tag{75}$$

we notice that, in equations (70), this relation selects only the close terms, with $x' \approx x$, while the farer terms in $x - x'$ are washed up by the uncertainty Δk in the oscillating functions under the x'-dependent integrals. By definition, these x-dependent integrals describe the attenuation of an electromagnetic wave squeezed in the x-domain, $\Delta x = 0$. We take into account a finite uncertainty Δx, in the vicinity of x, by extending these x'-integrals with the half-width $\Delta x/2$. We obtain the field equations

$$\frac{d}{dt}\mathcal{A}_+(x,t) = -\frac{1}{2}\omega_0 \vec{K}\vec{r}_{01}\mathcal{S}_+(x,t) - \frac{\Omega_D}{1_L}\int_0^{x-\Delta x/2} \mathcal{A}_+(x',t')e^{-\mathrm{i}[k(x-x')-\omega(t-t')]}dx' \quad \text{(76a)}$$
$$- \frac{\Omega_D}{1_L}\int_{x-\Delta x/2}^{x+\Delta x/2} \mathcal{A}_+(x',t')e^{-\mathrm{i}[k(x-x')-\omega(t-t')]}dx'$$
$$\frac{d}{dt}\mathcal{A}_-(x,t) = -\frac{1}{2}\omega_0 \vec{K}\vec{r}_{01}\mathcal{S}_-(x,t) + \frac{\Omega_D}{1_L}\int_{x+\Delta x/2}^{L_D} \mathcal{A}_-(x',t')e^{-\mathrm{i}[k(x'-x)-\omega(t-t')]}dx' \quad \text{(76b)}$$
$$+ \frac{\Omega_D}{1_L}\int_{x-\Delta x/2}^{x+\Delta x/2} \mathcal{A}_-(x',t')e^{-\mathrm{i}[k(x'-x)-\omega(t-t')]}dx'.$$

We notice that these are non-local in space equations including retarded contributions of the dissipation processes along the distance $|x - x'|$, and absorption processes in the vicinity Δx of the coordinate x. We consider the quantities under these integrals as spectral lines integrated over half-widths, $k\Delta x - \omega\Delta t = (k + \Delta k)\Delta x - (\omega + \Delta\omega)\Delta t = \Delta k\Delta x - \Delta\omega\Delta t = \frac{\pi}{6} + \frac{\pi}{6} > \frac{1}{2} + \frac{1}{2}$. By integrating the first integral of the first equation two times by parts, for a large distance $x - x'$ we obtain

$$\int_0^{x-\Delta x/2} \mathcal{A}_+(x',t')e^{-\mathrm{i}[k(x-x')-\omega(t-t')]}dx' = \frac{1}{\mathrm{i}k}\mathcal{A}_+(x',t')e^{-\mathrm{i}[k(x-x')-\omega(t-t')]}\Big|_0^{x-\Delta x/2}$$
$$- \frac{1}{\mathrm{i}k}\int_0^{x-\Delta x/2} \frac{d}{dx'}\mathcal{A}_+(x',t')e^{-\mathrm{i}[k(x-x')-\omega(t-t')]}dx'$$
$$= \left[-\frac{\mathrm{i}}{k}\mathcal{A}_+(x,t) + \frac{1}{k^2}\frac{d}{dx}\mathcal{A}_+(x,t)\right]\left[\cos\left(\frac{k\Delta x - \omega\Delta t}{2}\right) - \mathrm{i}\sin\left(\frac{k\Delta x - \omega\Delta t}{2}\right)\right] \quad \text{(77)}$$
$$= -\frac{1+\mathrm{i}\sqrt{3}}{2k}\mathcal{A}_+(x,t) + \frac{\sqrt{3}-\mathrm{i}}{2k^2}\frac{d}{dx}\mathcal{A}_+(x,t),$$

while, taking into account that on a vary short distance Δx the field amplitude is practically constant, the second integral of this equation takes a simple form

$$\int_{x-\Delta x/2}^{x+\Delta x/2} \mathcal{A}_+(x',t')e^{-\mathrm{i}[k(x-x')-\omega(t-t')]}dx' = \frac{1}{\mathrm{i}k}\mathcal{A}_+(x',t')e^{-\mathrm{i}[k(x-x')-\omega(t-t')]}\Big|_{x-\Delta x/2}^{x+\Delta x/2}$$
$$= \frac{2}{k}\mathcal{A}_+(x,t)\sin\left(\frac{k\Delta x - \omega\Delta t}{2}\right) = \frac{1}{k}\mathcal{A}_+(x,t). \quad \text{(78)}$$

These terms describe a slight variation of the wave-vector k, $k' = k + \kappa$, which means that the amplitude of the mean-value of the field operator takes a form

$$\mathcal{A}_+(x,t) = \tilde{\mathcal{A}}_+(x,t)e^{\mathrm{i}\kappa x}. \quad \text{(79)}$$

Taking into account that $\frac{\Omega_D}{ck^2 1_L} \ll 1$, while $\kappa = \frac{k}{1+\frac{2ck^2 1_L}{\sqrt{3}\Omega_D}} \ll k$, we get a field equation

$$\frac{\partial}{\partial t}\mathcal{A}_+(x,t) + c\left[\frac{\partial}{\partial x}\mathcal{A}_+(x,t) + \alpha'\mathcal{A}_+(x,t)\right] = -\frac{1}{2}\omega_0\vec{K}\vec{r}_{01}\mathcal{S}_+(x,t), \quad \text{(80)}$$

with an absorption coefficient

$$\alpha' = \frac{2\alpha c}{1_L}\sqrt{\frac{2M}{\hbar\omega}\frac{L_D}{1_L}}. \quad (81)$$

In the following calculations, we are interested in a form of this equation in a cavity with the length L_D,

$$\frac{d}{dt}\mathcal{A}_+(x,t) = -\gamma_F\mathcal{A}_+(x,t) - \frac{1}{2}\omega_0\vec{K}\vec{r}_{01}\mathcal{S}_+(x,t), \quad (82)$$

with a field decay rate

$$\gamma_F = c\alpha' = \frac{2\alpha c^2}{1_L^2}\sqrt{\frac{2M}{\hbar\omega}}L_D, \text{ or } 1_L\gamma_F = \frac{2\alpha c^2}{1_L}\sqrt{\frac{2M}{\hbar\omega}}L_D, \quad (83)$$

which we call field decay velocity. In section 6, it will be shown that the field decay velocity $1_L\gamma_F$ describes the field loss by dissipation, as the quantity $\mathcal{T}c$ describes the electromagnetic energy loss by radiation through the output mirror with the transparency $\mathcal{T}$. We notice that the decay rate of an electromagnetic field in a cavity is proportional to the length of this cavity. For a semiconductor chip with the thickness $L_D = 2\ mm$, we considered in our calculations in (3), from (83) we get a field decay rate $\gamma_F = 2.05 \times 10^7\ s^{-1}$, which is in agreement with the empirical values $\gamma_F = 10^7,\ 10^8\ s^{-1}$, we considered in these calculations. It is interesting that, in this model, the decay rate does not depend on the concentration of the dissipative clusters, since an increase of this concentration means a decrease of the density of states in every cluster, which, in this way, becomes smaller. These two variations cancel exactly one another in the final result. By taking into account the spreading of a dissipative electron wave-function beyond the the boundaries of its cluster due to the thermal motion, one obtains a lower value of the decay rate, but with an increase with the concentration of these clusters.

5. Optical equations for a system of quantum injection dots

From the quantum master equation (34), we derive optical equations for a two-level system. In the approximation of the slowly varying amplitudes, we consider the non-diagonal matrix elements

$$\rho_{10}(t) = \rho_{01}^*(t) = \frac{1}{2}\left[\mathcal{S}_+(t)e^{ikx} + \mathcal{S}_-(t)e^{-ikx}\right]e^{-i\omega t}, \quad (84)$$

and the population difference

$$w(t) = \rho_{11}(t) - \rho_{00}(t), \text{ with the normalization condition} \quad (85a)$$

$$1 = \rho_{11}(t) + \rho_{00}(t). \quad (85b)$$

Calculating the matrix elements of the two-level system, and averaging over the field states, from the master equation (34) we get:

$$\begin{aligned}\frac{d}{dt}\rho_{10}(t) = & -[\lambda_{01} + \lambda_{10} + \mathrm{i}(\omega_0 + \zeta_{11} - \zeta_{00})]\rho_{10}(t) \\ & +\vec{K}\left[(\langle a_+\rangle + \langle a_-^\dagger\rangle)e^{ikx} + (\langle a_+^\dagger\rangle + \langle a_-\rangle)e^{-ikx}\right]\omega_0\vec{r}_{10}[\rho_{00}(t) - \rho_{11}(t)] \\ & +(\zeta_{11} - \zeta_{00})^2\int_{t-\tau}^{t}\rho_{10}(t')e^{-\mathrm{i}[\phi_{10}(t') + \omega(t-t')]}dt' \end{aligned} \quad (86a)$$

$$\begin{aligned}\frac{d}{dt}\rho_{11}(t) = & -\frac{d}{dt}\rho_{00}(t) = 2[\lambda_{10}\rho_{00} - \lambda_{01}\rho_{11}] \\ & +\vec{K}\left[(\langle a_+\rangle + \langle a_-^\dagger\rangle)e^{ikx} + (\langle a_+^\dagger\rangle + \langle a_-\rangle)e^{-ikx}\right]\omega_0\vec{r}_{10}[\rho_{10}(t) + \rho_{01}(t)]. \end{aligned} \quad (86b)$$

From the expression (8) of the quantized electric field $\vec{E}$ in the plane-wave approximation, and mean-values of the annihilation operators of the form

$$\langle a_+ \rangle = \bar{a}_+(t)e^{-i\omega t} \tag{87a}$$
$$\langle a_- \rangle = \bar{a}_-(t)e^{-i\omega t}, \tag{87b}$$

we get the mean-value of this field,

$$\langle \vec{E} \rangle = \frac{1}{2}\left[\vec{\mathcal{E}}(t)e^{-i\omega t} + \vec{\mathcal{E}}^*(t)e^{i\omega t}\right], \tag{88}$$

with the time slowly-varying amplitude

$$\vec{\mathcal{E}}(t) = \vec{\mathcal{E}}_+(t)e^{ikx} + \vec{\mathcal{E}}_-(t)e^{-ikx}, \tag{89}$$

while the amplitudes of the two counter-propagating waves are

$$\vec{\mathcal{E}}_+(t) = 2i\frac{\hbar\omega}{e}\vec{K}\bar{a}_+(t) \tag{90a}$$
$$\vec{\mathcal{E}}_-(t) = 2i\frac{\hbar\omega}{e}\vec{K}\bar{a}_-(t). \tag{90b}$$

In this description we neglect the variation of the amplitudes inside the cavity, by taking into account these two amplitudes only as mean-values over the space coordinate, related by the boundary condition for the output mirror of transmission coefficient $\mathcal{T}$:

$$\vec{\mathcal{E}}_-(t) = -\sqrt{1-\mathcal{T}}\vec{\mathcal{E}}_+(t). \tag{91}$$

With the notations

$$\vec{g} = \frac{e}{\hbar}\vec{r}_{10} \tag{92}$$

for the coupling coefficient,

$$\gamma_\perp = \lambda_{01} + \lambda_{10} \tag{93}$$

for the dephasing rate,

$$\gamma_\parallel = 2(\lambda_{01} + \lambda_{10}) \tag{94}$$

for the decay rate,

$$\gamma_n = |\zeta_{11} - \zeta_{00}| \tag{95}$$

for the fluctuation rate of the self-consistent field, and

$$w_T = -\frac{\lambda_{01} - \lambda_{10}}{\lambda_{01} + \lambda_{10}}, \tag{96}$$

from (84)-(91) we obtain equations for the slowly-varying amplitudes

$$\frac{d}{dt}\mathcal{S}_+(t) = -[\gamma_\perp + i(\omega_0 + \gamma_n - \omega)]\mathcal{S}_+(t) + i\vec{g}\vec{\mathcal{E}}_+(t)w(t) + \gamma_n^2\int_{t-\tau}^{t}\mathcal{S}_+(t')e^{-i[\phi_{10}(t')+\omega(t-t')]}dt' \tag{97a}$$

$$\frac{d}{dt}w(t) = -\gamma_\parallel[w(t) - w_T] + (2-\mathcal{T})i\vec{g}\frac{1}{2}\left[\vec{\mathcal{E}}_+^*(t)\mathcal{S}_+(t) - \vec{\mathcal{E}}_+(t)\mathcal{S}_+^*(t)\right]. \tag{97b}$$

In equation (97b) we have taken into account that the term

$$\Phi_+(t) = \mathrm{i}\vec{g}\frac{1}{2}\left[\vec{\mathcal{E}}_+^*(t)\mathcal{S}_+(t) - \vec{\mathcal{E}}_+(t)\mathcal{S}_+^*(t)\right] \tag{98}$$

is a particle flow due to the forward electromagnetic wave propagating in the cavity, while

$$\Phi_-(t) = \mathrm{i}\vec{g}\frac{1}{2}\left[\vec{\mathcal{E}}_-^*(t)\mathcal{S}_-(t) - \vec{\mathcal{E}}_-(t)\mathcal{S}_-^*(t)\right] \tag{99}$$

is a particle flow due to the backward electromagnetic wave, which means that the two flows satisfy the boundary condition for the energy flow of the electromagnetic field

$$\Phi_-(t) = (1 - \mathcal{T})\Phi_+(t). \tag{100}$$

At the same time, calculating the mean-value of the field operator a, averaging over the states of the two-level system, and taking into account the relation

$$\langle c_i^+ c_j\rangle = \rho_{ji}(t), \tag{101}$$

from equation (34) we get the field equation

$$\frac{d}{dt}\langle a_+\rangle = -\mathrm{i}\omega\langle a_+\rangle + \vec{K}\omega_0\vec{r}_{10}[\rho_{10}(t) - \rho_{01}(t)]e^{-ikx}. \tag{102}$$

Thus, with (84), (87) and (90), we get a field equation for slowly-varying amplitudes

$$\frac{d}{dt}\vec{\mathcal{E}}_+(t) = -\mathrm{i}\omega_0\frac{\hbar\omega}{e}\vec{K}(\vec{K}\vec{r}_{10})\mathcal{S}_+(t). \tag{103}$$

We consider this equation for the components $u(t)$ and $v(t)$ of the polarization amplitude

$$\mathcal{S}_+(t) = u(t) - \mathrm{i}v(t), \tag{104}$$

and $\mathcal{F}(t)$ and $\mathcal{G}(t)$ of the electromagnetic field

$$\mathcal{E}_+(t) = \mathcal{F}(t) + \mathrm{i}\mathcal{G}(t), \tag{105}$$

and take into account the field dissipation described by the dissipation rate γ_F. We get

$$\frac{d}{dt}\mathcal{F}(t) = -\gamma_F\mathcal{F}(t) - g\frac{\hbar\omega_0}{2\varepsilon\mathcal{V}}v(t) \tag{106a}$$

$$\frac{d}{dt}\mathcal{G}(t) = -\gamma_F\mathcal{G}(t) - g\frac{\hbar\omega_0}{2\varepsilon\mathcal{V}}u(t). \tag{106b}$$

We consider these equations for the electromagnetic energy in the quantization volume $\mathcal{V}$, and introduce the energy flow through the surface $\mathcal{A}$ of this volume:

$$\frac{d}{dt}\left[\mathcal{V}\frac{1}{2}\varepsilon\mathcal{F}^2(t)\right] = -\mathcal{T}c\frac{1}{2}\varepsilon\mathcal{F}^2(t)\mathcal{A} - \gamma_F\mathcal{V}\varepsilon\mathcal{F}^2(t) - g\frac{\hbar\omega_0}{2}\mathcal{F}v(t) \tag{107a}$$

$$\frac{d}{dt}\left[\mathcal{V}\frac{1}{2}\varepsilon\mathcal{G}^2(t)\right] = -\mathcal{T}c\frac{1}{2}\varepsilon\mathcal{G}^2(t)\mathcal{A} - \gamma_F\mathcal{V}\varepsilon\mathcal{G}^2(t) - g\frac{\hbar\omega_0}{2}\mathcal{G}u(t). \tag{107b}$$

At the same time, from (97b) with (104) and (105), we derive the equation for the population difference (85a), and introduce the particle flow $\mathcal{I}$ in a two-level system, due to the electric current $I = eA_D N_e \mathcal{I}$ injected in the device:

$$\frac{d}{dt}w(t) = -\gamma_F[w(t) - w_T] + 2\mathcal{I} + (2 - \mathcal{T})g[\mathcal{F}(t)v(t) + \mathcal{G}(t)u(t)] \tag{108}$$

From (107) and (108) with (85), we get an equation of energy conservation:

$$\begin{aligned}\hbar\omega_0\mathcal{I} = \frac{d}{dt}\left\{\hbar\omega_0\rho_{11}(t) + (2-\mathcal{T})\mathcal{V}\frac{1}{2}\varepsilon[\mathcal{F}^2(t) + \mathcal{G}^2(t)]\right\} + \gamma_\parallel\left[\rho_{11}(t) - \frac{1+w_T}{2}\right]\hbar\omega_0 \\ +(2-\mathcal{T})(\mathcal{T}c\frac{\mathcal{A}}{\mathcal{V}} + 2\gamma_F)\mathcal{V}\frac{1}{2}\varepsilon[\mathcal{F}^2(t) + \mathcal{G}^2(t)].\end{aligned} \tag{109}$$

This equation describes the transition power $\hbar\omega_0\mathcal{I}$ of the active system providing the energy transfer processes involved in the dissipative super radiant decay: (1) the energy variation of the electron-field system, (2) the dissipative decay of the electron energy, proportional to $\gamma_\parallel$, (3) the radiation of the field energy, proportional to the light velocity c and the transmission coefficient $\mathcal{T}$ of the output mirror, and (4) the dissipation of the field energy, proportional to γ_F. In this equation, both waves leaving the quantum system and propagating in the cavity, the forward wave with an amplitude coefficient 1 and the backward wave with an amplitude coefficient $\mathcal{R} = 1 - \mathcal{T}$, are taken into account with the coefficient $1 + \mathcal{R} = 2 - \mathcal{T}$.

From the polarization equation (97a) with (104) and (105), the population equation (108), and the field equations (107), we obtain the equations of the slowly varying amplitudes of the system:

$$\begin{aligned}\frac{d}{dt}u(t) = & -\gamma_\perp[u(t) - \delta\omega v(t)] - g\mathcal{G}(t)w(t) \\ & +\gamma_n^2\int_{t-\tau}^{t}\{u(t')\cos[\phi_n(t') + (\omega-\omega_0)(t-t')] + v(t')\sin[\phi_n(t') + (\omega-\omega_0)(t-t')]\}dt'\end{aligned} \tag{110a}$$

$$\begin{aligned}\frac{d}{dt}v(t) = & -\gamma_\perp[v(t) + \delta\omega u(t)] - g\mathcal{F}(t)w(t) \\ & +\gamma_n^2\int_{t-\tau}^{t}\{v(t')\cos[\phi_n(t') + (\omega-\omega_0)(t-t')] - u(t')\sin[\phi_n(t') + (\omega-\omega_0)(t-t')]\}dt'\end{aligned} \tag{110b}$$

$$\frac{d}{dt}w(t) = -\gamma_\parallel[w(t) - w_T] + 2\mathcal{I} + (2-\mathcal{T})g[\mathcal{G}(t)u(t) + \mathcal{F}(t)v(t)] \tag{110c}$$

$$\frac{d}{dt}\mathcal{F}(t) = -\frac{1}{2}\mathcal{T}c\frac{\mathcal{A}}{\mathcal{V}}\mathcal{F}(t) - \gamma_F\mathcal{F}(t) - g\frac{\hbar\omega_0}{2\varepsilon\mathcal{V}}v(t) \tag{110d}$$

$$\frac{d}{dt}\mathcal{G}(t) = -\frac{1}{2}\mathcal{T}c\frac{\mathcal{A}}{\mathcal{V}}\mathcal{G}(t) - \gamma_F\mathcal{G}(t) - g\frac{\hbar\omega_0}{2\varepsilon\mathcal{V}}u(t), \tag{110e}$$

where $\phi_n(t') \equiv \phi_{01}(t') \equiv -\phi_{10}(t')$ is the phase fluctuation with a fluctuation time $\tau_n = 1/\gamma_n$, and

$$\delta\omega = \frac{\omega - \omega_0 - \gamma_n}{\gamma_\perp} \tag{111}$$

is the relative atomic detuning. In these equations, the coupling of the electron system to the electromagnetic field is described by a coupling coefficient for the dipole interaction $g = \vec{g}\vec{1}_E$. These equations also describe a dissipative decay of the electron system by the coefficients $\gamma_\parallel$

and $\gamma_\perp$, non-Markovian effects by time-integrals proportional to the fluctuation coefficient γ_n^2 in the polarization equations (110a) and (110b), a decrease of the electron-field coupling due to the field radiation by the term proportional to the coefficient $(2-\mathcal{T})$ in (110c), and a decrease of field by the radiation terms proportional to the product $c\mathcal{T}$ in (110d) and (110e), and by the terms proportional to the decay rate γ_F.

6. Superradiant quantum injection dots and heat conversion

The dynamic equations (110) take a simpler form in a stationary regime when the derivatives with time become zero and the polarization variables can be taken out from the integrals. Considering an integration over a fluctuation time $\tau_n = 1/\gamma_n$, we get a time oscillating term, generated by the fluctuations of the environment particles. In the Markovian approximation, when these oscillations are neglected, we get the steady-state equations:

$$-\gamma_\perp[u - \delta\omega v] - g\mathcal{G}w = 0 \tag{112a}$$

$$-\gamma_\perp[v + \delta\omega u] - g\mathcal{F}w = 0 \tag{112b}$$

$$-\gamma_\parallel(w - w_T) + 2\mathcal{I} + (2-\mathcal{T})g(\mathcal{G}u + \mathcal{F}v) = 0 \tag{112c}$$

$$-\Gamma_F\mathcal{F} - Gv = 0 \tag{112d}$$

$$-\Gamma_F\mathcal{G} - Gu = 0, \tag{112e}$$

where

$$G = g\frac{\hbar\omega_0}{2\varepsilon\mathcal{V}} \tag{113a}$$

$$\Gamma_F = \frac{1}{2}\mathcal{T}c\frac{\mathcal{A}}{\mathcal{V}} + \gamma_F. \tag{113b}$$

From the system of equations (112), for the resonance case ($\delta\omega = 0$), we calculate the flow density of the electromagnetic energy radiated by the device:

$$S = \mathcal{T}c\frac{1}{2}\varepsilon(\mathcal{F}^2 + \mathcal{G}^2). \tag{114}$$

We get

$$S = \frac{\frac{\hbar\omega_0}{(2-\mathcal{T})\mathcal{A}}}{1 + \frac{2\gamma_F\mathcal{V}}{\mathcal{T}c\mathcal{A}}}\left[\mathcal{I} - \left(-w_T\frac{\gamma_\parallel}{2} + \frac{\frac{1}{2}\mathcal{T}c\frac{\mathcal{A}}{\mathcal{V}} + \gamma_F}{\frac{g^2\hbar\omega_0}{\gamma_\perp\gamma_\parallel\varepsilon\mathcal{V}}}\right)\right]. \tag{115}$$

This expression of the flow density S has a nice physical interpretation, being proportional to the product of the transition energy $\hbar\omega_0$, divided to the radiation area of a quantum dot $\mathcal{A}$, with the difference between the particle flow $\mathcal{I}$ and a threshold value depending on coupling, radiation, and dissipation coefficients. This expression is valid when the quantization volume $\mathcal{V}$ of the field corresponds to the electromagnetic energy delivered by the whole system of $N_e N_t$ quantum dots to a volume unit, which means

$$\mathcal{V}[m^3] = \frac{1}{N_e[m^{-2}]N_t[m^{-1}]}, \tag{116}$$

where $N_e[m^{-2}]$ is the number of quantum dots per area unit, and $N_t[m^{-1}]$ is the number of super radiant junctions per length unit.

In a first approximation, we neglect the temperature variation due to the heat transfer through the semiconductor structure. To take into account this temperature variation, one has to make corrections of the parameters, to obtain the same transition frequency on the whole chain of super radiant junctions.

Such a device can be realized in two versions schematically represented in figure 4: (a) a longitudinal device, with the two mirrors M_1 and M_2 made on the two surfaces in the plane of the chip, of transmission coefficients $\mathcal{T}_0 = 0$ and $\mathcal{T} > 0$, which form a Fabry-Perot cavity coupling a super radiant mode that propagates in the x-direction of the injection current; (b) a transversal device, with the two mirrors M_1 and M_2 made on two lateral surfaces of the chip, of transmission coefficients $\mathcal{T}_0 = 0$ and $\mathcal{T} > 0$, which form a Fabry-Perot cavity coupling a super radiant mode that propagates in the y-direction, perpendicular to the injection current. While in version (a) the roles of the mirrors M_1 and M_2, and of the injection electrodes $\mathcal{E}_1$ and $\mathcal{E}_2$, are played by the same metalizations, made on the two surfaces in the plane of the chip, in version (b) the mirror metalizations M_1 and M_2, which are made on two lateral surfaces, are different from the electrode metalizations $\mathcal{E}_1$ and $\mathcal{E}_2$.

The two devices have the same structure, including layers of $GaAs$, with a narrower forbidden band and a heavier doping, for the quantum wells, and layers of $Al_xGa_{1-x}As$, with a larger forbidden band and a lighter doping, for the potential barriers. The margins of these bands are determined by the concentrations of the donors (acceptors) embedded in the semiconductor layers. For a longitudinal device (figure 4a), the $\bar{N}_t$ (dimensionless) quantum dots in the x-direction, radiate through an area $\frac{1}{N_e[m^{-2}]}$, which means

$$\mathcal{A}_L[m^2] = \frac{1}{N_e[m^{-2}]\bar{N}_t}, \tag{117}$$

while for a transversal device (see figure 4b), the $\sqrt{N_e[m^{-2}]A_D[m^2]}$ quantum dots in the y-direction, radiate through an area $\frac{L_D[m]}{\bar{N}_t}\frac{1}{\sqrt{N_e[m^{-2}]}}$, which means

$$\mathcal{A}_T[m^2] = \frac{L_D[m]}{N_e[m^{-2}]\bar{N}_t\sqrt{A_D[m^2]}}. \tag{118}$$

With the radiation area $\mathcal{A}_L$ ($\mathcal{A}_T$) of a quantum dot, from (115) we derive the flow density S_L (S_D), and the total flow of the electromagnetic field radiated by the device in the two versions:

$$\Phi_L = A_D S_L \tag{119a}$$
$$\Phi_T = L_D\sqrt{A_D}S_T. \tag{119b}$$

We obtain

$$\Phi_L = \frac{\bar{N}_t}{(2-\mathcal{T})(1+2\frac{1_L\gamma_F}{\mathcal{T}c})}\cdot\frac{\hbar\omega_0}{e}(I-I_{0L}) \tag{120a}$$

$$\Phi_T = \frac{\bar{N}_t}{(2-\mathcal{T})(1+2\frac{1_L\gamma_F}{\mathcal{T}c}\frac{A_D^{1/2}}{L_D})}\cdot\frac{\hbar\omega_0}{e}(I-I_{0T}), \tag{120b}$$

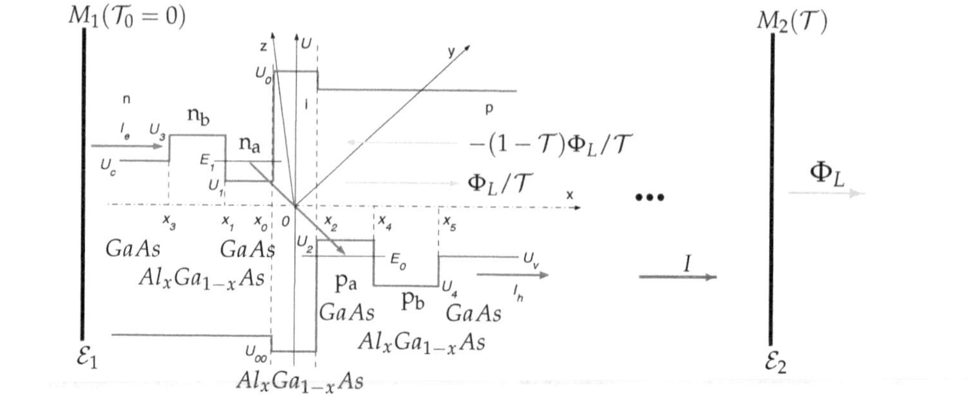

(a) Longitudinal super radiant device with the Fabry-Perot cavity oriented in the x-direction of the injected current $I = I_e = I_h$, i.e. perpendicular to the semiconductor layers.

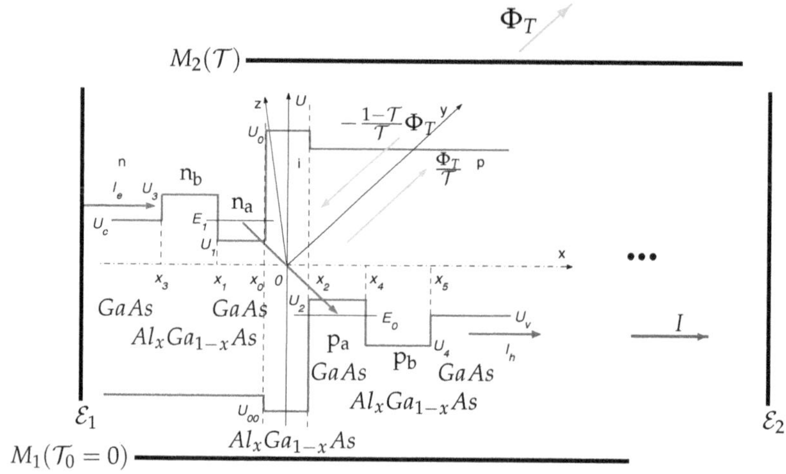

(b) Transversal super radiant device with the Fabry-Perot cavity oriented in the y-direction, perpendicular to the injected current $I = I_e = I_h$, i.e. in the plane of the semiconductor layers.

Fig. 4. Dissipative super radiant n-i-p device with two injection electrodes $\mathcal{E}_1$ and $\mathcal{E}_2$ and a Fabry-Perot cavity with the mirrors M_1 and M_2 of transmission coefficients $\mathcal{T}_0 = 0$ and $\mathcal{T}$, respectively, in two possible versions (a) and (b).

as a function of the injected current I and the threshold currents

$$I_{0L} = \frac{1}{2} e N_e A_D \gamma_{\parallel} \left[-w_T + \frac{\varepsilon \gamma_{\perp}}{g_L^2 \hbar \omega_0 N_e \bar{N}_t} (\mathcal{T} c + 2 \cdot 1_L \gamma_F) \right] \tag{121a}$$

$$I_{0T} = \frac{1}{2} e N_e A_D \gamma_{\parallel} \left[-w_T + \frac{\varepsilon \gamma_{\perp}}{g_T^2 \hbar \omega_0 N_e \bar{N}_t} \left(\mathcal{T} c \frac{L_D}{A_D^{1/2}} + 2 \cdot 1_L \gamma_F\right) \right], \tag{121b}$$

which depends on the field decay velocity (83). The threshold current is proportional to the threshold population, which includes three terms for the three dissipative processes that must be balanced by current injection for creating a coherent electromagnetic field: (1) a term $-w_T$, for a population inversion, (2) a term proportional to the light velocity c and the transmission coefficient $\mathcal{T}$, for the field radiation, and (3) a term proportional to decay rate γ_F, for the field dissipation.

From (112c) and (120), we notice that when the injection current $I = eN_eA_D\mathcal{I}$ is under the threshold value I_{0L} (I_{0T}), the radiation field is $\mathcal{F} + i\mathcal{G} = 0$, while the population difference w increases with this current. When the injection current I reaches the threshold current I_{0L} (I_{0T}), the population difference w reaches the radiation value

$$w_R = \frac{\mathcal{T}c\frac{A}{\mathcal{V}} + 2\gamma_F}{\frac{g^2\hbar\omega_0}{\gamma_\perp\varepsilon\mathcal{V}}}. \tag{122}$$

Increasing the injection current I beyond the threshold value I_{0L} (I_{0T}), the population difference keeps this value, while the super radiant field and the polarization ($u = -\frac{g}{\gamma_\perp}w_R\mathcal{G}, v = -\frac{g}{\gamma_\perp}w_R\mathcal{F}$) increases. However, the polarization (u, v) can not increase indefinitely, being constrained by the condition of the Bloch vector length $(2-\mathcal{T})(u^2 + v^2) + w^2 \leq w_T^2$. For the maximum value (u_M, v_M) of the polarization, while $u_M^2 + v_M^2 = (w_T^2 - w_R^2)/(2-\mathcal{T})$, the super radiant field reaches its maximum flow density

$$S_M = \frac{\mathcal{T}c\varepsilon}{2(2-\mathcal{T})}\left[w_T^2\frac{g^2\frac{\hbar^2\omega_0^2}{\varepsilon^2\mathcal{V}^2}}{\left(\mathcal{T}c\frac{A}{\mathcal{V}} + 2\gamma_F\right)^2} - \frac{\gamma_\perp^2}{g^2}\right]. \tag{123}$$

From this equation with equation (115) for $S = S_M$, we get the value $I_M = eN_eA_D\mathcal{I}_M$ of the injection current producing the maximum flow of the electromagnetic energy. Increasing the injection current beyond this value, the polarization (u, v) will not increase any more, but the population will increase, leading to a rapid decrease of the polarization. Neglecting the current increase from I_M to the value I_M' when the polarization vanishes, from equation (112c) with $w = -w_T$ and $u = v = 0$, we get a simple, approximate expression $I_M \approx I_M' = \frac{1}{2}eN_eA_D\gamma_\parallel(-w_T - w_T)$, which can be compared with (121). From the operation condition $I_{0L}, I_{0T} < I_M$, we get conditions for the coupling, dissipation, and radiation coefficients:

$$w_{IL} = \frac{\varepsilon_0\gamma_\perp}{g_L^2\hbar\omega_0N_e\bar{N}_t}(\mathcal{T}c + 21_L\gamma_F) < -w_T \approx 1 \tag{124a}$$

$$w_{IT} = \frac{\varepsilon_0\gamma_\perp}{g_T^2\hbar\omega_0N_e\bar{N}_t}(\mathcal{T}c\frac{L_D}{A_D^{1/2}} + 21_L\gamma_F) < -w_T \approx 1. \tag{124b}$$

From equations (46) and (53), with (47) and (43), we notice that the dephasing and decay rates (93) and (94) strongly depend on the i-layer thickness $x_2 - x_0$. In figure 5a, we represent the decay rates $\gamma_\parallel^P, \gamma_\parallel^E, \gamma_\parallel^{EM}$ for the three dissipative couplings, with the phonons of the crystal vibrations, the conduction electrons and holes, and the free electromagnetic field, $\gamma_\parallel = \gamma_\parallel^P + \gamma_\parallel^E + \gamma_\parallel^{EM}$. We also represent the fluctuation rate (95) with the components (48) for the two neighboring conduction regions n and p, $\gamma_n^2 = \left[\gamma_n^{(n)}\right]^2 + \left[\gamma_n^{(p)}\right]^2$. In figure

5b, we represent the coupling coefficients for a longitudinal and a transversal structure. For all these coefficients, we get quasi-exponential variations with the i-layer thickness $x_2 - x_0$. We notice that, in these structures, the phonon decay rate $\gamma_{\|}^{P}$, which is unavoidable, dominates the electric decay rate $\gamma_{\|}^{E}$, which depends on the separation barriers, while the electromagnetic decay rate $\gamma_{\|}^{EM}$ is negligible. It is remarkable that the decay rate of a quantum

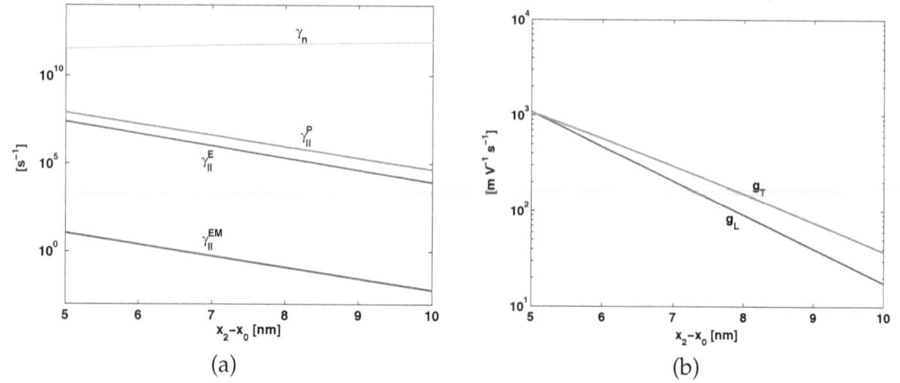

Fig. 5. (a) The dependence of the dissipative coefficients on the i-zone thickness; (b) The dependence of the coupling coefficients on the i-zone thickness, for a longitudinal and a transversal structure.

injection dot, with a value around $10^7 s^{-1}$, is significantly lower than the decay rates of other $GaAs$ structures, which are at least somewhere around $10^{12} s^{-1}$ (13). From figure 5b, we notice that, although the two coupling coefficients are calculated with completely different dipole moments, g_L with (42b)-(42c), and g_T with (42a), the values of these coefficients are approximately equal for small values of $x_2 - x_0$, and keep near values for thicker i-zones.

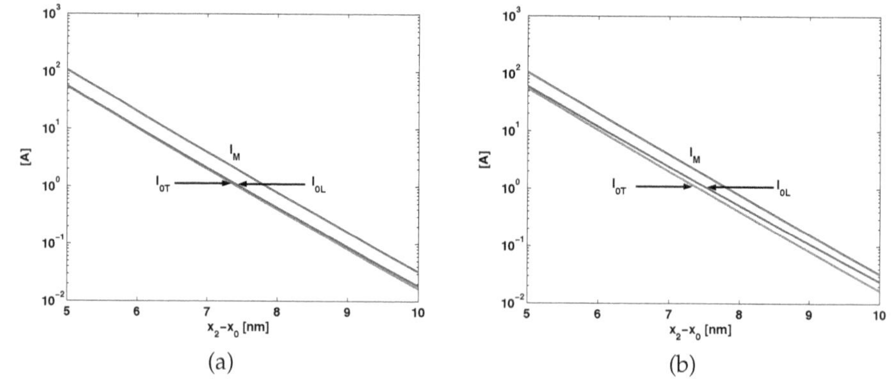

Fig. 6. The dependence of the threshold currents on the thicknesses of the i-zone for two values of the transmission coefficient of the output mirror: (a) $\mathcal{T} = 0.1$; (b) $\mathcal{T} = 0.5$.

From equations (121), we notice that the dissipative rates $\gamma_{\|}, \gamma_{\perp}$ and the coupling coefficient g_L (g_T), determine the threshold current I_{0L} (I_{0T}). In figure 6a we represent the dependence of these currents in comparison with the maximum current I_M, for two values of the output transmission coefficient $\mathcal{T}$. We notice that the operation condition $I_{0L}, I_{0T} > I_M$ is satisfied for both values of these coefficients. This property can be understood from the analytical expressions (121) or (124), having in view that the dephasing rate $\gamma_\perp$ and the square of the coupling coefficient g_L (g_T) are proportional to the square of the dipole moment, which means that the operation condition does not depend on this moment. Since the quantum

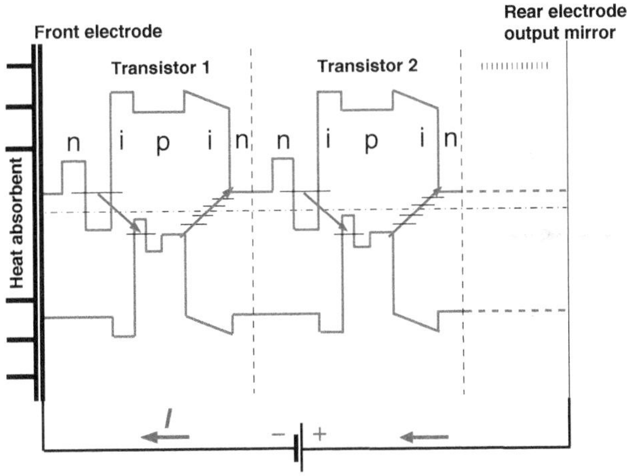

Fig. 7. Quantum heat converter, as a packet of super radiant transistors, thermally coupled to a heat absorbent. While a current I is injected in the device, an electromagnetic flow is obtained, mainly on the account of the heat absorption.

dot density N_e is determined by physical conditions, according to (25), the threshold current (121) can be controlled only by the number of super radiant transistors $\bar{N}_t$ in the structure. In our calculations we considered a number of superadiant transistors $\bar{N}_t = 1045$. While the heat propagates from the heat absorbent (see figure 7) throughout the semiconductor structure, a portion of this heat is absorbed by every super radiant transistor, producing a temperature decrease from the front electrode to the rear one. In figure 8a we represent the electric power and the radiation power as functions of the injected current, for a longitudinal and a transversal configuration of the device. A radiation power arises only when the injection current exceeds a threshold value. From (121a) and (121b), we notice that, due to the factor $\frac{L_D}{A_D^{1/2}}$ in the radiation term of the population inversion, the threshold current of a transversal device is lower than that of a longitudinal one. However, due the same factor at the denominator of (120b), the increase of the radiation power with the injection current is lower for a transversal device than for a longitudinal one. In figure 8b the total temperature variation in the semiconductor structure is represented. We notice that a rather high power of 200 W, that means 0.500 MW from an active area of 1 m^2, can be obtained at a rather low temperature difference of about 7 ^{0}C.

The radiation power of a transversal device becomes much higher by increasing the transmission coefficient from $\mathcal{T} = 0.1$ to $\mathcal{T} = 0.5$ and the transition dipole moment by diminishing the thickness of the i-zone from $x_2 - x_0 = 6.5\ nm$ to $x_2 - x_0 = 6\ nm$ as is represented in figure 9. In this case, the threshold current of the transversal device becomes significantly lower than that of the longitudinal one. The threshold current of the longitudinal device is sgnificantly lowered by decreasing the transmission coefficient from $\mathcal{T} = 0.5$ to $\mathcal{T} = 0.2$ as is represented in figure 10.

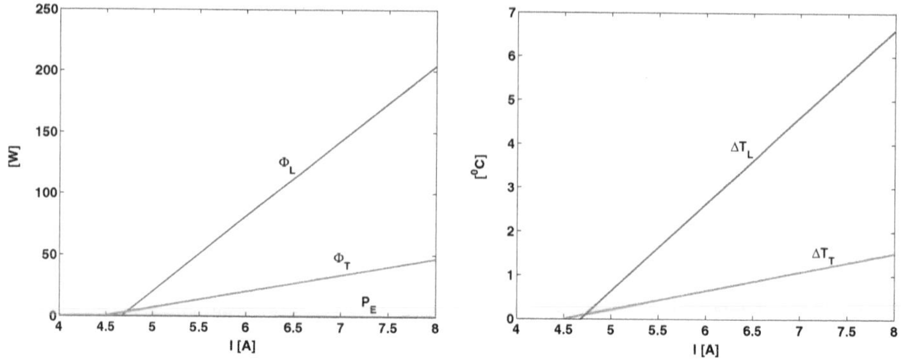

(a) (b)

Fig. 8. (a) The radiation powers Φ_L and Φ_T and the electric power P_E as functions of the injection current I, for $x_2 - x_0 = 6.5\ nm$, $\mathcal{T} = 0.1$, and $\gamma_F = 10^7\ s^{-1}$; (b) The temperature variations ΔT_L, ΔT_T as functions of the injection current I.

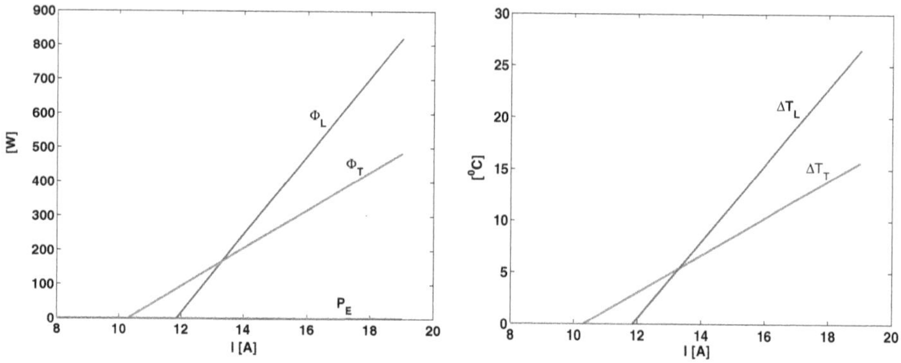

(a) (b)

Fig. 9. The radiation powers Φ_L and Φ_T, the electric power P_E, and the temperature variations ΔT_L, ΔT_T as functions of the injection current, for $x_2 - x_0 = 6\ nm$, $\mathcal{T} = 0.5$, and $\gamma_F = 10^7\ s^{-1}$.

It is remarkable that in the three cases represented in figures 8-10 the electric power dissipated in the device by the injection current I is much lower than the super radiant power. This is because, as one can notice also from (120), the super radiant power produced by the injected current corresponds to the high transition energy $\hbar\omega_0$ between the two zones n and p, while the power electrically dissipated by this current corresponds to a very low potential difference $U_c - U_{c1}$, necessary for carrying this current through the two rather thin highly conducting zones n and p (figure 7b). The difference between these two powers is obtained by heat absorption, when the electrons are excited from the lower potential of the p-zone to the higher

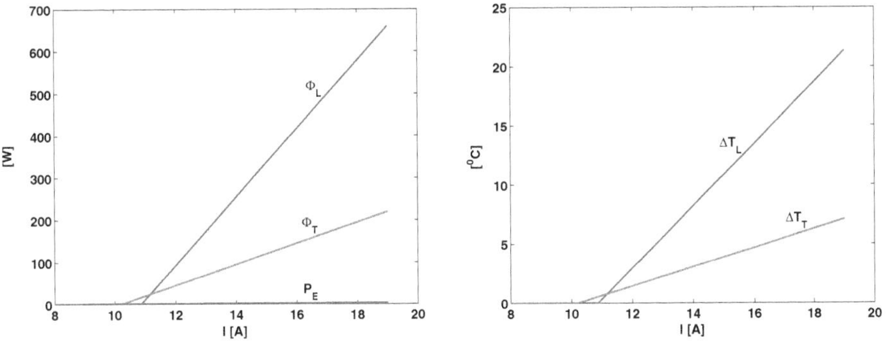

(a) (b)

Fig. 10. The radiation powers Φ_L and Φ_T, the electric power P_E, and the temperature variations ΔT_L, ΔT_T as functions of the injection current, for $x_2 - x_0 = 6\ nm$, $\mathcal{T} = 0.2$, and $\gamma_F = 10^7\ s^{-1}$.

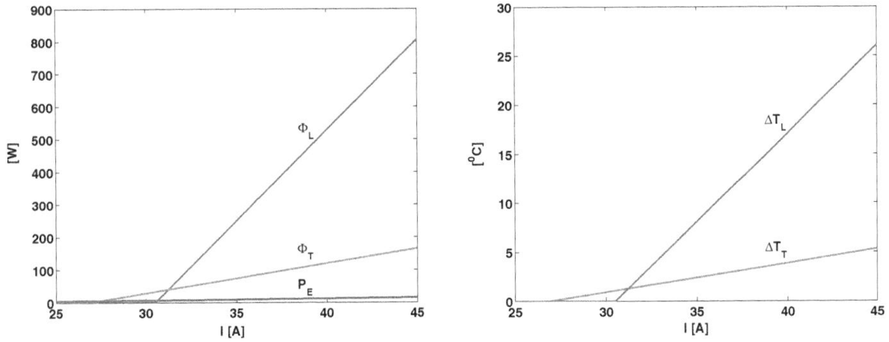

(a) (b)

Fig. 11. The radiation powers Φ_L and Φ_T, the electric power P_E, and the temperature variations ΔT_L, ΔT_T as functions of the injection current, for $x_2 - x_0 = 5.5\ nm$, $\mathcal{T} = 0.5$, and $\gamma_F = 10^8\ s^{-1}$.

potential of the n-zone of the base-collector junction. In figure 11 we consider a much larger decay rate of the electromagnetic field, $\gamma_F = 10^8\ s^{-1}$ instead of $\gamma_F = 10^7\ s^{-1}$, when the operation conditions (124) are also satisfied. In this case, we also obtain a high radiation power, but with a higher injection current, which, however, does not produce an important electrical power P_E, dissipated in the device.

We study the time evolution of a quantum heat converter, by solving the time dependent system of equations (110), for a step current injected at the initial $t = 0$, and a fluctuation that arises at a certain time $t > 0$. Non-Markovian fluctuations are time-evolutions of polarization, population and field due to the self-consistent field of the environment particles that, in our case, are the quasi-free electrons and holes in the conduction regions of the device. In figure 12, we represent the dynamics of a longitudinal device with a thickness of the i-zone $x_2 - x_0 = 5.5\ nm$ and a transmission coefficient of the output mirror $\mathcal{T} = 0.1$, while the threshold current is $I_{0L} = 24.1149\ A$ and the maximum current is $I_M = 46.0995\ A$. We consider a step current $I = 45\ A$ injected at time $t = 0$. In the Markovian approximation, the super radiant power $\Phi_L(t)$ of a longitudinal quantum heat converter is generated as in figure 12a, while the population $w(t)$ and polarization variables $u(t)$, $v(t)$ have the time-evolutions represented in figure 12b. The sudden jumps of the polarization variables in figure 12b, are detailed in

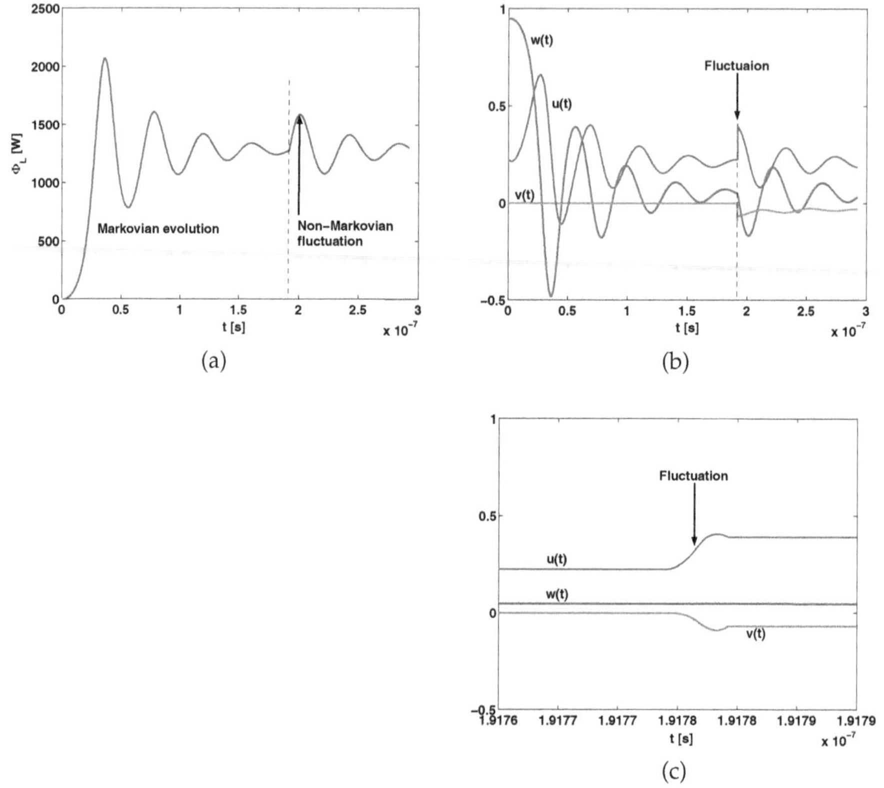

Fig. 12. Dynamics of a longitudinal super radiant device with $x_2 - x_0 = 5.5\ nm$ and $\mathcal{T} = 0.1$ when a step current of $I = 45\ A$ is injected in the device: (a) super radiant power; (b) polarization and population; (c) polarization fluctuation in a short timescale.

figure 12c, in a short timescale. At $t = 0$, the population increases from the equilibrium value w_T for the temperature T, to $w(0) = w_T + 2I/(eN_eA_D\gamma_{\parallel})$ and, after that, while the radiation field increases, the population decreases tending to an asymptotic value. With an appropriate choice of the phase of the initial polarization, $v(0) = 0$, $u(0)$ takes a value corresponding to the maximum value $-w_T$ of the Bloch vector, which is $u(0) = \sqrt{[w_T^2 - w^2(0)]/(2 - \mathcal{T})}$. In the Markovian approximation, the electromagnetic power is growing to a certain value, and after a short oscillation tends to the asymptotic value that according to (120a) is $\Phi_L = 1.2843 \times 10^3\ W$. However, in the non-Markovian approximation, random fluctuations of the polarization, population, and field arise. These fluctuations are described by the time integrals in the polarization equations (110a) and (110b) depending on the time-dependent phase term $\phi_n(t')$, with a mean-value of the fluctuation time $\tau_n = 1/\gamma_n$. From figure 5a, we notice that the fluctuation rate γ_n is four orders higher than the decay rate $\gamma_{\parallel}$, corresponding to the timescale of the Markovian processes. In equations (110a) and (110b), we take a positive fluctuation with a duration tn $\tau_n = 2.6305 \times 10^{-12}$s, followed by a negative one with the same duration. In figure 12c such a fluctuation is represented in a short timescale, specific to the non-Markovian fluctuations, while in figures 12a and 12b it is represented in a long timescale specific to the Markovian processes. We notice that, while the polarization variables $u(t)$ and $v(t)$, which depend on the transition elements of the density matrix, undertake considerable variations in a fluctuation time, for population and super radiant field these variations only initialize long time oscillations.

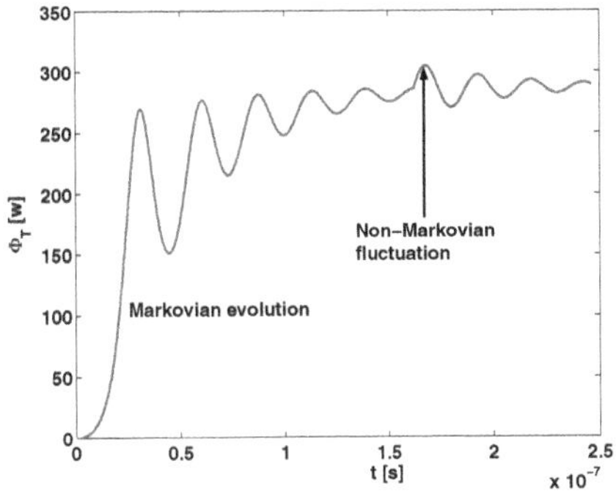

(a)

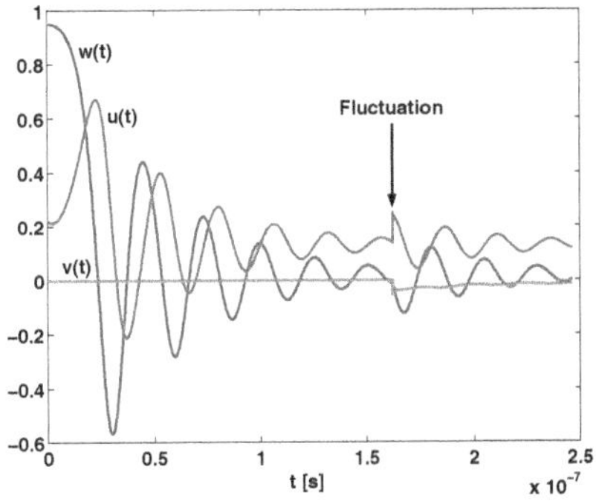

(b)

Fig. 13. Dynamics of a transversal super radiant device with $x_2 - x_0 = 5.5\ nm$ and $\mathcal{T} = 0.1$ when a step current of $I = 45\ A$ is injected: (a) Superradiant power; (b) Population and polarzation.

In figure 13, we represent the dynamics of the transversal device with the same semiconductor structure and injected current, while the threshold current takes a lower value $I_{0T} = 23.4528\ A$. This decrease of the threshold current for a transversal device, in comparison with a longitudinal one with the same semiconductor structure, is obtained due to the field amplification on the longer path of the field propagation in the plane of the quantum dot layers, which is described by the term $c\mathcal{T}\frac{L_D}{A_D^{1/2}}$ in equations (121). However, this small difference is not very significant, since, according to equation (83), a longer propagation path leads also to a higher decay rate of the field, i.e. to an increase of the dissipative term $1_L\gamma_F$. We notice that, while the radiation power is lower, this device is much less sensitive to the thermal fluctuations described by the non-Markovian term. This decrease of the radiation power for a transversal device, in comparison with a longitudinal one with the same semiconductor structure, is obtained due to the factor $\frac{A_D^{1/2}}{L_D}$ at the denominator of equation (120b). An essential advantage of a transversal quantum heat converter, in comparison with a longitudinal one, consists in injection electrodes as zero-transmission mirrors, i.e. these electrodes are thick metalizations, providing an uniform current injection in the device. For an uniform current injection, a longitudinal quantum heat converter needs a special output structure, as a high transmission output Fabry-Perot cavity (4). Although for a transversal device we obtained a lower radiation power than a longitudinal one, it could be advantageous for some applications: for instance to obtain a powerful radiation device, as a stack of many transversal quantum heat converters. Another application could be an electric generator with the three semiconductor devices of the system, transversal quantum heat converter, quantum injection system, and total quantum injection system, in the same plane, eventually stuck on the same pad.

In figures 12 and 13, we considered a positive fluctuation followed by a negative one, which means an integration over a first interval of time $\tau_n = 1/\gamma_n$ with a phase $\phi_n = 0$ followed by an integration over a second interval of time τ_n with a phase $\phi_n = \pi$ in the polarization equations (110a) and (110b).

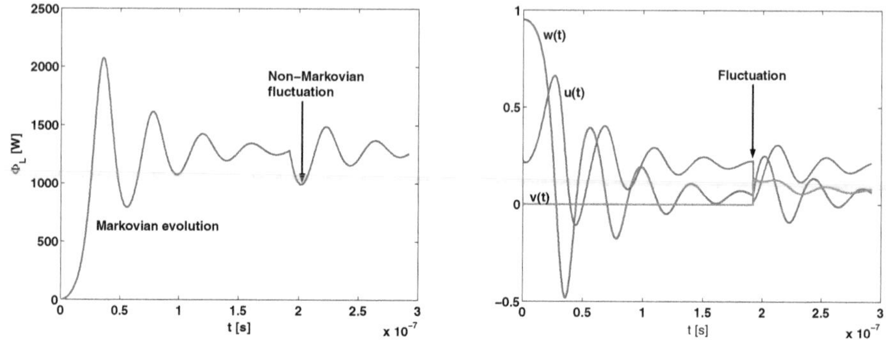

(a) (b)

Fig. 14. Dynamics of a longitudinal super radiant device with a negative fluctuation ($\phi_n = \pi$), followed by a positive one ($\phi_n = 0$).

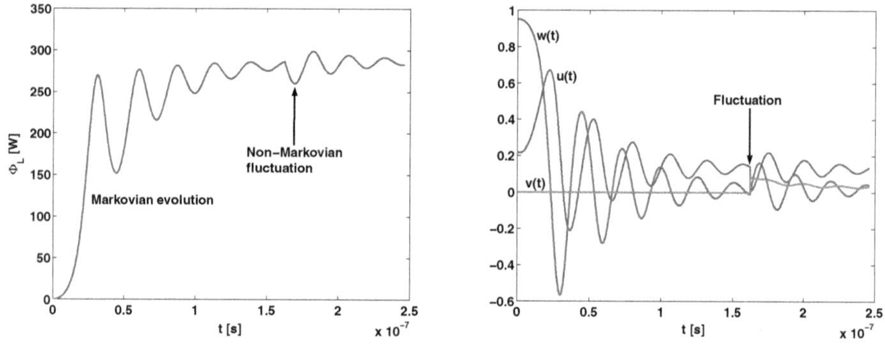

(a) (b)

Fig. 15. Dynamics of a transversal super radiant device with a negative fluctuation ($\phi_n = \pi$), followed by a positive one ($\phi_n = 0$).

Changing the phases of the fluctuations, i.e. taking a negative fluctuation followed by a positive one (figures 14 and 15), we get similar evolutions but with opposite signs. Obviously, the realistic evolution of a device is the result of the random phases ϕ_n, arising during the whole evolution of the system. Thus, the system dynamics takes a noisy form, with the polarization undertaking rapid variations during a fluctuation time, while the population and the super radiant field are only initialized into slow oscillations.

7. Conclusions

We presented a new kind of quantum dots, with a quantum well for electrons in the n-region of an n-i-p heterostructure, and a quantum well for holes in the p-region of this structure. These quantum wells are separated from the two n and p conduction regions by transparent potential barriers, and separated from one another by the potential barrier of the i-region. Such a quantum dot, we call "quantum injection dot", can be compared with a conventional quantum dot, as a small semiconductor region with a narrower forbidden band in a much larger i-region of an n-i-p semiconductor structure. Quantum injection dots have mainly been conceived for conversion of environmental heat into usable energy, while the conventional quantum dots are mainly used in information technology. A quantum injection dot differs in many respects from a conventional quantum dot, where the quantum well for holes is placed under the quantum well for electrons, as conduction and valence bands of the same semiconductor region: (1) while a quantum injection dot is supplied with electrons and holes from the two conduction regions by quantum tunneling through the n and p separation barriers, without any energy increase, a conventional quantum dot is supplied with electrons and holes only by providing a substantial energy, necessary to raise these electrons and holes from the n and p conduction regions to the conduction and valence bands of the i-region, from where they fall in the two potential wells of the quantum dot; (2) a quantum injection dot provides the electron transfer from the n-region to the p-region only by quantum transitions, while the electron transfer provided by a conventional quantum dot includes additional transport processes from the two n and p regions to the i-region where this quantum dot is located; (3) a quantum injection dot is a one-electron normalized two-level quantum system, while a conventional quantum dot is a confinement semiconductor region where many electrons and holes are simultaneously present to provide a larger probability for the super radiant transitions. In comparison with a conventional quantum dot, a quantum injection dot is much less dissipative, and, due to its simpler structure, enables a much higher packing degree in a semiconductor structure.

We studied a system of quantum injection dots by using the available means of quantum mechanics: (1) we calculated wave-functions, dipole moments, and eigenvalue equation for energy; (2) we derived equations for the dissipative super radiant dynamics of the system; (3) we obtained analytical coefficients depending only on physical characteristics and universal constants, without any phenomenological parameter. In the dynamics of a quantum dot system, we distinguish five dissipative processes: (1) correlated transitions with phonons of the crystal lattice vibrations, which is the dominant dissipation process (2) correlated transitions with quasi-free electrons and holes in the conduction regions, (3) correlated transitions with the quasi-free electromagnetic field, which are negligible, (4) transitions stimulated by the thermal fluctuations of the self-consistent field of the electrons and holes in the conduction regions, (5) non-Markovian processes induced by these fluctuations. However, we found that the fluctuation time is much shorter than the decay time, which means that the system is in fact quasi-Markovian, while the non-Markovian fluctuations manifest themselves only as a noise. For the propagation of the electromagnetic field throughout the semiconductor structure, by taking into account the dissipative interaction with the quasi-free electrons an holes in the conduction regions, we obtained an analytical expression of the field decay rate as a function of effective masses, frequency, and propagation path.

We studied a device converting environmental heat into coherent electromagnetic energy in two versions: (1) longitudinal quantum heat converter, with the electromagnetic field propagating in the direction of injected current, i.e. emerging from the surface the semiconductor structure, and (2) transversal quantum heat converter, with the electromagnetic field propagating in a perpendicular direction to the injected current, i.e. emerging from a lateral surface of the semiconductor structure. We found operation conditions for the physical characteristics of the semiconductor structure. We studied the dependence of the dissipative rates, coupling coefficients, and threshold currents as functions of the i-region thickness, which enables the control of these quantities in a large field of values. We found that the operation conditions do not depend on the i-layer thickness. When this thickness is decreased, the injected current and the corresponding super radiant power increase. However, these quantities of interest can not be indefinitely increased, especially due to the temperature variation induced by the heat propagation throughout the structure, which tends to produce an atomic detuning of the quantum dot layers. We highlighted the super radiant dissipative dynamics under a step current injection, and thermal fluctuations of the conduction electrons and holes.

8. References

[1] Kent D. Choquette and John F. Klem, Long wavelength vertical cavity surface emitting laser, US 6,931,042 B2 (US Patent Office, Aug. 16, 2005).

[2] Ashkan A. Arianpour, James P. McCanna, Joshua R. Windmiller, Semeon Y. Litvin, Photovoltaic device employing a resonator cavity, US 2008/0128023 A1 (US Patent Office, Jun. 05, 2008).

[3] E. Stefanescu, Master equation and conversion of environmental heat into coherent electromagnetic energy, Prog. Quantum Electron. 34 (2010) 349-408.

[4] Eliade Stefanescu, Lucien Eugene Cornescu, Longitudinal quantum heat converter, US 20090007950 (US Patent Office, 01-08-2009), http://www.freepatentsonline.com/y2009/0007950.html.

[5] Eliade Stefanescu, Lucien Eugene Cornescu, Transversal quantum heat converter, US 20100019618 (US Patent Office, 01-28-2010), http://www.freepatentsonline.com/y2010/0019618.html.

[6] Eliade Stefanescu, Lucien Eugene Cornescu, Quantum injection system, US 20090007951 (US Patent Office, 01-08-2009), http://www.freepatentsonline.com/y2009/0007951.html.

[7] E. Stefanescu, W. Scheid, and A. Sandulescu, Non-Markovian master equation for a system of Fermions interacting with an electromagnetic field, Ann. Phys. 323 (2008) 1168-1190.

[8] T. Fließbach, Statistische Physik, Lehrbuch zur Theoretischen Physik IV (Elsevier, München 2007).

[9] G. W. Ford, J. T. Lewis, and R. F. O'Connell, Master Equation for an Oscillator Coupled to the Electromagnetic Field, Ann. Phys. 252 (1996) 362-385.

[10] V. M. Axt and S. Mukamel, Nonlinear Optics of semiconductor and moleclar nanostructures, Rev.Mod..Phys. 70 (1998) 145-287.

[11] Eliade Stefanescu, Dynamics of a Fermi system with resonant dissipation and dynamical detailed balance, Physica A 350 (2005) 227-244.

[12] Haward Carmichael, an Open Quantum System Approach to Quantum Optics, in Lecture notes in Physics (Springer Verlag, Berlin 1993).

[13] Hartmut Haug and Antti-Pekka Jauho, Quantum Kinetics in Transport and Optics of Semiconductors (Springer-Verlag, Berlin, Heidelberg, New York, 1998).

[14] Günter Mahler and Volker A. Weberruß, Quantum Networks - Dynamics of Open Nanostructures (Springer-Verlag, Berlin, Heidelberg, New York, 1995).

2

Electron Transport Properties of Gate-Defined $GaAs/Al_xGa_{1-x}As$ Quantum Dot

Dong Ho Wu and Bernard R. Matis
Naval Research Laboratory, Washington, DC, USA

1. Introduction

In this chapter we explore transport properties of lateral, gate defined quantum dots in $GaAs/Al_xGa_{1-x}As$ heterostructures. The term "quantum dot" as defined here refers to small regions of charge carriers within a 2-dimensional electron gas (2DEG), established via electrically biased surface gates used to isolate the charge carriers from the rest of the 2DEG, which are confined to length scales on the order of nanometers. While there are several other forms of quantum dots, including colloidal and self-assembled dots, in this chapter, however, we consider only gate defined quantum dots.

Recent advancements in the research areas of quantum dot (QD) and single electron transistors (SET) have opened up an exciting opportunity for the development of nanostructure devices. Of the various devices, our attention is drawn in particular to detectors, which can respond to a single photon over a broad frequency spectrum, namely, microwave to infrared (IR) frequencies. Here, we report transport measurements of weakly coupled double quantum dots, fabricated on a GaAs/AlGaAs 2-dimensional electron gas material, under the influence of external fields at 110GHz. In this experiment, transport measurements are carried out for coupled quantum dots in the strong-tunneling Coulomb blockade (CB) regime. We present experimental results and discuss the dependence on quantum dot size, 2DEG depth, fabrication techniques, as well as the limitations in developing a QD photon detector for microwave and IR frequencies, whose noise equivalent power (NEP) can be as sensitive as 10^{-22} $W/Hz^{1/2}$.

The charging energy E_C of a quantum dot is the dominant term in the Hamiltonian and is inversely related to the self capacitance of the dot C_{dot} according to $E_C = e^2/C_{dot}$. The temperature of the charge carriers within the 2DEG must be kept below a certain value, namely K_BT, so that the thermal energy of the electrons does not exceed the charging energy E_C of the dot. Keeping the temperature below the K_BT limit prevents electrons from entering or leaving the dot at random, thereby allowing one to control the number of electrons in the dot. In order to raise the operating temperature T of the single photon detector we must also raise the charging energy E_C, which is accomplished by decreasing C_{dot}. Since C_{dot} is directly related to the dimensions of the quantum dot our focus was directed at decreasing the overall size of the quantum dots. For smaller gate-defined quantum-dots the inclusion of shallower 2DEG is necessary.

However the experiments that we carried out to determine the effect of 2DEG depth on lateral gate indicated that leakage currents within a GaAs/AlGaAs heterostructure increased dramatically as the 2DEG depth became shallower. At this moment the leakage current in shallower 2DEG materials is one of the most significant technical challenges in achieving higher operating temperature of the single photon detector.

2. Gate-defined quantum-dots

2.1 2-dimensional electron gas (2DEG)

In contrast to colloidal and self-assembled quantum dots, which are physically well defined small dots separated from other media, the gate-defined quantum dot means charge carriers (either electrons or holes) confined in a small region, which is formed by electrically biased gates surrounding the region. First the charge carriers are confined within the so-called 2-dimentional electron gas (2DEG) material, which is typically made of GaAs/AlxGa1-xAs heterostructure. Figure 1 shows an example of the vertical profile of a 2DEG heterostructure and the corresponding energy-band diagram.

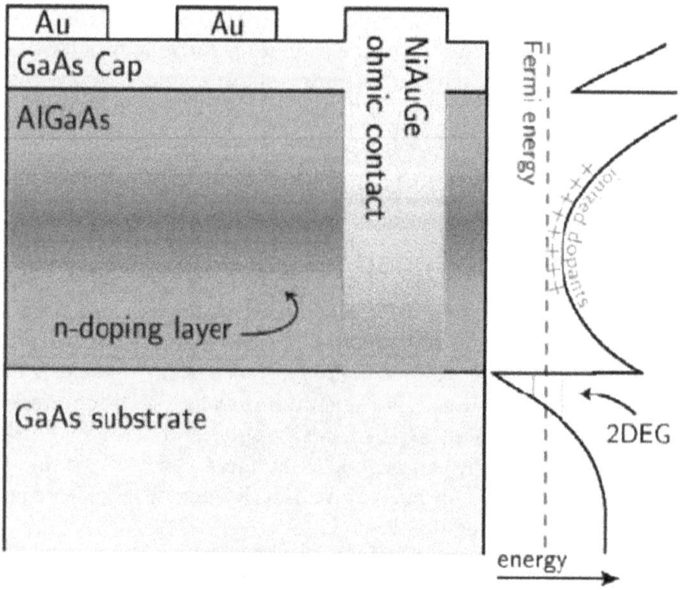

Fig. 1. An example of a the vertical profile of the GaAs/AlGaAs 2DEG heterostructure (Left) and the corresponding energy-band diagram (Right).[Ref.1] Included in the vertical profile are the patterned metallic surface gates (Au) that define the nanostructure devices and the ohmic contacts (NiAuGe), which when annealed penetrate through the top layers of the structure and make electrical contact to the 2DEG.

One layer of AlGaAs in particular contains a region of n-type dopants, either as a single layer (so called δ-doped layer) or homogeneously doped (modulation doping). In our case the dopants are Si atoms, which are deposited within an AlGaAs layer and are separated away from the 2DEG by an undoped AlGaAs spacer layer. The spacer layer which is typically 10 - 100 nm thick is to minimize the effect of scattering from the dopants; the various layer thicknesses can be modulated to vary the properties of the 2DEG.

At low temperature each Si atom produces a free electron as the electrons become thermally ionized [2, 3]. The offset in the conduction bands between GaAs and AlGaAs results in each free electron migrating toward the energetically favorable GaAs substrate layer. The charge carriers still feel the electrostatic attractive forces from the ionized donor atoms, however, and ultimately become trapped at the interface between the GaAs layer and an undoped AlGaAs layer. These trapped electrons are called 2-dimentional electron gas (2DEG). As the temperature decreases to very cold temperatures (< 1 K) the thermal smearing of the vertical "z" profile of the 2DEG becomes less pronounced as the electrons occupy only the lowest energy levels up to the Fermi Energy, resulting in a very clean glass of electrons confined within a 2-dimensional plane.

Because the lattice constants of GaAs and AlGaAs are only slightly different (~7% mismatch) the interface is essentially defect free. Because of this defect free interface and the separation of the 2DEG from the Si dopants 2DEG can have high electron mobility, $\mu_e \sim 10^5$ - 10^7 cm^2V/s, and long mean free paths, ℓ ~ 1 - 1000 nm. These properties are often exploited for quantum dot devices which require coherent and ballistic electron transport behavior.

2.2 Gate-defined quantum-dot

The local electron density within the 2DEG can be manipulated by placing electrodes on GaAs cap surface, as shown in Figure 1. When a negative bias voltage is applied to the electrodes the negatively charged gates repel electrons in the 2DEG. If the negative field strength is strong enough all electrons beneath the electrodes will be fully depleted. The electrodes can be lithographically arranged over an area with a certain geometric shape, such as a circular disk. An example is shown in Figure 2.

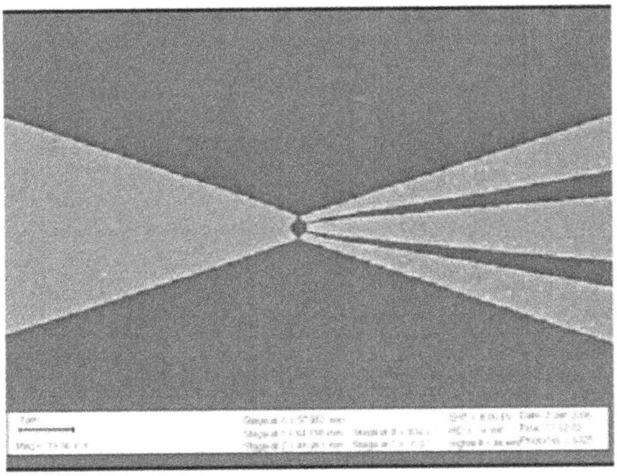

Fig. 2. An example of gate-defined quantum dot. The four gates are fabricated by the e-beam lithography, and surround a circular disk in the center, which becomes the quantum dot when the four gates are negatively biased. The lithographically defined circular-disk size is about 250 nm in diameter. However the actual size of the quantum dot depends on the strength of the bias voltage. Also the shape of the quantum dot depends on how the bias voltage is applied to each gate, and it can be deviated from the disk shape.

With strong enough negative bias voltages applied to the electrodes, electrons confined inside the area (e.g. circular disk) will be isolated from the rest of the electrons in the 2DEG. These isolated electrons in the area (e.g. disk) are called the gate-defined quantum-dot and the rest of the electrons in the 2DEG are called the reservoir. While the configuration of the gates influence the overall shape and determine the maximum size of quantum dot, actual shape and size of the isolated electron puddle (i.e. quantum dot) are dependent upon the strength of the negative bias voltage applied to each gate. The gap between gates is often called the quantum point contact (QPC) and is typically a few tens of nm. It pinches off electrons when the negative bias voltage is applied to the gates. The QPCs can individually tune the potential barriers between the dot and the reservoirs, and hence control the tunneling rate from the leads and the dot. The transport through a quantum dot can be divided into two categories, "open" and "closed," depending upon the conductance of the QPCs. For strong coupling, the conductance $G > e^2/h$, where each QPC passes one or more modes, the dot is considered "open." In an "open" dot electrons are classically allowed to travel through the dot from one reservoir to the other. For weak coupling, $G < e^2/h$, where each QPC is set to pass less than one fully transmitting mode, the dot is considered "closed." If the bias voltage is large enough the electrons near the quantum point contacts are completely pinched off, making the quantum dot to be "closed" or isolated from the reservoir. However electrons can tunnel through the "closed" quantum dot, allowing very small currents. Therefore the conductance is orders of magnitude lower than that of 2DEG.

Quantum Dots are often referred to as zero-dimensional systems, as the electronic motion is entirely restricted in all directions. The size of quantum dot is typically smaller than a few hundred nano-meters in diameter. As electrons are confined within such a length scale the spacing between each quantum energy level of the electron becomes very pronounced when the temperature of the quantum dot drops below 4.2 K and thermal smearing is very much reduced. Because of these well defined quantum energy levels of electrons within the quantum dot the tunneling currents through the quantum dot exhibit the characteristic Coulomb blockade effect.

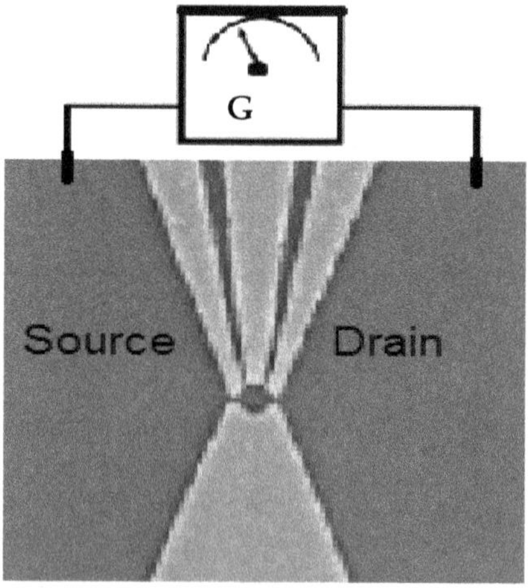

Fig. 3. The conductance G (tunneling current) is measured across a quantum dot device, from the source to drain reservoirs.

The Coulomb blockade occurs due to the fact that conduction through the dot is prevented for most settings of the electrostatic gates simply because the available energy levels within the dot are not in alignment with the Fermi levels in the source and drain (i.e. reservoir). An electron is unable to tunnel into the dot if the energy needed to add an additional electron (from N to N + 1 electrons) is above the Fermi Energy in the source. Similarly an electron is unable to tunnel out of the dot if the energy carried by that electron is less than the Fermi Energy in the drain. If electrons have enough energy to tunnel into the dot and then tunnel out of the dot, the measured conductance displays a large conductance spike, which indicates tunneling currents. This is known as a Coulomb blockade peak.

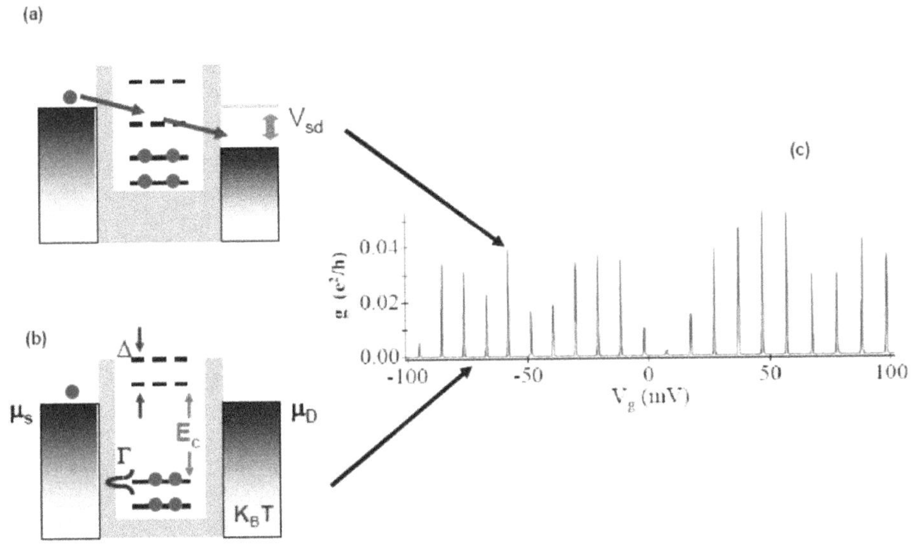

Fig. 4. (a) and (b): Coulomb blockade energy diagrams. (a) The energy level in the quantum dot allows the electron to tunnel through the dot. The tunneling currents produce a Coulomb blockade peak. (b) The energy level in the quantum dot is higher than the electron's energy so that the electron cannot tunnel through the quantum dot. Therefore the conductance is zero, and there is no Coulomb blockade peak in (c). As a gate voltage is swept the energy levels of the dot are raised and lowered, making the quantum energy levels to move in and out of alignment with the chemical potentials (μ_S and μ_D) of the source and drain, resulting in large spikes in the conductance. The interval and the sharpness of Coulomb blockage peaks are determined by the quantum energy level spacing Δ and the finite thermal broadening Γ for a given temperature T.

For the tunneling currents and the Coulomb blockade five separate energy parameters need to be considered, including the source-drain voltage V_{sd}, the chemical potentials of source μ_S and drain μ_D, the charging energy E_C and the thermal energy of charge carriers K_BT. For the conductance measurement a small source-drain voltage Vsd, which is typically limited to be less than a few μV so as not to impart energy to the electrons greater than the thermal energy, is held across the dot. The source-drain voltage results in the chemical potential difference between the chemical potentials of source and drain so that $eV_{sd} = \mu_S - \mu_D$. The charging energy E_C is an additional Coulomb energy that is needed to add an additional electron to the quantum dot, and can be expressed as

$$E_C = e^2/2C_{dot}. \quad (1)$$

Here C_{dot} is the self-capacitance of the quantum dot. At temperature T an electron has the thermal energy K_BT. If the thermal energy becomes comparable or larger than the charging energy it causes the electron randomly to tunnel through the quantum dot, and also results in a thermal broadening larger than the energy level spacing Δ. Then the quantum dot will not be functional, as the electron is no longer controllable by the gate bias voltage. Hence it is very important to keep the quantum dot at very low temperatures so that its thermal energy is well below the charging energy (i.e. $E_C > K_BT$).

The quantum dot can manipulate the flow of an individual electron by controlling the gate bias voltage. As shown in Figure 5, a quantum dot with a capacitively coupled gate can be used as a single electron transistor. The bias voltage applied to the gate raises or lowers the energy level of the dot so that each single electron can tunnel through the quantum dot. Such a device is called a single electron transistor. We utilized the single electron transistor for our quantum dot single photon detector.

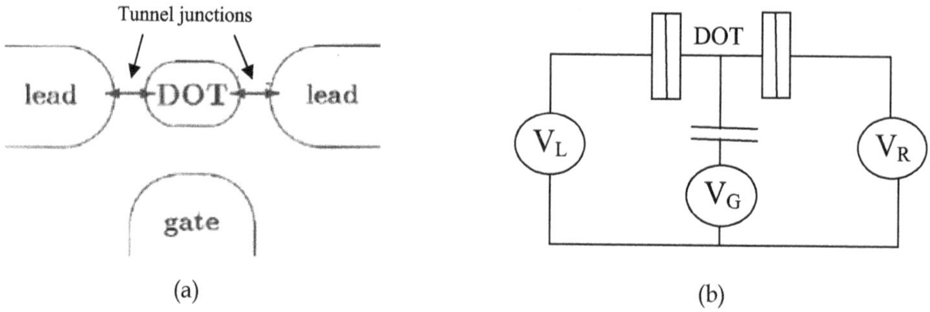

Fig. 5. (a) A schematic diagram of a single electron transistor made of a quantum dot. (b) An equivalent circuit for a single electron transistor.

3. A single photon detector based on coupled double quantum dots

The quantum energy levels as well as the level spacing Δ can be adjusted by controlling the physical parameters of the quantum dot. A photon can change the energy level of a quantum dot, which leads to electron tunneling through the quantum dot. This is known as photon assisted tunneling in a quantum dot. In 2000 Komiyama and his coworkers exploited this property and developed a detector, which can detect a single photon at far-infrared frequencies. The quantum dot size that they used in the experiments was about 500 nm in diameter fabricated on a 100 nm thick 2DEG substrate. Their large quantum dots resulted in a large self capacitance C_{dot} and a small charging energy $E_C = e^2/2C_{dot}$. Hence their detectors had to be operated at 100 mK or below, which made the detector less practical. We attempted to adopt their quantum dot detector technology and raise the charging energy and the operating temperature by reducing the quantum dot size.

3.1 A shallow 2-dimentional electron gas for quantum dot single photon detector

For the design of our quantum dot detector we have performed numerical calculations. The calculations indicate that our detector should be fabricated on a shallow 2 dimensional electron gas (2DEG) substrate in order to achieve an operating temperature above 4 K. [4] As shown in Figure 6 the depth of the 2DEG that we have used for the detector is about 40 nm. This is much shallower in comparison with the Komiyama group's 2DEG, which was buried approximately 100 nm beneath the un-doped GaAs cap layer.

The high-mobility GaAs/$Al_{0.24}Ga_{0.76}As$ heterostructure crystal was grown by molecular beam epitaxy in the [001] direction. The heterostructure layers were deposited on an n-type GaAs substrate, carried a 5000Å thick GaAs buffer layer, a non-inverted heterostructure (500 Å thick GaAs/ 140Å thick $Al_{0.24}Ga_{0.76}As$), a δ-doped barrier layer (250 Å thick $Al_{0.24}Ga_{0.76}As$), and a δ-doped GaAs cap layer (10 Å thick). The silicon n-type dopants (level $6\times10^{18}/cm^3$) provide the excess charge carriers (target value was $6\times10^{11}/cm^2$ at room temperature), which constitute a 2 dimensional electron gas (2DEG) at the hetero-interface 400 Å below the wafer surface and 140 Å from the dopant atoms.

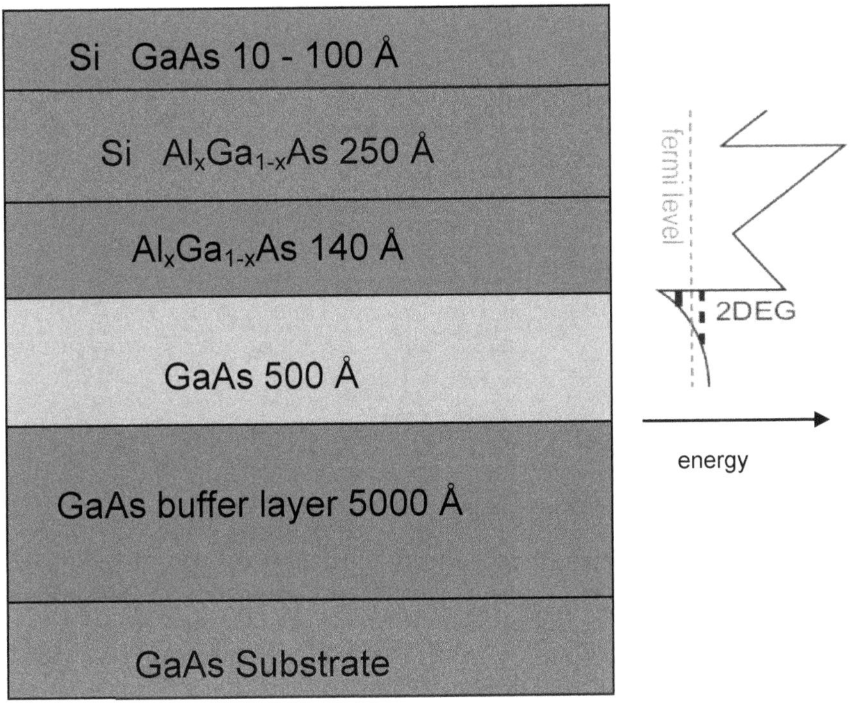

Fig. 6. The 2DEG structure we have used for our double-dot detector. For this work the 2DEG depth was a shallow 40 nm beneath the surface of the wafer. Here, x = 0.24.

For the characterization of 2DEG as well as for the quantum dot device good ohmic contacts should be made on the GaAs cap layer, as illustrated in Figure 1. A good ohmic contact has a non-zero internal resistance R_c that obeys Ohm's law for all current densities of interest. The contact should work at the lowest temperatures reached in quantum dot experiments where thermionic currents are negligible, but tunnel currents are allowed [5-7]. Fabrication of good ohmic contacts is not always trivial. The standard process includes depositing metals onto the surface and then annealing them into the wafer in order to make electrical contact to the 2DEG. We have used GaAs/AlGaAs heterostructures with several different 2DEG depths ranging from a shallow 40 nm to a deeper 160 nm. In each case a separate ohmic recipe had to be developed.

After several steps of cleaning procedures we carried out acid wet etch in order to remove any GaAs oxide layer that has formed on the surface. [8] Then we performed metal deposition on the 2DEG substrate. With the following recipes we have achieved low resistance, good ohmic contacts at cryogenic temperatures.

For shallow 2DEG	For 160nm deep 2DEG
1. 5 nm Ni	1. 5 nm Ni
2. 40 nm Ge	2. 125 nm Ge
3. 80 nm Au	3. Wait 30 min.
4. Wait 30 min.	4. 250 nm Au
5. 35 nm Ni	5. 50 nm Ni
6. 30 nm Au	6. Wait 30 min.
	7. 50 nm Ge
	8. 100 nm Au
	9. 35 nm Ni

Table 1. Ohmic contact recipes for two different 2DEG depths that have given low resistance (<100 Ω) at low temperature when annealed above the eutectic point.

The first Ni layer acts as a wetting layer and enhances the uniformity of the contacts; 5 nm is enough as this layer should not be thick. Otherwise it may prevent the other elements from penetrating into the wafer. The 2:1 ratio of Au:Ge forms a eutectic mixture, which is the ratio of two substances with the lowest melting point (a 2:1 ratio is essentially 88% Au and 12% Ge by weight with the melting point of this eutectic at ~ 380 °C). Each metal was evaporated one at a time. The second Ni layer acts as a barrier for the top layers of metals. The metalized 2DEG substrate is then submersed in Acetone for liftoff, and then rinsed with IPA and DI-H_2O. Finally it was dried by blowing dry N_2 gas.

In order to make electrical contact to the 2DEG the metals must be annealed into the substrate after the liftoff process. For the annealing we used AS-One 150 Rapid Thermal

Annealer (RTA) from ANNEALSYS with the following annealing procedure: Start with a ramp to 260 °C at 10% power to drive off any moisture from the chip. The pyrometer target is set at 510 °C for 100 seconds for the 160 nm deep 2DEG's, and 450 °C for 100 seconds for the 40, 43, and 90 nm deep 2DEG's.

The resulting R_c resistances for each contact are on the order of tens of kΩ at room temperature and decrease to a value on the order of kΩ at 4.2 K for the 160 nm deep 2DEG. For the shallower 2DEGs the contact resistances are even lower; they are on the order of kΩ at room temperature.

Low temperature measurements for the ohmic contacts and the 2DEG mobility characterizations are carried out using a Physical Property Measurement System (Quantum Design). The system is also equipped with a 9 T superconducting magnet and is capable for carrying out our Hall Effect measurements. Figure 7 shows the temperature dependence of both good and poor Ohmic contacts that were measured at zero field.

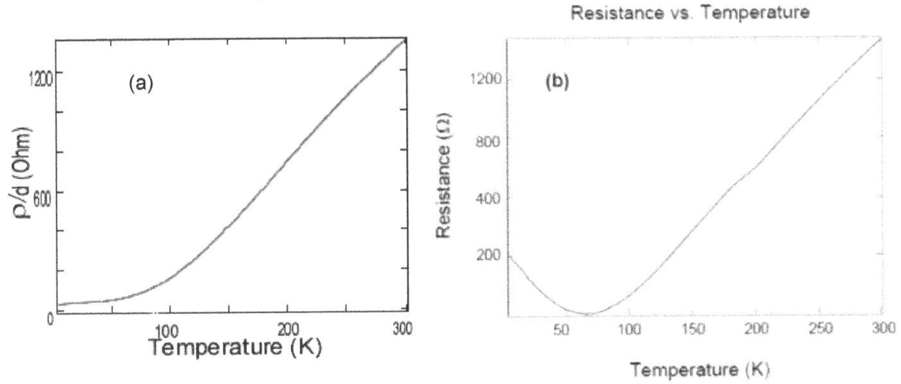

Fig. 7. Temperature dependence of resistance for (a) a good ohmic contact and (b) a poor ohmic contact. Preferably, the contact resistance is as low as possible at low temperatures.

After the success in Ohmic contact fabrication the 2DEG was characterized by measuring the Hall properties of micron-size Hall bars, which were fabricated on the 2DEG material. A standard Hall bar geometry, which is shown in Figure 8, is defined by wet etching and the metallic electrodes and ohmic contacts are patterned via optical lithography. Hall measurements reported in this paper were taken on a 50 μm wide Hall bar with a 700 μm distance between longitudinal taps. Electrical contact is made with the 2DEG by lithographically patterned Ni-Au-Ge Ohmic contacts, which when annealed at temperatures above 400 degrees Celsius provide for low resistive transport into and out of the 2DEG at cryogenic temperatures.

Two different Hall bars were fabricated, with and without an overlaying Si_3N_4 (silicon nitride) dielectric layer, which was tested to shield the 2DEG along the mesa edge from unwanted field effects caused by voltage biased leads. For the characterization of ohmic contacts we used a standard Van der Pauw experimental configuration. As shown in Figure 6, the resistivity decreased with temperature monotonically indicating the correct Ohmic contact behavior.

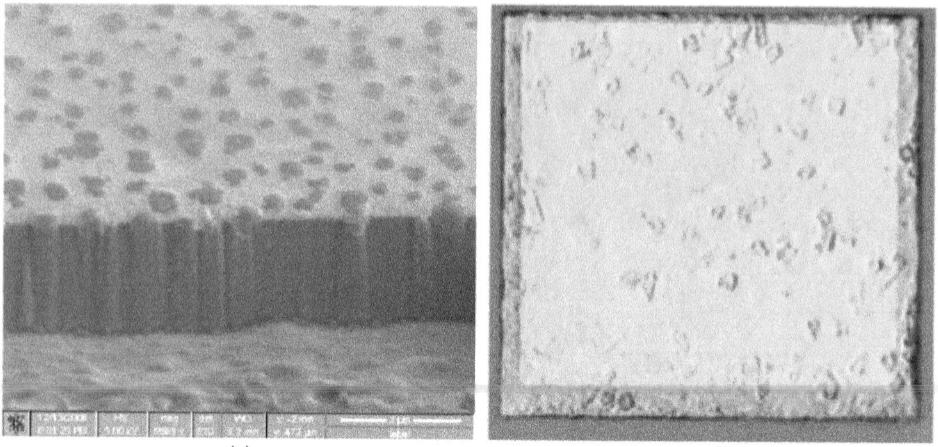

(a) (b)

Fig. 8. SEM and optical images taken of an annealed, low resistance ohmic contact. (a) The cross-sectional SEM image, taken after a Focused Ion Beam (FIB) cut into the contact, shows the puncturing of the deposited metal into the host GaAs/AlGaAs wafer. (b) An optical micrograph of a contact measured to have less than 1 kΩ of resistance at 4.2 K. The dimensions of the contact are 200 μm x 200 μm. The smaller gold square is additional metal deposited during the last optical lithography step (large gate pads) to help in wire-bonding.

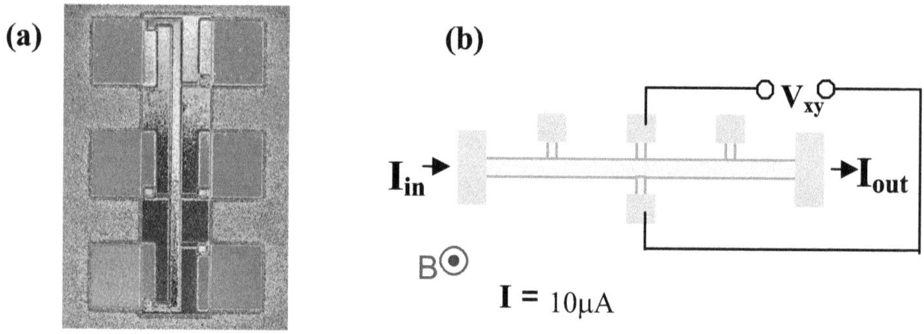

Fig. 9. (a) Optical micrograph of a micron-size Hall bar fabricated on the 2DEG material, and (b) the schematic diagram of the Hall bar and its characterization. Note that a magnetic field *B* perpendicular to the Hall bar is applied during the characterization, and a 10 μA current was used.

When a magnetic field is applied to 2DEG, electrons moving within the 2-dimensional system experience a Lorentz force that pushes them into circular orbits. Since in the 2-dimensional system only certain orbits (or energy states) are quantum mechanically allowed, the energy levels of the circular orbits are quantized, just as in the discrete set of allowed energy levels in an atom. These quantized energy states, or *Landau levels*, can be expressed as

$$E_j = (j - \tfrac{1}{2})\, h\, e\, B \,/\, (2\, \pi\, m) \tag{2}$$

Here, j is an integer, h the Planck's constant, e the fundamental electron charge, and m the electron mass. Assuming a fixed electron density n for a 2-dimensional system, at low temperatures all electrons occupy the lowest allowable energy state, or Landau level, filling it only partially. As the field B is swept toward zero the capacity for each Landau level to hold each electron decreases according to

$$N = e\,B\,/\,h \tag{3}$$

where N is the number of orbits that can be packed per Landau level into each cm^2 of the system. At various points along the magnetic field all electrons fill up an exact number of Landau levels with all higher energy states remain empty. When this occurs the B-field is quantized and can be expressed as

$$B = (n\,h\,/\,e)\,/\,j, \tag{4}$$

where n is the electron density for a given state. Then the magneto-resistance -- resistance measured along the initially supplied current path -- drops and the Hall resistance R_H becomes quantized as

$$R_H = B\,/\,(n\,e) = h\,/\,(j\,e^2). \tag{5}$$

The first expression is just the classical Hall resistance while the second expression comes from substituting the values for B into the first expression. From this equation it is possible to extract the charge carrier density of the material by examining the periodicity of the plateaus in the quantum Hall effect measurement.

Our Hall resistance measurements were carried out on a patterned Hall bar shown in Figure 9. A drive current of 10 μA, which was the minimum current setting available on our Physical Properties Measurement System at a frequency of 30 Hz was supplied across the length of the Hall-bar, and a magnetic field B was applied along the direction perpendicular to both the current path and the measured V_H direction. A 9 Tesla superconducting magnet was used to generate the field, though for safety purposes the magnet was only ramped to 7 T in each direction. The measurements performed at 1.7 K

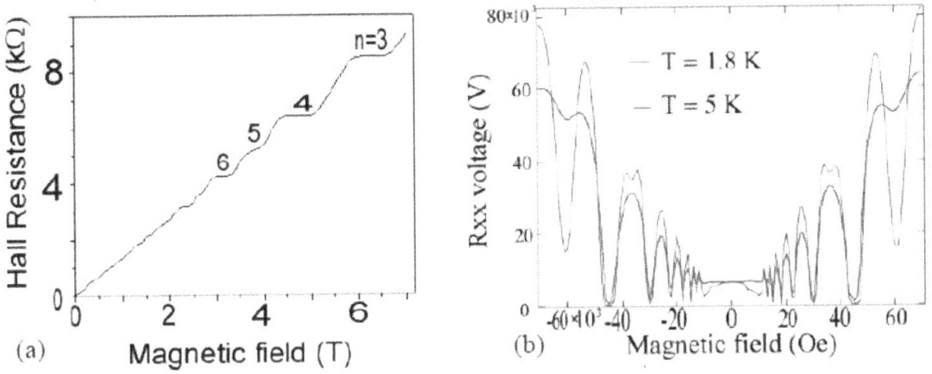

Fig. 10. (a) The magnetic field dependence of the Hall resistance R_{xy} measured at T=1.7 K. Note the quantized Hall resistance. (b) The Shubnikov-de Haas oscillations in the longitudinal resistance R_{xx}.

show that the magnetic field dependence of the Hall resistance is quantized to $R_{xy} = h/je^2$ in our 2DEG material. Also, the longitudinal resistance R_{xx} measured as a function of magnetic field exhibits the characteristic Shubnikov-de Haas oscillations, as shown in Figure 10.

2DEG Property	Symbol	Value	Units
Charge carrier density	n	5.0×10^{11}	cm^{-2}
Charge carrier mobility	μ	3.0×10^{5}	cm^2/Vs
Effective mass	m^*	0.067	$m_e = 9.1\times10^{-28}g$
Spin degeneracy	g_S	2	
Valley degeneracy	g_V	1	
Density of states	$\rho(E) = g_S g_V m^*/(h^2/2\pi)$	2.8×10^{10}	$cm^{-2}meV^{-1}$
Landau level spacing	$1/\rho(E)$	3.57	$\mu eV \mu m^2$
Fermi wave vector	$k_F = (4\pi n/g_S g_V)^{1/2}$	1.8×10^{6}	cm^{-1}
Fermi Energy	$E_F = (hk_F/2\pi)^2/2m^*$	17.88	meV
Fermi wavelength	$\lambda_F = 2\pi/k_F$	35	nm
Fermi velocity	$v_F = (hk_F/2\pi)/m^*$	3.07×10^{7}	cm/s
Scattering time	$\tau = m^* \mu/e$	11	ps
Mean free path	$\ell = v_F \tau$	3.5	µm
Cyclotron radius	$r_C = (hk_F/2\pi)/eB$	26	$nm/B^{1/2}$

Table 2. Typical parameters of our 40 nm deep 2DEG formed in a GaAs/AlGaAs heterostructure. The unit of B is in Tesla.

From the periodicity of the plateaus in Figure 10 (a) and (b), the 2DEG charge carrier density n was estimated to be about 5.0×10^{11} charges/cm^2 while the charge carrier mobility μ was estimated to be about 3.0×10^5 cm^2V/s. These two parameters were then used to obtain for example the Fermi Energy E_F, mean free path ℓ, Fermi wavelength λ_F, and effective mass m*. Table 2 lists the various properties that were calculated for one of our shallower (40 nm thick) 2DEGs.

3.2 Fabrication of gates on a shallow 2DEG and gate-defined double quantum dots

As our heterostructure material showed the typical 2DEG behavior, we fabricated quantum-point-contact (QPC) devices to see further 2DEG behavior in another nano-device form. Moreover, we did this to test our device fabrication technique. The gap on the QPC was set at 250 nm, shown in Figure 11. We tested the device at 4.2 K using AC lock-in techniques and found that the device did indeed exhibit quantized resistance behavior on account of the quantized transverse electron momentum through the QPC.

A Quantum Point Contact is defined as a short one-dimensional channel that is connected adiabatically to large source and drain reservoirs and that supports one or more wave modes. Our QPCs were made by electron beam lithography where two small metallic electrodes are patterned to form a small gap between them (100 nm – 1 µm in a typical QPC experiment). When the device is very cold and the negative bias voltage applied to it is strong enough to fully deplete electrons in the local 2DEG underneath, the electrons within the 2DEG are forced through a narrow constriction having now been permitted to move in only one direction. The width of the channel can be controlled by adjusting the gate voltages and can be made small enough to be comparable to the Fermi wavelength of the electrons (~40 nm). When the wavelength of the electrons is on the order of or greater than the characteristic size of the system quantum effects become pronounced. Here, since the Fermi wavelength is comparable to the width of the QPC's narrow constriction quantum effects are observable. Figure 11 shows examples of QPC's while Figure 12 shows a quantized resistance obtained from a QPC shown in Figure 11 (a), which indicates the quantization of the conductance in the QPC.

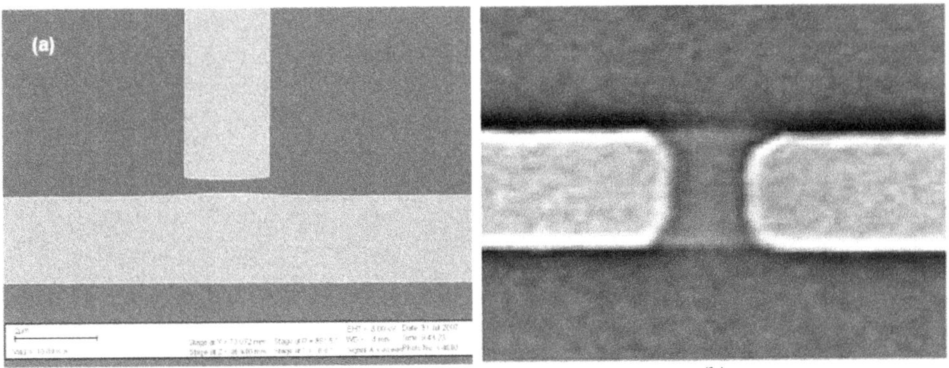

(a) (b)

Fig. 11. Two Scanning Electron Micrographs (SEM) of QPC's used to restrict the electrons in the 2DEG to motion in one direction. (a) QPC with a ~250 nm wide gap. (b) QPC with a ~300 nm wide gap. Note the slight bend at the edges of the electrodes in (a), which is due to the *proximity effect*, a result of secondary backscattering electrons in the electron beam lithography writer.

If the negative voltages on the QPC electrodes are made sufficiently strong so that the first subband is above the Fermi level, then the electrons can only tunnel across from one reservoir to the other, and the QPC then acts as a tunnel barrier. A good review of the theory regarding quantum point contacts can be found in [9]. The conductance is calculated starting with the simple Hamiltonian

$$H = p_x^2/2m^* + eV(x) + p_y^2/2m^* \tag{6}$$

Here, V(x) is the confining potential from the gate electrodes in the lateral direction. In this Hamiltonian V(y), the potential in the longitudinal direction that describes the transition from the 2DEG reservoirs to the constriction, is not included, assuming the one-dimensional electron's motion in the x-direction. V(x) takes on a parabolic form in the lateral direction

$$V(x) = \tfrac{1}{2}\, m^* \omega_o^2 x^2. \tag{7}$$

The solutions to the Schrödinger equation with this V(x) can be written in the Energy eigenvalue form

$$E_n = (n-1/2)\hbar\omega_o + \hbar^2 k_y^2/2m^*. \tag{8}$$

Here n is an integer (n = 1, 2,...). The conductance of the QPC can be calculated using Landauer-Buttiker formalism if the transmission probability are known, and is given by (see ref.[9] for details)

$$G = \frac{2e^2}{h} \sum_{n=1}^{N} T_n(E) \tag{9}$$

Essentially, the summation is over all modes of the QPC and $T_n(E)$ represents the transmission probability of each individual mode. For small V_{sd} values this can be simplified by making the approximation $T_n(E) = T_n(E_F)$. If there is no backscattering from the QPC (although this assumption is not realistic), $\Sigma T_n = N$, where N is an even number integer (N = 0, 2, 4,...) for the case of no applied magnetic field, representing each fully occupied subband. Then the conductance of a QPC can be written as

$$G = \frac{2e^2}{h} N \tag{10}$$

The conductance G of a QPC is quantized in units of $2e^2/h$ depending on the number of modes accessible in the device.

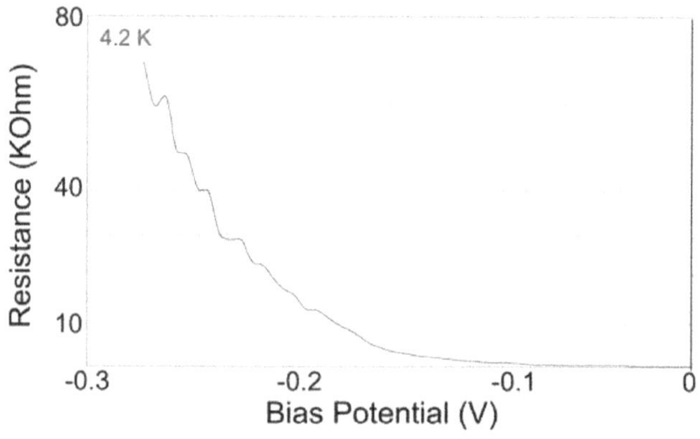

Fig. 12. Quantized resistance obtained from the quantum point contact shown in Figure 11 (a), which indicates the quantized conductance of the QPC at 4.2K.

The quantized conductance also can be seen from a quantum point contact formed by a gap between the gates, which are fabricated to define the quantum dot. An example of a single quantum dot is shown in Figure 2 and also in the inset of Figure 13. The conductance measurements at 80 mK through a quantum point contact formed by the gates 4 and 8 exhibits a well defined quantized conductance.

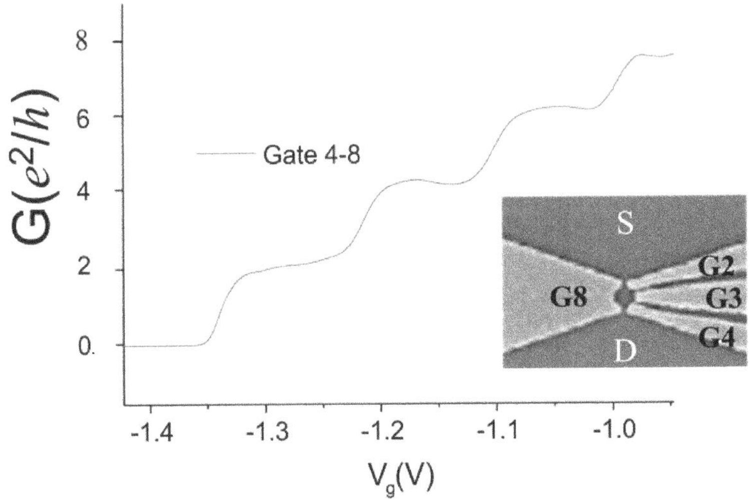

Fig. 13. The conductance G through a quantum point contact between the gates G4 and G8 is quantized in units of e^2/h, with the multiple of 2 arising from spin degeneracy. Inset: a single quantum dot.

3.3 Weakly coupled double quantum dot for a single photon detection application

For a small number of electrons in the quantum dot it is possible to calculate many-electron wave-functions and energy states. The many-body spectrum at zero magnetic field is then governed by the quantum confinement energy E_q and the charging energy E_c. For the simplest case of two parabolically confined electrons these parameters may be expressed in terms of l_o, which is related to a characteristic frequency Ω_0 determined by the electrostatic environment, as

$$l_o = (\hbar / m\, \Omega_0)^{1/2}. \tag{11}$$

The confinement length of the harmonic oscillator can be expressed as

$$E_q = \hbar^2 / (m^* l_o^2)$$

and the charging energy as

$$E_c = e^2 / 2C \propto e^2 / (4\pi\mu\varepsilon\, l_o).$$

If one uses the quantum dot as a photon detector, the characteristic frequency Ω_0 is related to the frequency of the photon absorbed by the quantum dot. This means that the photon frequency of the quantum dot detector can be tuned by adjusting the electrostatically defined quantum dot size. When a photon is absorbed by the dot its energy level is shifted resulting in a pair of excited electrons and holes. The excited charge can tunnel to the electron reservoir (i.e. outside of the quantum dots), resulting in the conductance-resonance peak shift. The variation of conductance can be detectable when the quantum dot absorbs even a single photon. As demonstrated by Komiyama and his coworkers [10, 11], such photon detection can be achieved using a single quantum dot or weakly-coupled double quantum-dots. Since the photon detection using double quantum-dots seemed to be more practical than that of a single quantum dot, which can be achieved by applying a considerable magnetic field (3.4 - 4.15 Tesla) to the quantum dot, we adopted the double quantum-dot technique.

3.4 Double quantum dot photon-detector

Similar to Komiyama's double quantum dot detector, our single photon detector consists of double quantum dots in a parallel geometry that is defined by metallic electrodes deposited on the 2DEG substrate surface. In this experiment we have used several different 2DEG substrates with the 2DEG depth ranging from 40 nm to 160 nm, and fabricated more than several hundred devices. Figure 14(a) shows the gate electrodes and the Ohmic contacts with the quantum dots located at the center of the white frame. Figure 14 (b) shows another SEM picture of the double quantum dots, which is a magnified view of the center part of Figure 14(a). The lower quantum dot (QD1) acts as a photon absorber and the upper quantum dot (QD2) functions as a single electron transistor.

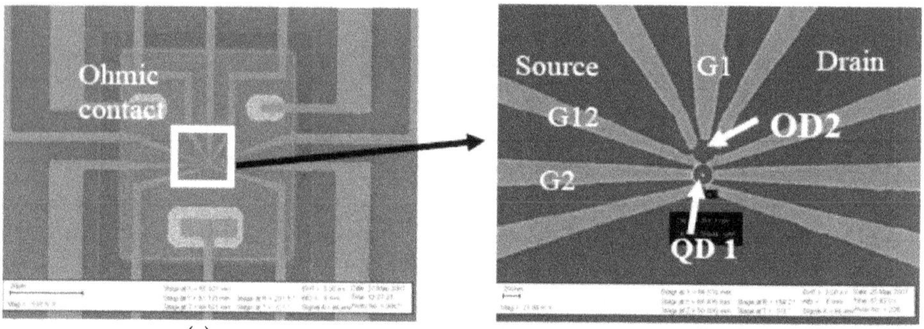

(a) (b)

Fig. 14. (a) Overall SEM view of the NRL double quantum dot detector. (b) SEM picture of the electrode defined double quantum dots.

The gate electrodes were defined via e-beam lithography after which we deposited a 50 Å thick Cr layer (acts as a wetting layer) on the surface of a GaAs/AlGaAs heterostructure and then a 150 Å thick Au layer on top of the Cr layer. The diameters of QD1 and QD2, as defined by the surrounding electrodes, are roughly 250 nm with the diameter of the SET dot (QD2) slightly smaller than that of the absorber dot (QD1). As mentioned earlier, the actual size of the quantum dot is dependent upon the strength of the negative bias voltage applied to the gates: the stronger the bias voltage the smaller the quantum dot. As the capacitance and the electrochemical potential of the quantum dot are closely related to the number of isolated electrons, one can control the capacitance and the electrochemical potential by adjusting the gate voltage. The plunger gate G_1 shown in Figure 14 provides experimental control of the SET dot's self capacitance (C_1) and electrochemical potential (μ_1), and the pair of gates labeled G_2 control the absorber dot's self capacitance (C_2) and electrochemical potential (μ_2). The electrodes labeled G_{12} control the potential barrier that couples the SET dot and the absorber dot.

As the absorber (QD1) and the SET (QD2) are weakly coupled by the voltage on gates G12, the excited energy level of the absorber alters the energy levels of the SET. If an energy level of the SET aligns within the energy levels of the source and the drain, electrons begin to flow through the SET. These excited energy levels of the absorber and the SET are in what are referred to as meta-stable states, which survive typically on the order of or less than a few milli-seconds. This short meta-stable state is due to the fact that a finite probability exists that an electron from one of the large 2DEG reservoirs "hops" onto the absorber dot. This results in a change in the energy of the absorber dot, which can affect the energy level matching between the SET dot and source and drain, since the two dots are electrostatically coupled. Since the electron mobility is very high (3.0×10^5 cm^2/Vs) and the electron density is very large ($5.0\text{x}10^{11}$ cm^{-2}) in the 2DEG, a significantly large number (~10^6 - 10^7) of electrons can flow through the SET within the short time of a meta-stable state, resulting in electric currents on the order of a pico-ampere. By employing a lock-in technique one can readily measure such currents. This operating principle is somewhat analogous to the photomultiplier tube as a single photon triggers a measureable electron flow in the detector.

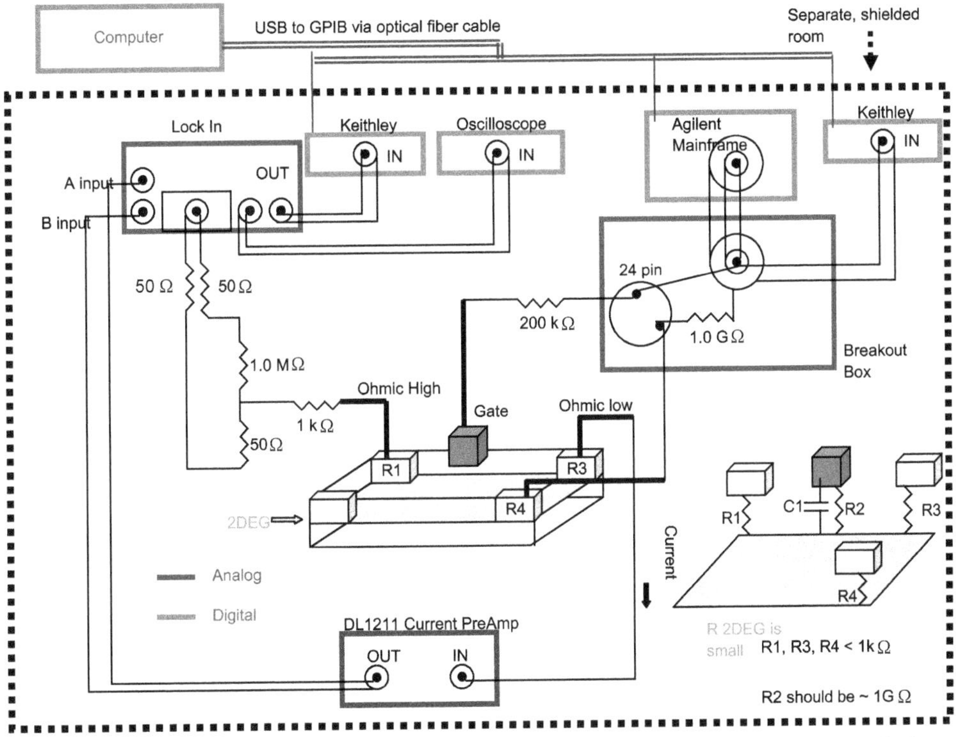

Fig. 15. A schematic diagram of our experimental setup to measure low current signals from a device. This diagram shows the wiring between the electronic equipment and the gate and Ohmic pads on a device. A lock-in sources an AC signal (~ 0.1 V) and a simple voltage divider circuit 10^5:1 is connected to an Ohmic contact on the high side and is used to supply the μV drop across the device. An Agilent E5270B Mainframe supplies the gate voltages through a homemade breakout box. Final data values can be read off of the Keithley digital multi-meters (DMM) and oscilloscope. The computer that collects data is housed in a separate room and is connected to the equipment via an optical fiber.

The electrons (~10^6 -10^7) flowing through the SET result in an electric current, which is on the order of a pico-Ampere (10^{-12}) or less. [1, 3, 12-14] In order to measure this weak current one should carefully design the experimental set up. Since electrical noises can induce currents much larger than pico-Ampere, it is necessary to minimize ambient electrical noises, which usually can be achieved by carrying out the measurements within a shielded room, and also by employing a lock-in technique. [15-17] An example is shown in Figure 15. A measurement includes the application of a source-drain voltage (or current source) over the device, or part of the device, and measuring the resulting current or voltage signal as a function of various parameters, such as the negative voltages applied to the depletion gates, temperature, electromagnetic fields, etc.

During the measurements it is important to keep the current and voltage across a device small enough in order to maintain the device temperature sufficiently cold. The energy associated with the voltage drop across the source and the drain, V_{sd}, should not exceed the electron thermal energy, K_BT_e, within the 2DEG. If $eV_{sd} > K_BT_e$, then the electrons within the Fermi reservoirs may enter or leave the quantum dots at random and/or the higher energy charge states may become allowed within the dots. Then the electron flow through the quantum dot cannot be controlled. Therefore, it is necessary to limit the voltage drop across a device such that $eV_{sd} < K_BT$, where K_B is the Boltzmann's constant. For example, V_{sd} should be less than 345 μV for $T_e = 4$ K, or 8.62 μV for $T_e = 100$ mK.

3.5 Photon detection

A schematic diagram of our photon detection setup is shown in Figure 16. For this demonstration we used a double quantum-dot detector fabricated on 100 nm 2DEG, and employed an HP 85105 millimeter-wave controller and an HP W85104A test-set module, which were attached to an HP8510C Vector Network Analyzer. The millimeter-wave signal was sent through micro-coaxial cable (Lakeshore Type C cable). To modulate the signal we split the micro-coaxial cable and made two sets of dipole antennae, which face each other across an optical beam chopper (Stanford Research SR540) set to produce 1 - 2 Hz modulation. However, at the low modulation frequency, the chopper's blade did not rotate smoothly resulting in irregular modulation so much that the modulation interval was highly irregular. Also, we note that the millimeter-wave signal was highly attenuated through the micro-coaxial cables, as well as through the dipole-antenna to dipole-antenna coupling. We estimate the attenuation rate was much more than 5dB per foot for the micro-coaxial cable and the coupling efficiency through the dipole-antenna coupling to be less than 10%. Since the initial mm-wave input from the W85104A was approximately 50 μW, and the transmission efficiency of the millimeter-wave photon through the coaxial cable and the dipole-antenna coupling was extremely poor, we think that the millimeter-wave signal radiated onto the double dot detector was sub-microwatts.

We measured the temperature dependent conductance as well as the bias voltage dependent conductance of our double dot device. The experiments indicate that although the millimeter-wave signal power was very weak our double dot device could detect the signal (Figure 17). However, the results may not indicate single photon detection. We think that our double dot detector could detect a few millimeter-wave photons at 100 mK.[18]

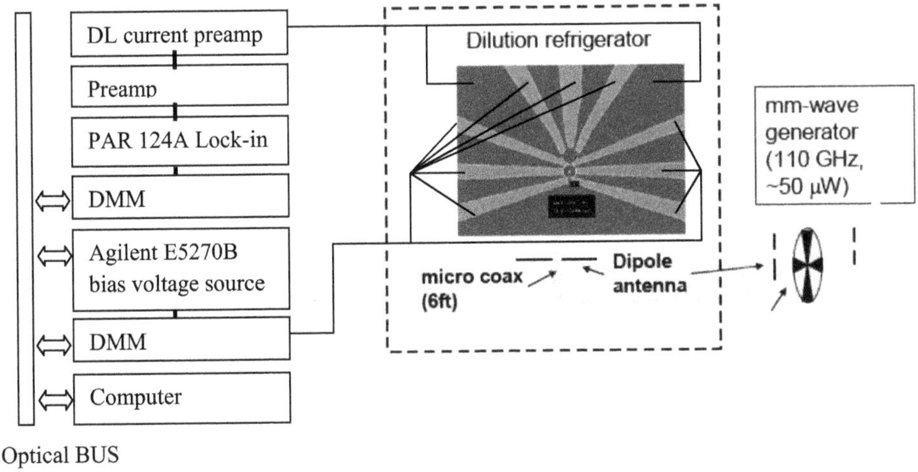

Fig. 16. Experimental set up for the millimeter-wave photon detection with our double dot detector.

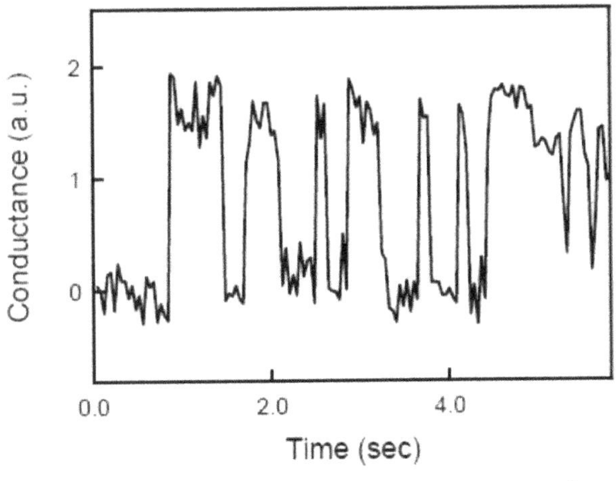

Fig. 17. The conductance variation as the double dot device detects millimeter-wave photons

While experiments indicate that it was possible to detect some photons at 110 GHz with the double quantum-dot structure shown in Figure 14, the detection efficient was very poor. We suspect that the inefficiency was largely due to the gate G2, which was supposed to function as an antenna. With its improper shape as an antenna, it did not efficiently pick up photons. Later we modified the double quantum dot detector, implementing bow-tie antenna geometry for the gate G2, as shown in Figure 18.

Also we attempted to reduce the quantum dot size in order to detect photons at an elevated temperature. As discussed earlier, a quantum dot detector should be operated at a temperature T that has a thermal energy K_BT below the charging energy of the quantum dot. The charging energy E_C is given as $E_C = e^2/C_{dot}$ where C_{dot} is the self capacitance of the quantum dot, which is proportional to the size of quantum dot. Hence, by reducing the quantum dot size, one can raise the charging energy E_C, as well as the operating temperature T of the quantum dot detector.

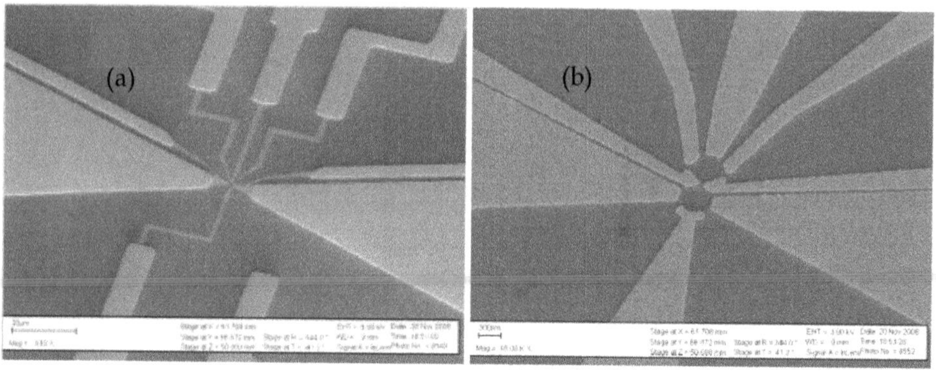

Fig. 18. SEM images of a modified double quantum-dot detector. (a) Overall view. (b) Magnified view of the center of picture (a), showing double quantum dots and a bow-tie antenna. The lower quantum dot is coupled to a dipole (bow-tie) antenna fabricated to absorb THz frequencies, while the upper dot acts as a single electron transistor.

As a rule of thumb the depletion length that is the lateral depletion of the electrons around the gates is roughly equal to the depth of the 2DEG. This means that, for a smaller quantum dot, we need to use a shallower 2DEG substrate. For our experimental demonstration of a photon detector at temperatures considerably higher than 100 mK, we have fabricated about a hundred double quantum-dot detectors on 2DEG substrates with thickness ranging from 40 nm to 160 nm.

4. Leakage currents in GaAs/AlGaAs heterostructures

Our experiments revealed that the quantum dot detectors fabricated on a shallow 2DEG suffered from problems associated with overwhelming leakage currents. Within a GaAs/AlGaAs heterostructure the leakage currents increased dramatically as the 2DEG depth became shallower. Since the leakage currents dominate, it was not possible to obtain any discernible signal from the quantum dot detector. Also the leakage currents caused severe damage to the quantum dot gates, often resulting in a short circuit on the gates.

Measurements were performed to determine the currents that flow between the 2DEG and a laterally defined depletion gate on the wafer's surface. Current is measured as a function of the voltage applied to the gate when the gate is biased with respect to the 2DEG underneath. While we expect that the current should be zero ideally or much less than pico-Amperes, the actual leakage current measured is orders of magnitude larger than anticipated. Even our numerical calculation, which was performed along with our experimental efforts, indicates that the leakage current is substantially larger than previously expected for shallow 2DEG wafers. Unless we find a way to prevent this large leakage it may lead to a limit for the maximum operating temperature obtainable for our quantum-dot photon detector.

Figure 19 shows a simple setup used to test the leakage currents within our GaAs/AlGaAs 2DEG substrate. The leakage current measurements were performed for several different 2DEGs with depths of 40 nm, 43 nm, 90 nm, and 160 nm. As shown in Figure 19, the voltage potential is applied directly to a gate and to an Ohmic contact in the reservoir. Any leakage current between the gate and the 2DEG is measured by a current amplifier (DL1211). The DL1211 converts the measured current signal into a voltage signal, which is read by the Keithley multimeter.

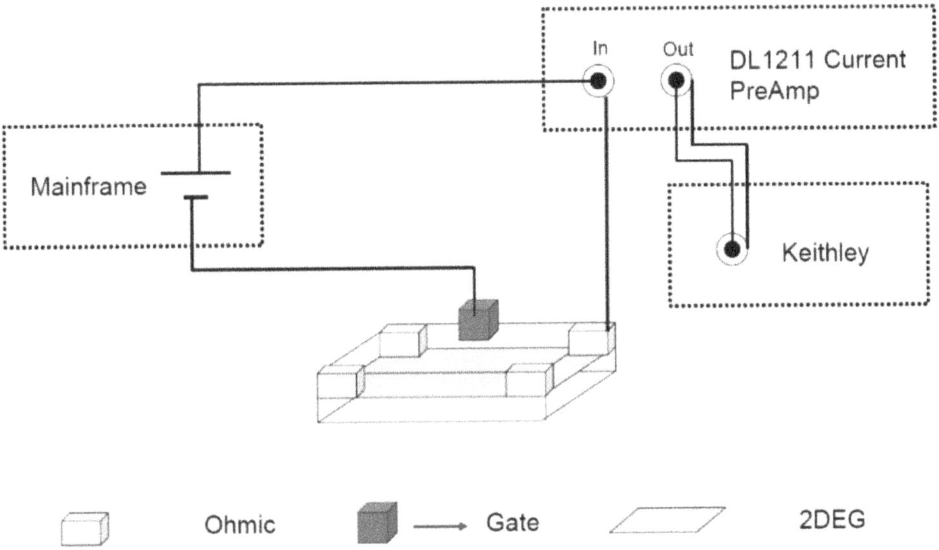

Fig. 19. A schematic experimental setup for the measurement of leakage currents in our 2DEG wafers. The voltage potential is supplied by an Agilent E5270B Precision Measurement Mainframe.

The leakage currents increased 6 orders of magnitude when the 2DEG depth was varied from 160 nm to 40 nm. Some of the shallowest (40 nm) 2DEG substrates generated leakage currents as high as tens of micro-Amperes, while the leakage current from the thickest one was much smaller, less than a pico-Ampere. When leakage currents are as high as several micro-Amperes the electron flow in and around the quantum dots cannot be controlled, and it is impossible to obtain any meaningful signal from the quantum dot detector. Our experiments further revealed that the strength of leakage currents vary depending on the individual 2DEG substrate. In other words, when we measure the leakage currents from two different 40 nm thick 2DEG substrates, we obtain very inconsistent results. This suggests us that the leakage current problem may not be entirely due to the intrinsic property of a shallow 2DEG, rather it may suggest that the problem is related to the defects in the 2DEG substrate.

Figure 20 shows an example of our leakage current data obtained from a 43 nm deep 2DEG at 4.2 K. The leakage currents increased linearly with the gate potential, and could reach as high as several hundred nano-Amperes. In order to investigate the origin of this leakage current we performed numerical simulations, in which we assumed the leakage currents were through a 1-dimensional barrier. In this simplified model only the tunneling across a Schottky barrier from a metal to a GaAs layer under a reverse gate bias was considered. The model is so simplified that we did not include the charge carrier interaction with the heterostructure, such as how the charge carriers, after having passed the Schottky barrier, travel through several regions of GaAs and AlGaAs, including a heavily doped AlGaAs layer, before reaching the 2DEG. The inclusion of these would make the simulations more realistic. However it will require implementing a lot more difficult calculations in the simulations. So we only considered the problem only the top layer of GaAs. We know that this simulation is not realistic, but we think that it will give us some insights about the leakage currents.

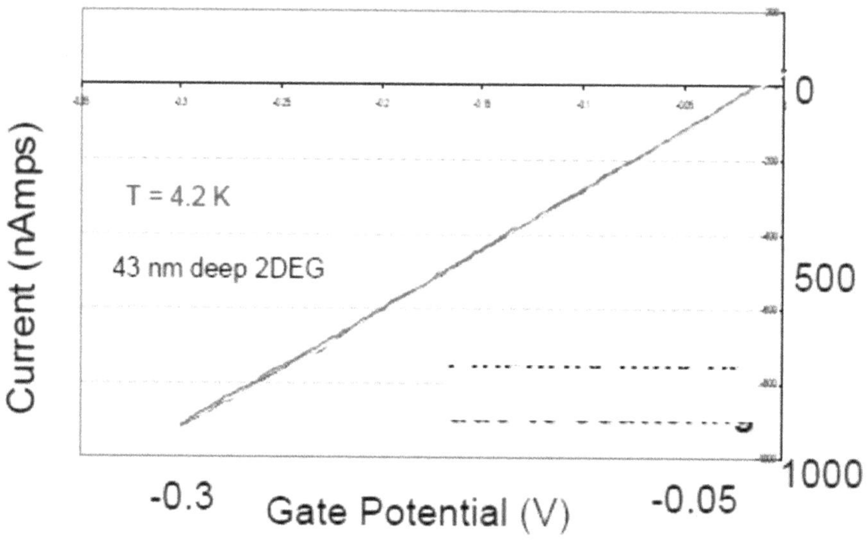

Fig. 20. Leakage current data taken from a 43 nm deep 2DEG at 4.2 K. Leakage currents for this shallow 2DEG structure can be as high as hundreds of nano-Amperes.

The simulations indicate that a shallower 2DEG leads to a larger leakage current, and the leakage current can exceed a thousand nano-Amperes. These results are at least qualitatively consistent with our experimental results. A simulated result that shows the leakage current as a function of gate voltage is presented in Figure 21. Apparently the exponential relationship between the leakage current and the bias voltage is not consistent with our experiments. (See Figure 20) The discrepancy may be due to the fact that our model is too simple and does not reflect realistic conditions, for instance the scattering that the charge carriers experience, due to the Si dopants, as they pass from the lateral surface gates to the 2DEG, and the effect of lattice mismatching between GaAs and AlGaAs.

In order to construct a device using quantum dots, one should minimize the leakage current since it not only prevents the proper control of the quantum dot but also sometimes leads to physical damage to the quantum dot. The gates surrounding the quantum dot are very small, typically less than 20 nm thick and a few tens of nm wide. Around the quantum point contacts formed by the gates the cross-sections of these metal-structures can be as small as a few tens of nm^2. When the leakage current exceeds several hundred nano-Amperes, the current density near the quantum point contacts can exceed 10^{10} - 10^{11} A/m^2. The current density may become high enough to fuse metallic structures near the quantum point contacts. This is what presumably happened to some of our quantum dot devices. An example is shown in Figure 22.

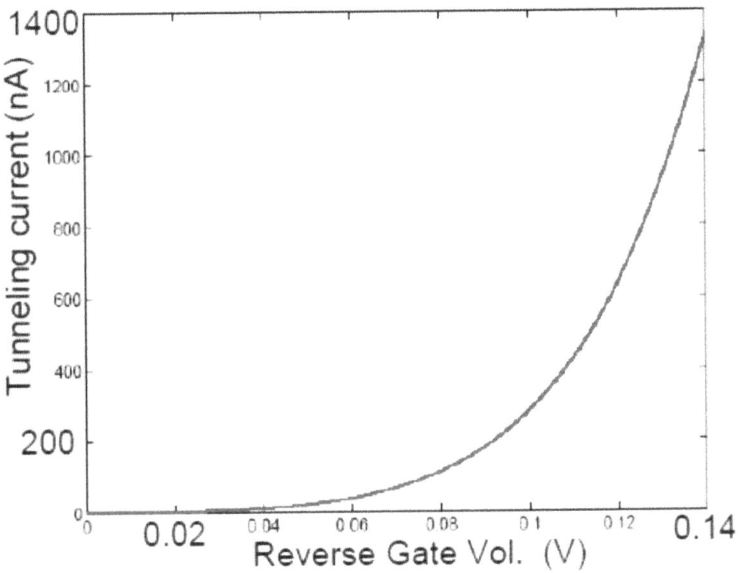

Fig. 21. An example of a numerical calculation showing the leakage current across a Schottky barrier from a metal to a GaAs layer under a reverse bias.

Our attempt to raise the operating temperature of a quantum dot photon detector did not succeed. The leakage current in shallower 2DEG materials remains one of the most significant technical challenges in achieving higher operating temperatures for single photon detectors. The origin of leakage currents in 2DEG substrates and a method to avoid them are topics for future research.

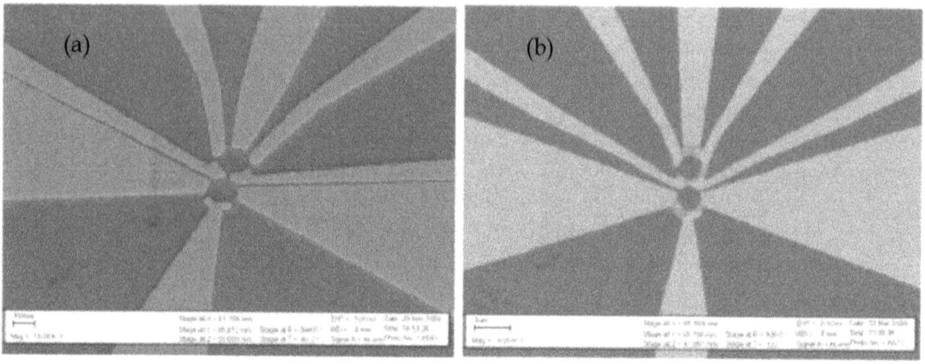

Fig. 22. SEM images of (a) BEFORE and (b) AFTER testing the quantum dots, which involves a large leakage current flowing through the device. The damage to the thin nanostructure gates near the center of the picture in (b) is due to the large leakage current that passes through the very small nanostructure (e.g. quantum point contacts between the gates).

5. References

[1] W. G. van der Wiel, S. DeFranceschi, J. M. Elzerman, T. Fujisawa, S. Tarucha, L. P. Kouwenhoven, Electron transport through double quantum dots, *Reviews of Modern Physics* 75 (2003).

[2] T. J. Thornton, M. Pepper, H. Ahmed, D. Andrews, G. J. Davies, One-Dimensional conduction in the 2D electron gas of a GaAs/AlGaAs Heterojunction, *Phys. Rev. Lett.* 56, 11 (1986).

[3] NATO ASI Series, Quantum Transport in Semiconductor Submicron Structures, *Series E: Appl. Sci.* 326 (1996)

[4] Vurgaftman, J. R. Meyer, D. H. Wu, K. Bussmann, and B. T. Jonker, Spectral simulation of GaAs and InAs quantum-dot terahertz detectors designed for higher-temperature operation, *Journal of Applied Physics* 100, 064509 (2006).

[5] P. O'connor, A. Dori, M. Feuer, R. Vounckx, Gold-germanium-based ohmic contacts to the two-dimensional electron gas at selectively doped semiconductor heterointerfaces, *IEEE Transactions on Electron Devices* ED-34, 4 (1987).

[6] M. Furno, F. Bonani, G. Ghione, Transfer matrix method modeling of inhomogeneous Schottky barrier diodes on silicon carbide, *Solid-State Electronics* 51, 466-474 (2007).

[7] J. Crofton, P. A. Barnes, A comparison of one, two, and three band calculations of contact resistance for a GaAs ohmic contact using the WKB approximation and a numerical solution to the Schrodinger equation, *J. Appl. Phys.* 69, 11 (1991).

[8] D. A. Muller, T. Sorsch, S. Moccio, F. H. Baumann, K. Evans-Lutterdot, G. Timp, The electronic structure at the atomic scale of ultrathin gate oxides, *Nature* 399, 758 (1999)

[9] H. van Houten, C. W. J. Beenakker, B. J. van Wees, Quantum point contacts *Nanostructured Systems*, M. A. Reed, Ed. (Academic Press, San Diego, 1992) 35, pp. 9-112 (1992)

[10] O. Astafiev, S. Komiyama, T. Kutsuwa, V. Antonov, Y. Kawaguchi, K. Hirakawa, Single-photon detector in the microwave range, *Applied Physics Letters* 80, 22 (2002).
[11] S. Komiyama, O. Astafiev, V. Antonov, T. Kutsuwa, H. Hirai, A single-photon detector in the far-infrared range, *Nature* 403, 405 (2000).
[12] L. P. Kouwenhoven, C. M. Marcus, P. L. McEuen, S. Tarucha, R. M. Westervelt, N. S. Wingreen, Electron transport in quantum dots, *Proceedings of the Advanced Study Institute on Mesoscopic Electron Transport* (1997).
[13] C. Livermore, C. H. Crouch, R. M. Westervelt, K. L. Campman, A. C. Gossard, The Coulomb Blockade in Coupled Quantum Dots, *Science* 274, 1332 (1996)
[14] L. P. Kouwenhoven, C. M. Marcus, P. L. McEuen, S. Tarucha, R. M. Westervelt, N. S. Wingreen, Electron transport in quantum dots, *Mesoscopic Electron Transport*, Kluwer Academic, Dordrecht (1997)
[15] F. Pobell, Matter and Methods at Low Temperatures, 2nd Edition (2009)
[16] H. W. Ott, Noise Reduction Techniques in Electronic Systems, 2nd Edition (1988)
[17] R. C. Richardson, E. N. Smith, Experimental Techniques in Condensed Matter Physics at Low Temperatures, Advanced Book Classics (1998)
[18] D. H. Wu, B. R. Matis, K. Bussmann, Scalable single photon detector for terahertz and infrared applications, *International Journal of High Speed Electronics and Systems* 18, 4 (2008)
[19] B. J. van Wees, H. van Houten, C. W. J. Beenakker, J. G. Williamson, L. P. Kouwenhoven, D. van der Marel, C. T. Foxon, Quantized conductance of point contacts in a two-dimensional electron gas, *Phys. Rev. Lett.* 60, 9 (1988).
[20] T. H. Oosterkamp, T. Fujisawa, W. G. van der Wiel, K. Ishibashi, R. V. Hijman, S. Tarucha, L. P. Kouwenhoven, Microwave spectroscopy of a quantum-dot molecule, *Nature* 395, 873 (1998).
[21] L. P. Kouwenhoven, S. Jauhar, J. Orenstein, P. L. McQueen, Y. Nagamune, J. Motohisa, H. Sakaki, Observation of photon-assisted tunneling through a quantum dot, *Phys. Rev. Lett.* 73, 25 (1994).
[22] K. Ikushima, Y. Yoshimura, T. Hasegawa, S. Komiyama, T. Ueda, K. Hirakawa, Photon-counting microscopy of terahertz radiation, *App. Phys. Lett.* 88, 152110 (2006).
[23] S. M. Cronenwett, S. M. Maurer, S. R. Patel, C. M. Marcus, C. I. Duruoz, J. S. Harris Jr., Mesoscopic Coulomb blockade in one-channel quantum dots, *Phys. Rev. Lett.* 81, 26 (1998).
[24] S. Gustavsson, M. Studer, R. Leturcq, T. Ihn, K. Ensslin, D. C. Driscoll, A. C. Gossard, Frequency-selective single-photon detection using a double quantum dot, *Phys. Rev. Lett.* 99, 206804 (2007).
[25] S. R. Patel, S. M. Cronenwett, D. R. Stewart, A. G. Huibers, C. M. Marcus, C. I. Duruoz, J. S. Harris Jr., K. Campman, A. C. Gossard, Statistics of Coulomb Blockade Peak Spacings, *Phys. Rev. Lett.* 80, 20 (1998)
[26] M. H. Devoret, R. J. Schoelkopf, Amplifying quantum signals with the single-electron transistor, *Nature* 406, 1039 (2000)
[27] H. L. Stormer, *Rev. Mod. Phys.* 71, 875 (1999)
[28] M. Pioro-Ladriere, J. H. Davies, A. R. Long, A. S. Sachrajda, L. Gaudreau, P. Zawadzki, J. Lapointe, J. Gupta, Z. Wasilewski, S. Studenikin, Origin of switching noise in GaAs/$Al_xGa_{1-x}As$ lateral gated devices, *Phys. Rev. B* 72, 115331 (2005)

[29] H. Hashiba, V. Antonov, L. Kulik, A. Tzalenchuk, P. Kleinschmid, S. Giblin, S. Komiyama, Isolated quantum dot in application to terahertz photon counting, *Phys. Rev. B* 73, 081310(R) (2006)

[30] S. Kim, M. S. Sherwin, J. D. Zimmerman, A. C. Gossard, P. Focardi, D. H. Wu, Room temperature terahertz detection based on electron plasma resonance in an antenna-coupled GaAs MESFET, *Appl. Phys. Lett.* 92, 253508 (2008)

3

Tunneling Atomic Force Microscopy of Self-Assembled In(Ga)As/GaAs Quantum Dots and Rings and of GeSi/Si(001) Nanoislands

Dmitry Filatov[1], Vladimir Shengurov[1], Niyaz Nurgazizov[2], Pavel Borodin[2] and Anastas Bukharaev[2]
[1]Technical Physics Research Institute, N.I. Lobachevskii University of Nizhny Novgorod,
[2]E.K. Zavoisky Kazan' Physical-Technical Institute, Kazan' Scientific Centre, Russian Academy of Sciences
Russia

1. Introduction

Scanning Tunnelling Microscopy (STM) has been being used for the investigation of the morphology and of the atomic structure of the semiconductor nanostructures extensively since early 1990-s (Medeiros-Ribeiro et al., 1998). More recently, STM in Ultra High Vacuum (UHV) has been applied also to the investigation of the spatial and energy distributions of the local density of states (LDOS) in the quantum semiconductor heterostructures. For example, Cross-Sectional STM (X-STM) has been applied to the visualization of the envelope wavefunctions of the quantum confined states in the GaSb/InAs(001) quantum wells (QWs) (Suzuki et al., 2007) and in the self-assembled InAs/GaAs(001) quantum dots (ODs) (Grandidier et al., 2000). Also, the surface InAs/GaAs(001) QDs grown by Molecular Beam Epitaxy (MBE) have been investigated by UHV STM *in situ* (Maltezopoulos et al., 2003). The peaks related to the quantum confined states in the QDs have been observed in the differential conductivity $\sigma_d = dI_t/dV_g$ spectra of the STM tip contact to the QDs (here I_t is the tip current and V_g is the gap voltage). The $\sigma_d(x, y)$ images of the ODs (x and y are the tip coordinates on the sample surface) recorded at the values of V_g corresponding to the peaks in the $\sigma_d(V_g)$ spectra correlated with the probability density patterns $|\chi(x, y)|^2$ where $\chi(x, y)$ are the lateral components of the electron quantum confined states envelope wavefunctions calculated for the pyramidal InAs/GaAs(001) ODs defined by the (101) facets (Stier et al., 1999).

The present chapter is devoted to the investigation of the electronic states in the self-assembled semiconductor nanostructures [namely, the InAs/GaAs(001) QDs, the InGaAs/GaAs(001) quantum rings (QRs), and the GeSi/Si(001) nanoislands] by Tunnelling Atomic Force Microscopy (AFM). The samples with the surface self-assembled semiconductor nanostructures were scanned across by a conductive Si AFM probe covered by a conductive coating (Pt, W_2C, or a diamond-like film) in the contact mode. The bias voltage V_g was applied between the AFM probe and the sample. Simultaneously with the acquisition of the topography $z(x, y)$, the $I-V$ curves of the probe-to-sample contact $I_t(V_g)$ were acquired in each point in the scans.

Earlier, Tunnelling AFM has been applied mainly to the characterization of the local electrical properties of the thin dielectric films on the conductive substrates (Yanev et al., 2008). Also, Tunnelling AFM in UHV has been applied to the tunnel spectroscopy of individual Au nanoclusters in the ultrathin SiO_2/Si films (Zenkevich et al., 2011). The present chapter summarizes a series of the original studies where the authors have applied Tunnelling AFM to the mapping of the LDOS in the self assembled semiconductor nanostructures for the first time (Filatov et al., 2010, 2011, Borodin et al., 2011). The main advantage of Tunnelling AFM compared to UHV STM is that the former allows the *ex situ* investigation of the surface semiconductor nanostructures covered by a native oxide layer during the sample transfer from the growth setup to the AFM one through the ambient air. This makes the STM studies of these samples hardly possible.

Another distinctive feature of the studies present in this chapter is that the samples with InGaAs/GaAs(001) QDs and QRs have been grown by Atmospheric Pressure Metal Organic Vapour Phase Epitaxy (AP-MOVPE). In most studies reported in the literature, the QDs grown by MBE or by Low Pressure (LP) MOVPE have been investigated. Nevertheless, the investigations of the electronic properties of the InAs/GaAs(001) QDs grown by AP MOVPE are of a considerable interest because this growth method is more promising for the commercial device production due to its lower cost and higher productivity as compared to MBE and LP-MOVPE.

Also, the structures with the self-assembled GeSi/Si(001) nanoislands studied in the present chapter were grown by a novel technique of Sublimation MBE (SMBE) in GeH_4 ambient. In this method, the Si layers are grown from an ordinary sublimation source in UHV. To deposit Ge, GeH_4 is introduced in the growth chamber at the pressure of $\sim 10^{-2} \div 10^{-4}$ Torr and undergo pyrolysis on the heated substrate. So far, this method is some hybrid between the conventional MBE from the sublimation source and LP VPE. Again, in the majority of works, the GeSi/Si(001) nanoislands grown by MBE have been studied (Berbezier & Ronda, 2009). The main advantage of the hybrid technique of SMBE in GeH_4 ambient as compared to the ordinary VPE of Si and Ge from silanes and germanes, respectively is that SMBE allows growing the Si layers of high crystalline quality and purity at relatively low temperatures (450 $\div$ 500°C) keeping the high enough growth rates. In addition, SMBE offers a broader choice of the doping impurities as well as a wider range of their concentrations achievable than VPE. On the other hand, the deposition of Ge from a gaseous precursor provides higher uniformity of the nominal thickness of the deposited Ge layer d_{Ge} over the substrate surface.

Having applied Tunnel AFM to the investigation of the LDOS in the self assembled semiconductor nanostructures described above, we have observed the patterning of the probe current images $I_t(x, y)$ of the InAs/GaAs(001) QDs and of the InGaAs/GaAs(001) QRs as well as the peaks in the $\sigma_d(x, y)$ spectra of the contact of the AFM probe to the sample surface attributed respectively to the spatial and energy distributions of the LDOS in the quantum heterostructures. The results of the LDOS mapping by Tunnelling AFM have been compared to the results of the calculations of the probability density patterns $|\chi(x, y)|^2$ reported in the literature. The Tunnel AFM data allowed the identification of the quantum confined states in the InAs/GaAs(001) QDs grown by AP MOVPE the interband optical transitions between which are manifested in the photosensitivity (PS) spectra of the QD structures. Finally, the direct measurements of the LDOS spectrum in the self assembled $Ge_xSi_{1-x}/Si(001)$ nanoislands demonstrated the type I conduction band alignment in them at $x < 0.45$.

2. Self assembled InAs/GaAs(001) quantum dots

2.1 Growth and characterization

In this subsection, the details of the growing the samples for the Tunnelling AFM investigation by AP MOVPE and of their characterization are presented.

The InAs/GaAs(001) QD structures for the tunnel spectroscopy of the LDOS in the conduction band were grown on the n^+-GaAs(001) substrates by Dr. B. N. Zvonkov in Research Institute for Physics and Technology, N. I. Lobachevskii University of Nizhny Novgorod, Russia using a homemade setup for AP MOVPE from trimethylgallium, trimethylindium, and AsH_3. The schematic of the QD structures for the Tunnelling AFM investigations is shown in Fig. 1, *a*. The substrates were misoriented from (001) by 3 ÷ 5° towards <110>. The donor concentration in the substrate material was ~10^{18} cm^{-3}. The GaAs buffer layers with the thickness $d_b \approx 200$ nm were doped by Si heavily up to the donor concentration ~ 10^{18} cm^{-3} using pulsed laser sputtering of a bulk Si target placed in the MOVPE reactor. More detains on this growth technique can be found elsewhere (Karpovich et al., 2004a). The 3 nm thick intentionally undoped GaAs spacer layers were grown between the n^+-GaAs buffers and the InAs ODs. The latter were grown at the substrate temperature T_g = 530°C, the nominal thickness of the InAs layers d_{InAs} was ≈ 5 monolayers (ML).

The morphology of the grown samples was first examined by ambient air AFM using NT MDT® Solver Pro™ instrument in Contact Mode. The NT MDT® CSG-01 silicon AFM probes were used. The curvature radii of the probe tips R_p were < 10 nm (according to the vendor's specifications).

The morphology of the surface QD arrays was characterized quantitatively by the following parameters:

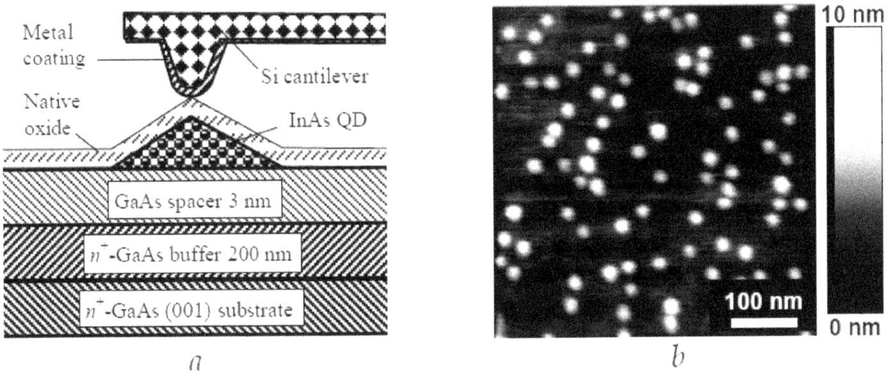

Fig. 1. The schematic (*a*) and an ambient air AFM image (*b*) of a structure with the InAs/*n*-GaAs/ n^+-GaAs(001) surface QDs.

- The average height of the QDs <h>
- The averaged diameter of the QDs <D>. The value of D for given QD was defined as $D = P/2\pi$ where P was the perimeter of the AFM image of the QD measured at the level of 0.1 <h> above the surface of the wetting layer
- The surface density of the QDs N_s

The morphological parameters of the QD arrays listed above were determined by the digital processing of the AFM data. In order to identify the QDs on the surface of the wetting layer, a threshold particle recognition algorithm has been applied.

The key issue in the investigations of the morphology of the surface InAs/GaAs(001) QDs by AFM was the effect of convolution (Bukharaev et al., 1999) originating from a relatively large AFM probe tip radius R_p (~ 10 nm) as compared to the typical sizes of the InAs/GaAs(001) QDs defined by the (101) facets ($D = 12 \div 18$ nm, $h = 5 \div 6$ nm). In order to extract the actual size and shape of the InAs/GaAs(001) QDs grown by AP MOVPE from the AFM data, we have applied the digital processing of the AFM images using an original software for the correction of the convolution artifacts (so called "deconvolution"). This software utilizes the "virtual AFM" algorithm (Bukharaev et al., 1998).

In order to apply this algorithm to the correction of the convolution artifacts, one needs to know the exact geometry of the actual probe tip used in the experiment. The specifications provided by the AFM cantilevers' vendors appear to be insufficient often. The actual probe tip shape may be determined by the measuring of a standard sample with the geometry known *a priori*. However, the theory of the convolution artifacts sets up an essential requirement on the geometrical parameters of the standard sample to be close to these ones of the samples under study [i. e. the InAs/GaAs(001) QDs in our case]. In order to evaluate the AFM tip shape, we have used a structure with the surface self assembled GeSi/Si(001) pyramidal-shaped nanoislands (Medeiros-Ribeiro et al., 1998) as a standard sample. The structure was grown by Dr. A. V. Novikov, Institute for Physics of Microstructures, Russian Academy of Sciences (Nizhny Novgorod, Russia) using standard MBE. The self assembled GeSi/Si(001) pyramidal-shaped nanoislands are defined by the (105) facets, their heigt can be extracted just from the AFM data directly. Note that the convolution artifacts do not affect the acuracy of the measurements of the heights h of the GeSi/Si(001) nanoislands as well as of the InAs/GaAs(001) QDs. So far, all the parameters needed to determine the actual probe shape can be determined from a single AFM scan of a single GeSi/Si(001) pyramidal island.

Another ulitmate requrement is that the hieght of the topographic elements on the standard sample should exceed the heihgts of the invesitgated objects [namely, the InAs/GaAs(001) QDs]. The self assembled GeSi/Si(001) pyramid islands satisfy this requirement as thier height h may reach ~ 10 nm when grown in the appropriate conditions, that is well enough for the InAs/GaAs(001) QDs (typically, $h = 5 \div 6$ nm). So far, the GeSi/Si(001) pyramid nanaoislands appear to be a good natural standard for the measurement of the probe shape in the particular case of the deconvolution of the AFM images of the self-assembled InAs/GaAs(001) QDs.

The raw images of the QDs seemd round, their lateral sizes D were 30 to 50 nm (Fig. 2, *a*). In the AFM images after the deconvolution (Fig. 2, *b*) the (101) faceting of the InAs/GaAs(001) QDs grown by AP MOVPE is seen clearly.

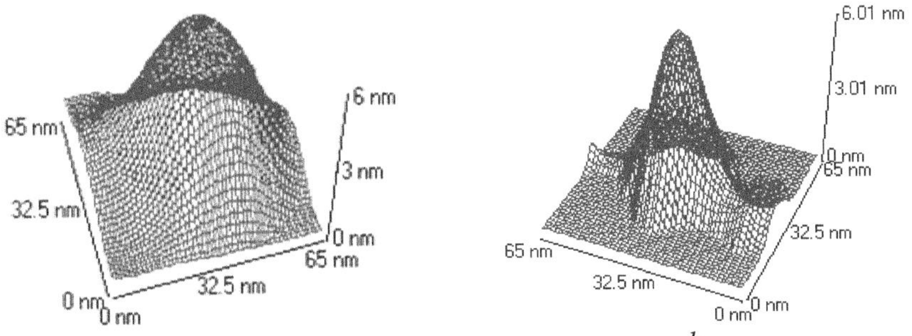

a *b*

Fig. 2. The AFM images of a surface InAs/GaAs(001) QD grown by AP MOVPE: as measured (*a*) and after the deconvolution (*b*).

The base sides were directed along <110>, their length $b = 14 \div 18$ nm like the ones grown by MBE (Ledentsov et al., 1999, Maltezopoulos et al., 2003). The top of the pyramid was truncated slightly (Fig. 2, *b*) that could be explained noting that the top of the pyramid is a concentrator of the tensile strain. So far, the truncation of the pyramid reduces the overall elastic energy of the QD. Similar shape of the QD tops has been observed by UHV STM (Maltezopoulos et al., 2003). Therefore, the aspect ratio A_R of the InAs/GaAs(001) QDs is less than 1:2 slightly.

The size quantization energy spectrum of the InAs/GaAs(001) QDs grown by AP MOVPE has been examined by photoluminescence (PL) spectroscopy at 77K and by the photoelectric spectroscopy at 300K. The structures for the PL spectroscopy have been grown on the semi-insulating GaAs(001) substrates, the buffer layers have been undoped intentionally. The InAs QDs in these structures have been capped by a 30 nm thick GaAs cladding layer. The structures for the photoelectric measurements had the same design except the substrates were form n^+-GaAs. The InAs in all three types of the structures (for Tunnelling AFM, for PL, and for the photoelectric spectroscopy) have been grown in the same conditions (T_g = 530°C, d_{InAs} = 5 ML). The details of the measurement techniques as well as the analysis of the experimental results can be found elsewhere (Karpovich et al., 2004b).

2.2 The conduction band states

In this subsection, the results of the Tunnelling AFM investigations of the quantum confined electron states in the surface InAs QDs are presented and discussed. Also, the results of the studies of the laterally coupled surface InAs QDs are presented.

The Tunnelling AFM studies were carried out at 300K using Omicron® MultiProbe P™ UHV system equipped by Omicron® UHV VT AFM/STM. A typical topographic image $z(x, y)$ and the probe current one $I_t(x, y)$ of an InAs/GaAs(001) QD sample are presented in Fig. 3, *a* & *b*, respectively. An increasing of I_t every time the AFM tip encounters the QD surface had been observed. This observation has been attributed to the electron tunnelling between the AFM tip and the conductive buffer layer through the quantum confined states in the QDs (a qualitative band diagram of a contact of a Pt coated AFM tip to an InAs/GaAs/n^+-GaAs biased negatively is presented in Fig. 4, *b*).

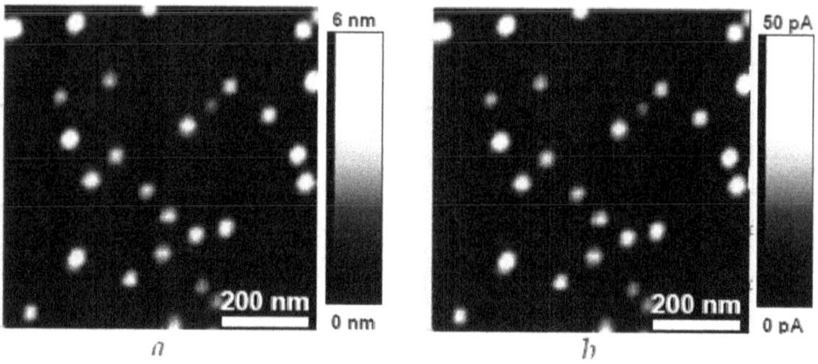

Fig. 3. The topographic (*a*) and the probe current (*b*) images of the surface InAs/GaAs/n^+-GaAs(001) QDs. V_g = -3.7 V. Reproduced from (Filatov et al., 2010) under license by IoP Publishing Ltd.

The differential conductivity spectra $\sigma_d(V_g) = dI_t/dV_g$ of the tunnel contact of a Pt coated AFM tip to different points of the surface an InAs/GaAs/n^+-GaAs(001) QD are presented in Fig. 4. , *a*. The $\sigma_d(V_g)$ spectra have been calculated from the measured $I_t(V_g)$ curves of the probe-to-sample contact by the numerical differentiation with the nonlinear smoothing. The points on the QD surface where the respective $I_t(V_g)$ curves had been measured are marked in Fig. 5, *a*. The peaks observed in the $\sigma_d(V_g)$ spectra were attributed to the tunnelling of the electrons between the metallic tip coating and the conductive substrate through the quantum confined states in the QDs (Fig. 4. , *b*). The native oxide on the sample surface formed a potential barrier, the second triangle potential barrier was formed by the depletion layer of the contract of the metal tip coating to the GaAs/n^+-GaAs. In the UHV STM studies (Maltezopoulos et al., 2003) the first potential barrier was formed by the vacuum gap between the metal STM probe and the sample surface.

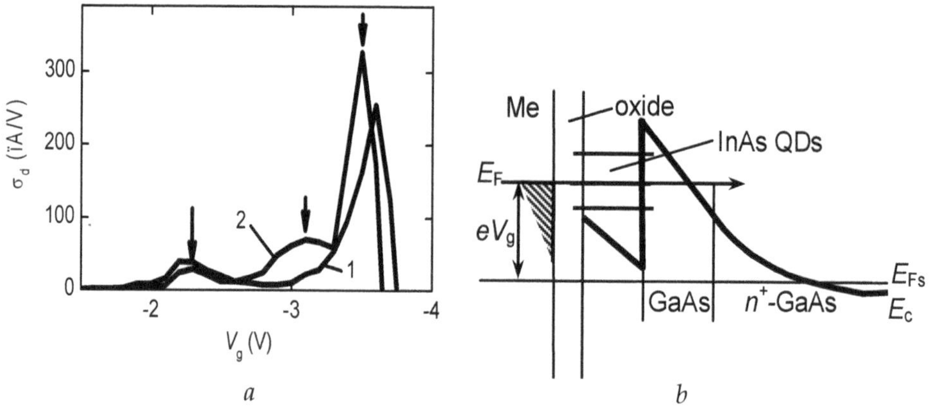

Fig. 4. The differential conductivity spectra $\sigma_d(V_g)$ (*a*) and the qualitative band diagram (*b*) of a negatively biased contact of a metal coated AFM tip to an InAs QD on the *n*-GaAs/n^+-GaAs (001) substrate. The curve numbers denote the points of the $I - V$ curves' measurements marked in Fig. 5, *a*. Reproduced from (Filatov et al., 2010) with permission from ©Pleiades Publishing, Ltd.

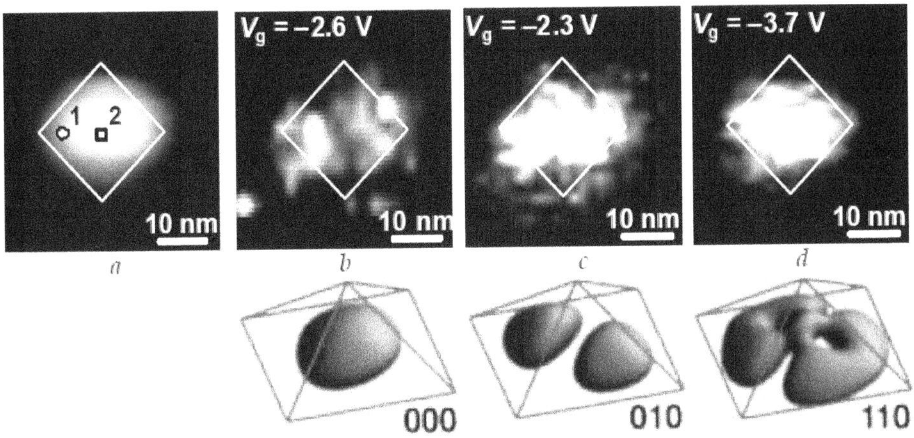

Fig. 5. The UHV AFM (a) and the probe current (b – d) images of an InAs QD on the n-GaAs/n^+-GaAs(001) substrate. Reproduced from (Filatov et al., 2010) with permission from ©Pleiades Publishing, Ltd. Below, the surfaces of the equal probability density $|\chi(x, y, z)|^2 = 0.65$ calculated for several quantum confined electron states $|n_1n_2n_3>$ in a pyramidal InAs/GaAs(001) QD with the base side length b = 16 nm are presented. Reproduced partly from (Stier et al., 1999) with permission from ©American Physical Society.

The AFM and the probe current images of an individual InAs/n^+-GaAs QD are presented in Fig. 5. Note that because of the convolution effect, the AFM image of the QD is rounded and enlarged as compared to the actual QD size revealed using the deconvolution (b = 14 ÷ 16 nm, see Sec. 2.1 above).

Following (Maltezopoulos et al., 2003), the patterns of the $I_t(x, y)$ images at certain values of V_g corresponding to the maxima of the $\sigma_d(V_g)$ spectra in Fig. 4, a have been related to the spatial distribution of the LDOS in the (x ,y) plane:

$$\rho_E(x,y) \propto \sum_{n_1n_2n_3=0}^{N_1N_2N_3} \left|\chi_{n_1n_2n_3}(x,y)\right|^2 . \qquad (1)$$

Here the envelope wavefunctions $\chi(x, y)$ were considered to be spin-independent and twofold spin-degenerated. The summation in (1) was taken over the states below the Fermi level in the probe coating material E_F. In other words, in the case of the QDs grown on the n-GaAs/n^+-GaAs(001) substrate the energies of the respective quantum confined states must satisfy the following condition:

$$E_{n_1n_2n_3} < E_{Fs} + eV_g , \qquad (2)$$

where E_{Fs} is the Fermi level in the n^+-GaAs buffer (see Fig. 4. , b). Condition (2) defines the upper limits of the summation N_1, N_2, and N_3 in (1).

Again, following (Maltezopoulos et al., 2003), the tunnel current images of the QDs $I_t(x, y)$ were compared to the probability density patterns $|\chi(x, y, z)|^2 = const$ calculated for the quantum confined electron states in the pyramidal InAs/GaAs(001) QDs (Stier et al., 1999).

The probe current images of the QDs obtained by Tunnelling AFM were more noisy than the ones obtained by STM in UHV (Maltezopoulos et al., 2003) that was attributed to the nonuniformity of the thickness of the native oxide covering the QDs. Nevertheless, several electron quantum confined states in the QDs were identified, the respective images of the $|\chi(x, y, z)|2 = 0.65$ surfaces (Stier et al., 1999) are shown in Fig. 5.

In order to associate the spectral positions of the peaks in the $\sigma_d(V_g)$ spectra with the quantum confined level energies $E_{n_1 n_2 n_3}$, one must take into account the partial voltage drop on the depletion layer between the QDs and the n^+-GaAs buffer layer as well as the one on the surface states at the semiconductor/native oxide interface. Following (Suzuki et al., 2007), we have applied a simple one-dimensional model based on the solution of one-dimensional Poisson's equation (Feenstra & Stroscio, 1987) to account for the voltage drop on the depletion layer of the probe-to-sample contact. In order to account for the surface charge density on the surface states on the interface between the sample surface and the native oxide, we have applied Hasegawa's model (Hasegawa & Sawada, 1983). The calculations have shown that approximately ½ of V_g drops on the surface states.

The InAs/GaAs (001) surface QD structures grown by AP MOVPE are featured by the presence of a considerable number of the QDs arranged in the pairs along the growth steps in a close proximity to each other (see Fig. 1, *b*) that may result in a considerable overlap of the envelope wavefunctions of the quantum confined states in the adjacent QDs. An example of the UHV AFM image as well as a series of the probe current images of a pair of the laterally coupled surface InAs/GaAs(001) QDs are presented in Fig. 6.

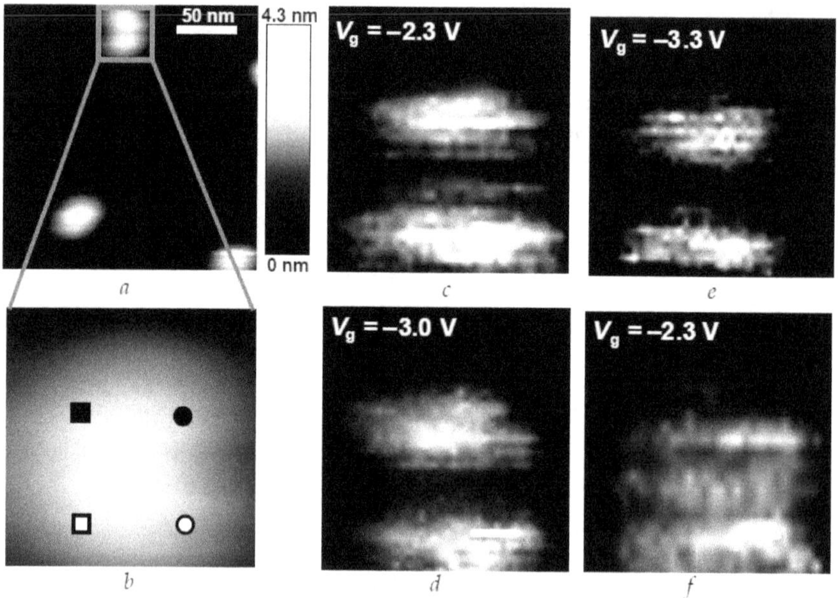

Fig. 6. The UHV AFM (*a*, *b*) and the probe current (*c* – *f*) images of the laterally coupled surface InAs/GaAs(001) QDs. The symbols in fig. (*b*) mark the points of the measurement of the respective tunnel spectra presented in Fig. 7.

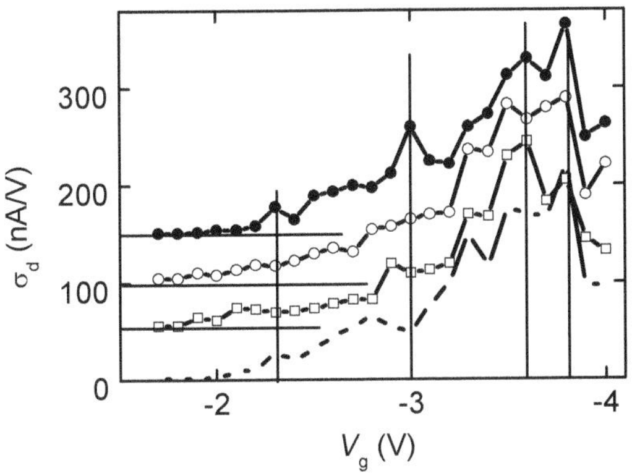

Fig. 7. The differential conductivity spectra $\sigma_d(V_g)$ of the tunnel contact of a Pt coated AFM tip to the laterally coupled InAs/n-GaAs/n^+-GaAs(001) QDs. The points of measurement of the initial $I-V$ curves are denoted in Fig. 6, *b* by the respective symbols.

The differential conductivity spectra $\sigma_d(V_g)$ of the tunnel contact of a Pt coated AFM tip to the laterally coupled InAs/n-GaAs/n^+-GaAs(001) QDs are presented in Fig. 7. The asymmetry of the current images in Fig. 6, *c*–*f* (as compared to the ones of the single QDs, see Fig. 5, *b*–*d*), which had varied with increasing V_g along with the splitting of the peaks related to the quantum confined states in the coupled QDs (Fig. 7) were attributed to the hybridization of the quantum confined states in the laterally coupled QDs. Similar patterns of the tunnel current images and tunnel spectra as well as of their dynamics with varying V_g have been observed while studying the hybridization of the quantum confined states in the GaSb/InAs(001) double symmetric QWs by X-STM in UHV (Suzuki et al., 2007).

2.3 The valence band states

In this subsection, the results of the Tunnelling AFM studies of the hole quantum confined states in the InAs/GaAs(001) QDs are presented. It is worth noting that the authors of the present chapter had applied Scanning Probe Microscopy technique for the studying of the valence band states in the InAs/GaAs(001) self assembled QDs for the first time.

The InAs/GaAs(001) QD samples for the investigation of the valence band states in the InAs QDs by Tunneling AFM have been grown on the p^+-GaAs(001) substrates. The GaAs buffer layers were also doped heavily by Zn from diethylzinc up to the acceptor concentration N_A ~ 10^{18} cm^{-3}. The intentionally undoped 3 nm thick GaAs spacer layers were grown prior to the deposition of InAs, as in the InAs/GaAs/n^+-GaAs(001) QD structures for the investigations of the electron states described in the previous subsection. The technological parameters of the process of growing the InAs QDs were the same, as in the case of the structures grown on the n^+-GaAs substrates: T_g = 530°C, d_{InAs} = 5 ML.

The differential conductivity spectra $\sigma_d(V_g)$ of the tunnel contact of a Pt coated AFM tip to an InAs QD on the GaAs/p^+-GaAs (001) substrate are presented in Fig. 8, *a*. The oscillations of the $\sigma_d(V_g)$ spectra have been attributed to the tunneling from the valence band states in the p^+-GaAs buffer to the free states above the Fermi level in the metal tip coating through the quantum confined hole states in the InAs QD (Fig 8, *b*).

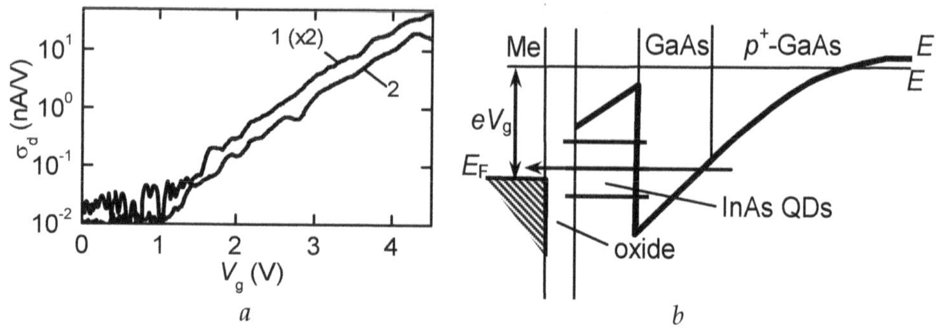

Fig. 8. The differential conductivity spectra $\sigma_d(V_g)$ (*a*) and the qualitative band diagram (*b*) of the positively biased contact of a Pt coated AFM tip to an InAs QD on the GaAs/p^+-GaAs (001) substrate. The curve numbers denote the points of the initial $I - V$ curves' measurements shown in Fig 9, *a*. Reproduced partly from (Filatov et al., 2010) under license from ©IoP Publishing, Ltd.

Comparing the $\sigma_d(V_g)$ spectra of the QDs on the p^+-GaAs (Fig. 8, *a*) with the ones of the QDs grown on n^+-GaAs (Fig 4, *a*), one can note that the peaks related to the tunnelling via the quantum confined electron states in the QDs were well resolved in the case of the InAs/n+-GaAs QDs while a nearly exponential curves with the weak oscillations only have been observed in the InAs/p^+-GaAs QDs(Fig. 8, *a*). This observation could be explained noting that according to (Stier et al., 1999), the energy spacing between the electron levels in the InAs/GaAs(001) QDs with the base size length $b = 15 \div 20$ nm is ~100 meV while the one for the hole levels is $10 \div 20$ meV only. As a result, the peaks in the $\sigma_d(V_g)$ spectra related to the quatum confined hole states in the InAs QDs were resolved poorly because of the thermal and structural broadening. Note also that the spectral spacing between the peaks in the $\sigma_d(V_g)$ spectra ascribed to the quantum confined hole states in Fig. 8, *a* was much less than the one for the conduction band states (Fig. 4, *a*), that agrees with proposed interpretation of the QDs' tunnel spectra as well.

The AFM and the probe current images of an individual InAs QD on the GaAs/p^+-GaAs(001) substrate are presented in Fig. 9. The probe current images $I_t(x, y)$ of the QD measured at different values of V_g were related to the lateral spatial distribution of the LDOS of the quantum confined hole states in the QD $\rho_E(x, y)$ at the respective vaules of E. In the case of the InAs QDs grown on the GaAs/p^+-GaAs(001) substrate, the hole quantum confined states, the energy of which satisfies the following condition:

$$E_{Fs} - eV_g < E_{n_1 n_2 n_3} < E_{Fs}\,, \tag{3}$$

where E_{Fs} is the Fermi level energy in the p^+-GaAs buffer (see the band sketch in Fig. 8, *a*) can manifest themselves in the probe current images $I_t(x, y)$.

Again, the maps of the probe current $I_t(x, y)$ have been compared to the equal probability density patterns $|\chi(x, y, z)|^2 = const$ calculated for several lower quantum confined hole states (Stier et al., 1999). Although the quality of the probe current images was not so good, several quantum confined states have been identified (Fig. 9).

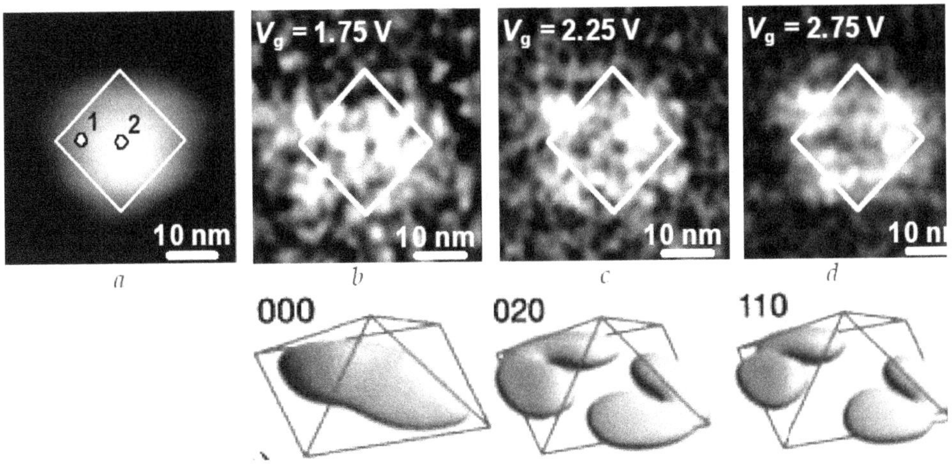

Fig. 9. The AFM (a) and the probe current ($b-d$) images of an InAs QD on the GaAs/p^+-GaAs(001) substrate. Reproduced partly from (Filatov et al., 2010) under license from ©IoP Publishing, Ltd. Below the surfaces of equal probability density $|\chi(x, y, z)|^2 = 0.65$ calculated for several quantum confined hole states $|n_1n_2n_3>$ in a pyramidal InAs/GaAs(001) QD with the base side length b = 16 nm are presented. Reproduced partly from (Stier et al., 1999) with permission from ©American Physical Society.

Having identified the quantum confined electron and hole states in the InAs/GaAs(001) QDs grown by AP MOVPE, we succeeded to identify the states the interband optical transition between which are manifested in the PS spectra of the QD structures grown by AP MOVPE.

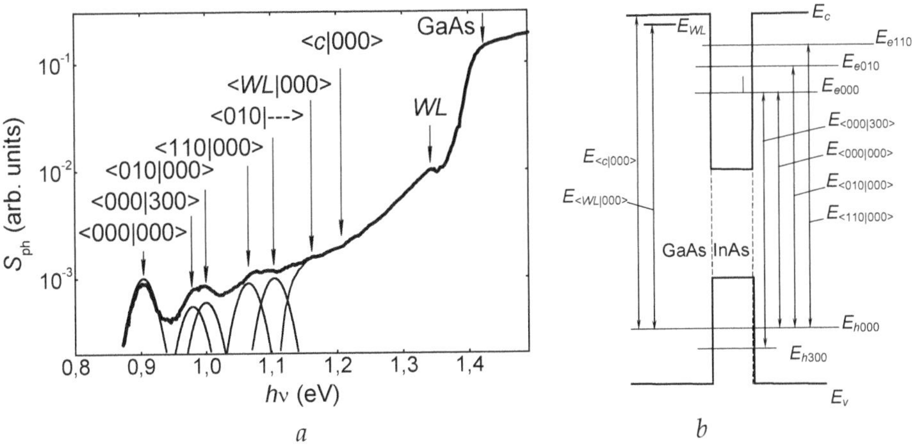

Fig. 10. The photosensitivity spectrum (a) and the band diagram (b) (300K) of an InAs/GaAs(001) QD structure grown by AP MOVPE. Reproduced from (Filatov et al., 2010) with permission from ©Pleiades Publishing Ltd.

An example of the PS spectrum $S_{ph}(h\nu)$ of an InAs/GaAs(001) QD structure measured by the photovoltage spectroscopy in a liquid electrolyte (Karpovich et al., 2004b) is presented in Fig. 10, *a*. The QDs in this particular structure were grown in the same conditions as in the samples for the Tunnelling AFM investigstions described above. Several peaks related to the interband optical transitions between the quantum confined electron and hole states in the QDs are present in the PS spectrum. Also, the PS bands related to the transitions from the ground hole state in the QD |000> to the ground electron states in the InAs wetting layer |WL> and to the conduction band states in GaAs have been observed. The band diagram of an InAs/GaAs QD structure with the transitions manifested in the PS spectrum (Fig. 10, *a*) shown schematically is presented in Fig. 10, *b*. This diagram was based on the best fit between the interband transition energies extracted from the PS spectrum in Fig. 10, *a* and the ones calculated from the data on the quantum confined level energies in the pyramidal InAs/GaAs(001) QDs as the function of the QD base size *b* (Stier et al., 1999). The best fit was found at $b \approx 16$ nm that is consistent with the ambient air AFM data obtained using the deconvolution (see Sec. 2.1). Note that many possible transitions are not manifested in the PS spectrum, because the overlap integrals between the envelopes of the respective electron and hole states are close to zero (Stier et al., 1999). Only the transitions between the states the overlap integrals for which are close to unity are manifested in the PS spectrum.

We have failed to identify one of the higher energy hole states marked as |---> in Fig. 10 because the data for the respective hole energy band are not presented in the publication by (Stier et al,. 1999). In the other aspects, the data on the morphology, on the electronic structure, and on the energy spectrum of the InAs/GaAs(001)QDs grown by AP MOVPE provided by ambient air and Tunnelling AFM, by the PL spectroscopy, and by the PS one appeared to be consistent with each other as well as with the results of the theoretical calculations reported in the literature (Stier et al., 1999).

3. InGaAs/GaAs(001) quantum rings

3.1 Growth and characterization

In this subsection, the details on the growth of the InAsGa/GaAs(001) QRs by AP MOVPE as well as of their characterization are presented.

The procedure of growing the self-assembled InAsGa/GaAs(001) QRs by AP MOVPE was different from that used in the standard MBE. Usually, in order to grow the InGaAs/GaAs(001) QRs by standard MBE, the InAs/GaAs(001) QDs are capped by a thin GaAs cladding layer with $d_c \approx \langle h \rangle$, and then the structures are annealed at the temperature $T_A \approx 600 \div 630$°C for several tens of minutes (Lorke et al., 2002). However, the QRs grown by AP MOVPE have been proven to form just during capping of the InAs QDs by the GaAs cladding layer (Baidus' et al., 2000). This phenomenon had been attributed to the enhanced surface diffusion of In adatoms in the AP MOVPE process as compared to MBE because in the former case the surface dangling bonds are passivated by hydrogen residual from the cracking of the metal organic compounds and/or arsine.

The InGaAs/GaAs(001) QR structures for the Tunnelling AFM investigations described in the present chapter have been grown on the n^+-GaAs(001) substrates misoriented by 5° towards <110>.

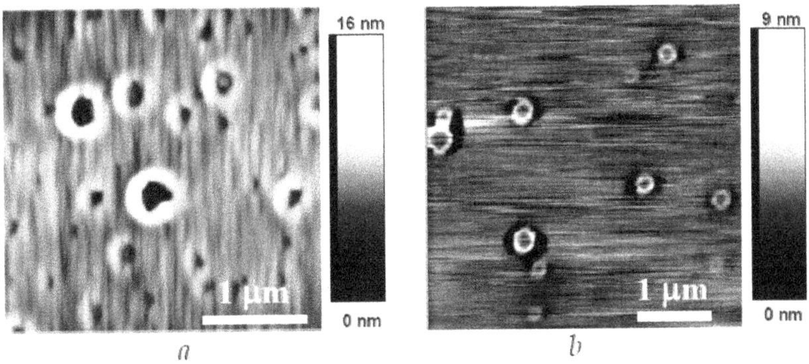

Fig. 11. The ambient air AFM image of an InGaAs/GaAs(001) QR structure (*a*); a liquid AFM image of an InGaAs/GaAs(001) QR structure with a 30 nm thick GaAs cladding layer removed by selective wet etching. Reproduced from (Filatov et al., 2011) with permission from ©Pleiades Publishing, Ltd.

The 3 nm thick intentionally undoped GaAs spacer layers have been grown on the n^+-GaAs buffer layers. Then, the InAs/GaAs QDs were grown at T_g = 530°C and capped by a GaAs cladding layer with the thickness d_c = 10 nm at T_g = 600°C. The nominal thickness of the InAs layerd_{InAs} was ≈ 5 ML. The ambient air AFM measurements have demonstrated the InGaAs QRs to form during capping the InAs QDs at elevated temperatures (Fig. 11, *a*). The QRs of various diameters D = 150 ÷ 400 nm have been observed on surface (Fig. 11, *a*). As it had been found earlier (Karpovich et al., 2004b), the structures with the surface InAs/GaAs(001) QDs grown by AP MOVPE are features by a number of the relaxed InGaAs clusters present on the surface (Fig. 12, *b*). These clusters are formed via the coalescence of the smaller InAs coherent QDs during growth, which, in turn, has been attributed to the enhanced surface diffusion in the AP MOVPE process. Within this paradigm, the formation of the smaller QRs has been attributed to the transformation of the smaller coherent InAs/GaAs QDs during capping while the formation of the larger QRs has been attributed to the transformation of the larger relaxed InGaAs clusters.

Also, the structure with d_c ≈ 30 nm was grown on a semi insulating GaAs(001) substrate in order to examine the optical properties of the QRs by PL spectroscopy. In order to reveal the morphology of the overgrown QRs, we have employed the removal of the GaAs cladding layer by wet selective etching followed by AFM investigation in liquid (Karpovich et al., 2004a). The application of liquid AFM was to avoid the probe fadeouts due to the residual etching solution contamination (Fig. 11, *b*).

The PL (77 K) and PS (300 K) spectra of the structures with the thickness of the GaAs cladding layer d_c ≈ 30 and 10 nm, respectively, are presented in Fig. 12, *a*. A peak of the edge PL of GaAs with the maximum energy $h\nu_m$ ≈ 1.51 eV (GaAs) and a peak with $h\nu_m$ ≈ 1.41 eV (WL) originating from the interband radiative transitions between the ground quantum confined states in the InAs/GaAs(001) wetting layer are present in the PL spectrum. Also, a broad PL band with $h\nu_m$ ≈ 1.2 eV attributed to the ground state transitions in the InGaAs/GaAs(001) QRs have been observed. This band has a longer wavelength shoulder and can be decomposed into two Gaussian components with $h\nu_m$ ≈ 1.2 and $h\nu_m$ ≈ 1.1 eV related to the smaller and larger QRs, respectively.

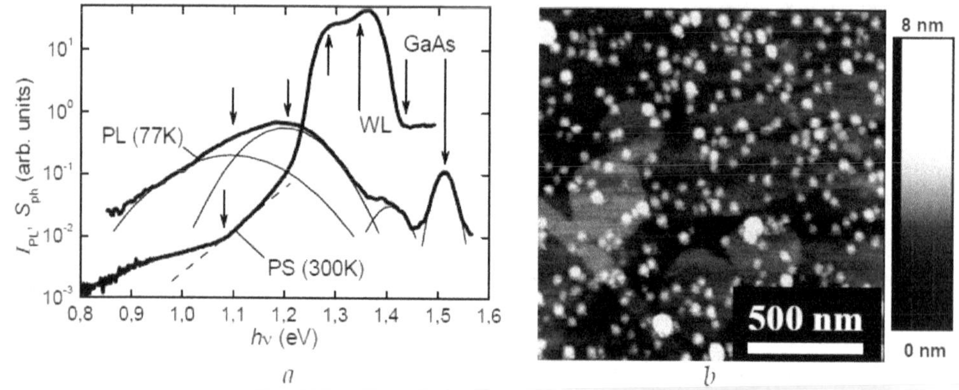

Fig. 12. The PL (77 K) and photosensitivity (300K) spectra of the InGaAs/GaAs(001) QR structures with the 30 nm and 10 nm cladding layers, respectively (*a*). Reproduced from (Filatov et al., 2011) with permission from ©Pleiades Publishing Ltd.; an AFM image of a surface InAs/GaAs(001) QDs (*b*).

Also, the PS bands with the edges at $h\nu \approx 1.36$ eV (WL), 1.29 eV, and 1.09 eV corresponding to the above PL peaks have been observed in the PS spectrum (Fig. 12, *a*). The difference in the values of the PL peak energies and the band edges ones in the PS spectrum could be related to the different in d_c, and, in turn, to the difference in the morphology, in the composition, and in the elastic strain of the QR material in these structures. Note that since the structure was photoexcited through the substrate, playing a role of a low-pass optical filter, the PS spectrum was truncated at $h\nu \approx 1.43$ eV that is the optical gap energy of n^+-GaAs. The PS band with the edge at $h\nu \approx 0.8$ eV was related, most likely, to the impurity PS from the defect complexes in GaAs (Karpovich et al., 2004b).

The In molar fraction x in the QR material ($In_xGa_{1-x}As$) was estimated from the PL and PS spectra to be 0.35 ÷ 0.4. The estimates were made by the best fit between the energies of the PL peak and of the PS band edges related to the ground state transitions in the QRs and the calculated values of the ground state transitions in the QRs E_0, x being the fitting parameter. The values of E_0 as a function of the QRs' sizes determined form the AFM images and of x were calculated by solving the Schrödinger's equation in the effective mass approximation for a model potential of a flat gasket-shaped ring (rectangular in the cross-section) with potential wells of finite height (Barticevic et al., 2000). The ring material was assumed to be pseudomorphic to the GaAs matrix. The materials parameters (the effective masses of electron and holes masses, the band offsets, etc.) were taken from (Stier et al., 1999).

3.2 Tunneling AFM investigations

In this subsection, the results of the investigations of the electronic structure of the self assembled InGaAs/GaAs(001) QRs by Tunnelling AFM are presented. Again, it is worth noting that the authors of the present chapter had applied Scanning Probe Microscopy technique for the studying of the LDOS in the self assembled semiconductor QRs for the first time.

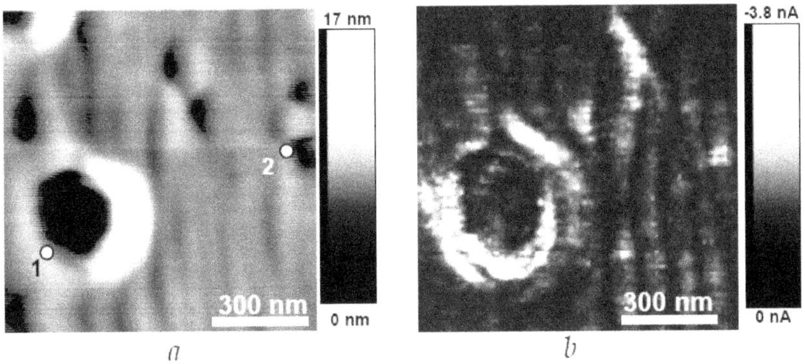

Fig. 13. The UHV AFM (*a*) and the probe current (*b*) images of an InGaAs/GaAs(001) QR structure. V_g = -1.37 V. The figures (1) and (2) mark the points of measurement of the $I - V$ curves (the respective $\sigma_d(V_g)$ spectra are presented in Fig. 14). Reprinted from (Filatov et al., 2010) under license by IoP Publishing Ltd.

The UHV AFM and the probe current images of an InGaAs/GaAs(001) QR structure are presented in Fig. 13. The QRs are manifested in the probe current image $I_t(x, y)$ by the increased probe current I_t that was related to the tunnelling of the electrons from the probe coating material into the n^+-GaAs buffer through the quantum confined states in the InGaAs/GaAs(001) QRs (Fig 4, *b*). The differential conductivity spectra $\sigma_d(V_g)$ of the contact of a Pt coated AFM tip to the surface of the InGaAs QRs on the n-GaAs/n^+-GaAs (001) substrate are presented in Fig. 14. The peaks related to the quantum confined states in the QRs are well expressed in the $\sigma_d(V_g)$ spectrum of the smaller QR (Fig. 14, *b*). However, the features, which could be related to the quantum size effect are present in the $\sigma_d(V_g)$ spectrum of the larger QR as well (Fig. 14, *a*). So far, the larger QRs could be classified as the quantum size structures as well in spite of their relatively large sizes. The calculations have shown that the size quantization in z direction (normal to the substrate) affects the energy spectrum of the electrons and holes in QRs most.

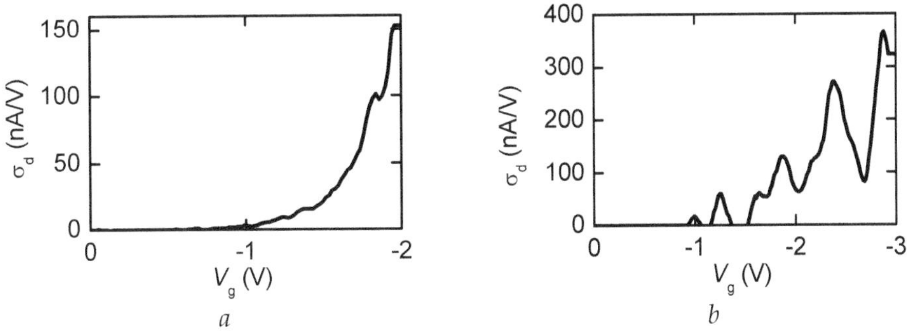

Fig. 14. The differential conductivity spectra $\sigma_d(V_g)$ of the contact of a Pt coated AFM tip to the surface of the InGaAs QRs of larger (*a*) and smaller (*b*) size on the n-GaAs/n^+-GaAs (001) substrate. The points of measurement of the initial $I - V$ curves are marked in Fig. 13, *a*. Reproduced from (Filatov et al., 2011) with permission from ©Pleiades Publishing, Ltd.

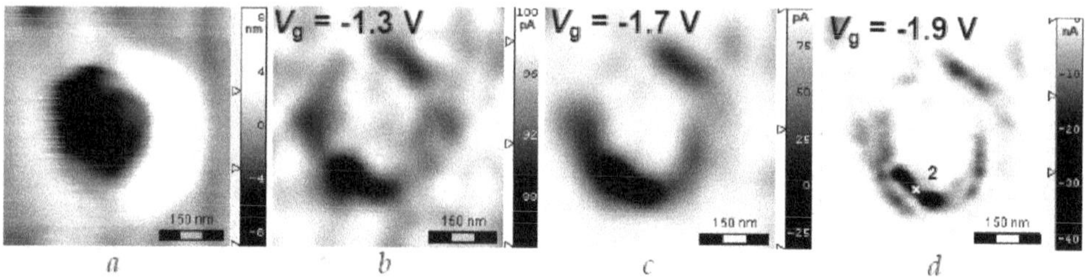

Fig. 15. The AFM (a) and the inverted probe current ($b-d$) images of an InGaAs QR on the n-GaAs/n^+-GaAs(001) substrate. Reproduced partly from (Filatov et al., 2011) with permission from ©Pleiades Publishing, Ltd.

The AFM images and the inverted probe current images of an individual InGaAs/GaAs(001) QR measured at different values of V_g are presented in Fig. 15. The probe current images of the QR change with increasing V_g. The most important result of the studies presented in this section was the observation of the angular patterning in the probe current images of the QRs. Ideally, in the circularly symmetric potential (Curie group of symmetry C_∞), the angular dependence of the envelope eigenfuntions of the quantum confined states is expressed in the polar coordinates (r, φ) by the term exp($-il\varphi$) where i is the imaginary unity and $l = 0, \pm 1, \ldots$ is the angular quantum number. As $|\exp(-il\varphi)|^2 = 1$ for any l and φ, no angular dependence of I_t should be observed.

The observation of the angular patterning in the current images of the QRs has been attributed to the asymmetry of the QRs' shape. In addition, one should take into account the effect of the piezoelectric field in the strained InGaAs, which also reduces the potential symmetry from C_∞ downto C_{2v} even in a perfectly round QR (Stier et al., 1999).

In order to account for the effect of the effect of the strain-induced piezoelectric field on the angular pattern of the LDOS in the InGaAs/GaAs(001) QRs, we have applied a simple model in the framework of the first-order perturbation theory. The lateral part of the spin-independent eigenfunction for the model flat gasket finite well height potential (Barticevic et al., 2000) can be written in the form:

$$\chi_{\mathrm{ml}}(\mathrm{r},\varphi)=\chi_{\mathrm{ml}}(\mathrm{r})\exp(-\mathrm{i}\mathrm{l}\varphi), \tag{4}$$

where $\chi_{ml}(r)$ is the radial part of $\chi_{ml}(r, \varphi)$ expressed via the Bessel functions and m is the radial quantum number. As the perturbation piezoelectric potential has the C_{2v} symmetry, an additional integral of motion, parity of states, appears. So, one can write the angular components of the correct envelopes of the zero-order approximation as

$$\begin{aligned}
\chi_0^{(0)}(\varphi) &= \frac{1}{\sqrt{2\pi}} \\
\chi_{\mu+}^{(0)}(\varphi) &= \frac{1}{\sqrt{\pi}}\cos(\mu\varphi), \\
\chi_{\mu-}^{(0)}(\varphi) &= \frac{1}{\sqrt{\pi}}\sin(\mu\varphi)
\end{aligned} \tag{5}$$

where $\mu = |l| = 1, 2, \ldots, |m-1|$.

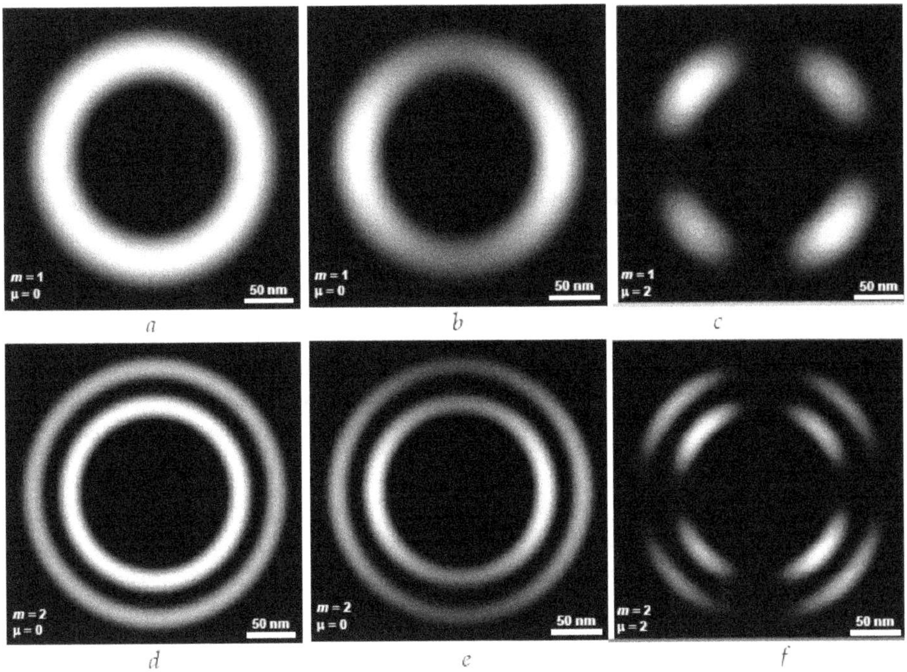

Fig. 16. The model maps of the spatial distribution of the lateral component of the probability density of the envelope wavefunction of the quantum confined states $|\chi_{nl}(r, \varphi)|^2$ in the plane of an $In_{0.35}Ga_{0.65}As$/GaAs(001) for different values of m and μ. (*a, d*): no perturbation potential; (*b, c, e, f*): the perturbation potential $V_0\cos(2\varphi)$ is imposed. Reproduced from (Filatov et al., 2011) with permission from ©Pleiades Publishing Ltd.

The model perturbation potential to account for the effect of the piezoelectric field has been selected in the form $V(\varphi)= V_0\cos(2\varphi)$. Such a model potential satisfies the symmetry C_{2v} and is convenient for the calculations of the matrix elements which appeared to be equal to zero except the $\langle ml | ml\pm2\rangle$ ones.

The model maps of $|\chi_{nl}(r, \varphi)|^2$ calculated for the $In_{0.35}Ga_{0.65}As$/GaAs(001) QRs with the outer and the inner radii of 70 and 140 nm, respectively, are presented in Fig. 16. When no perturbation potential is imposed, no angular patterning of the probability density maps takes place (Fig. 16, *a*&*d*). However, the model $|\chi_{nl}(r, \varphi)|^2$ maps calculated in the first order of the perturbation theory for V_0 = 5 meV demonstrate well expressed angular patterning. Note that the probability density images keep the C_{2v} symmetry regardless to the number of knots in the circle, in accordance with Curie's theorem.

Comparing the calculated $|\chi_{nl}(r, \varphi)|^2$ patterns with the the experimental probe current images of the QRs (cf, for example, Fig. 16, *c* & Fig. 15, *b*; Fig. 16, *f* & Fig. 15, *d*) demonstrate that in general the model presented above describes the qualitative features of the experimental current images more or less accurately. In turn, this could be considered as an evidence for the asymmetry of the quantum confining potential in the QRs as the origin of the angular patterning of these ones.

4. GeSi/Si(001) nanoislands

4.1 Growth and characterization

In this section, the results of the investigation of the LDOS in the self-assembled GeSi/Si(001) self assembled nanoislands by Tunnelling AFM are presented. Again, it should be stressed here that the authors of the present chapter had applied Scamming Probe Microscopy to the investigation of the LDOS in the self-assembled GeSi/Si(001) nanoislands for the first time. (Borodin et al., 2011).

The heterostructures with the surface GeSi/Si(001) nanoislands for the Tunnelling AFM investigations had been grown on the p^{+}-Si(001) substrates by SMBE in GeH_4 ambient using a home-made UHV setup (Svetlov et al., 2001). The qualitative band diagrams of the structures for the Tunnelling AFM investigations are presented in Fig. 19, *b* & *c*. The buffer layers of $\approx$ 200 nm in thickness were doped heavily by boron up to the hole concentration ~ 10^{18} cm^{-3}. More details on the growth technique can be found elsewhere (Filatov et al., 2008a, 2008b).

The topography of the samples destined to the Tunnelling AFM investigations was examined first by ambient air AFM. The AFM images of the structures with the surface GeSi/Si(001) nanoislands grown at various growth temperatures T_g are presented in Fig. 17. A dense array of the pyramid-shaped nanoislands ($N_s \sim 10^{11}$ cm^{-2}) has been observed on the surface of the sample grown at T_g = 500°C (Fig. 17, *a*). The (105) faceting of the pyramid nanoislands was resolved poorly, again, due to the convolution effect. A bimodal size distribution of the nanoislands took place on the surface of the sample grown at T_g = 600°C (Fig. 17, *b*). Along with the dome-shaped GeSi islands, a small number of larger so called super dome islands (Kamins et al., 1999) has been observed on the surface. The super dome islands originate from the coalescence of the smaller dome ones. It should be noted that in the structures grown by MBE the formation of the super dome islands has been observed in the process of the post-growth annealing (Kamins et al, 1999). In contrary, in SMBE in GeH_4 ambient the formation of the super dome islands has been observed just during growth (Filatov et al., 2008a, 2008b), that was attributed to Ostwald ripening.

The surface density of the islands N_s decreased and their sizes increased with increasing T_g while the bimodal size distribution remained in the structures grown at T_g = 700°C and 800°C (Fig. 17, *c* & *d*), in accordance with Lifshits-Sloyzov-Wagner theory.

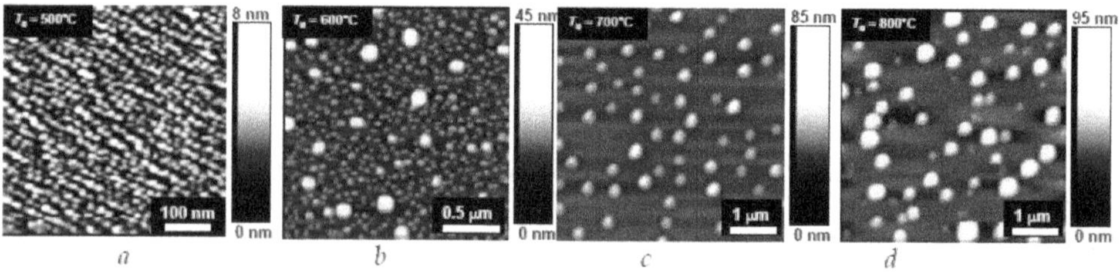

Fig. 17. The ambient air AFM images of the surface GeSi/Si(001) nanoislands grown by SMBE in GeH_4 ambient at various growth temperatures T_g.

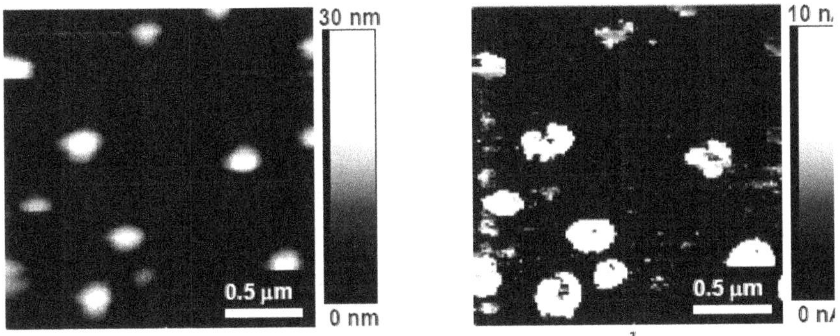

a *b*

Fig. 18. The UHV AFM (*a*) and the probe current (*b*) images of the GeSi/Si(001) nanoislands. V_g = 2.0 V. Reproduced from (Borodin et al., 2011) with permission from ©Pleiades Publishing, Ltd.

The composition of the GeSi/Si(001) nanoislands' material has been examined by PL spectroscopy (Filatov et al., 2008a, 2008b) and by Confocal Raman Microscopy (CRM, Mashin et al., 2010). The GeSi/Si(001) structures grown on the *p*-Si substrates with the nanoislands grown in the same conditions as the surface ones destined to the Tunnelling AFM studies but capped with the 40 nm thick cladding Si layers had been used for the optical investigations. It has been found that the nanoislands grown within T_g = 600 ÷ 800°C consisted of Ge_xSi_{1-x} alloy with the Ge molar fraction x decreasing from ≈0.55 downto ≈ 0.25 with T_g increasing from 600°C up to 800°C (Filatov et al., 2008b, Mashin et al., 2010).

4.2 Tunneling AFM investigations

The UHV AFM and the probe current images of the surface GeSi/Si(001) nanoislands grown at T_g = 700°C are presented in Fig. 18, *a* and *b*, respectively. The GeSi islands of various sizes and shapes have been observed on the sample surface (Fig. 18, *a*). The smaller islands were dome shaped while the larger ones (super dome islands) were shaped as the truncated pyramids defined by the (101) facets.

The spots of increased probe current I_t in the probe current image (Fig. 18, *b*) were located at the places corresponding to GeSi islands. The increased probe current has been related to the tunnelling of the electrons from the electronic states in the valence band of the GeSi islands into the free states above the Fermi level E_F in the Pt AFM tip coating through the native oxide layer covering the island surface. The band diagram of a positively biased contact between the Pt coated AFM tip and a surface GeSi/*p*-Si/*p*+-Si island is shown in Fig. 19, *c*. It should be noted that the sizes of the spots of increased I_t in the probe current image (Fig. 18, *b*) are larger than the sizes of the topographic images of the respective islands in Fig. 18, *a*. This has been attributed to the convolution effect due to the relatively large radius of the curvature of the AFM tip R_p. Typical values of R_p for NT MDT® NSG-01 Pt coated AFM probes are ≈ 35 nm, according to the vendor's specifications.

The differential conductivity spectra $\sigma_d(V_g)$ of the contact of a Pt coated AFM tip to the structure with the surface pyramid Ge/Si(001) nanoislands grown at T_g = 500°C are presented in Fig. 19, *a*.

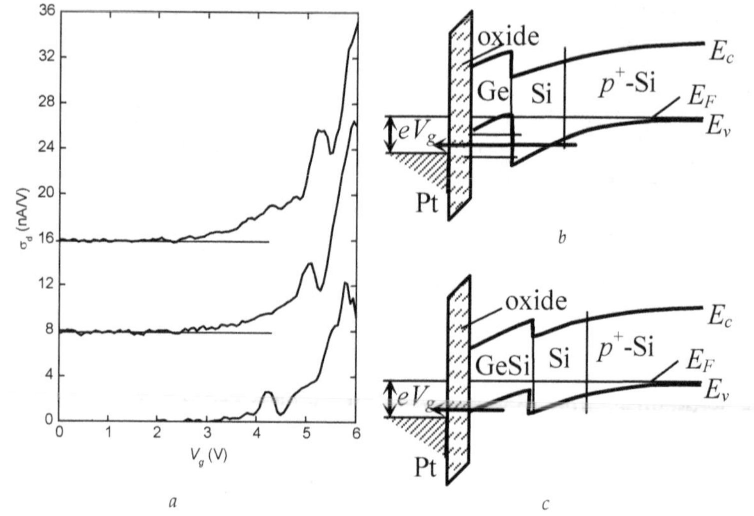

a *c*

Fig. 19. The differential conductivity spectra $\sigma_d(V_g)$ of the contact of a Pt coated AFM tip to the structure with the surface pyramid Ge/Si(001) nanoislands grown at T_g = 500°C (*a*), the qualitative band diagrams of the positively biased contact s of the Pt coated AFM tip to the pyramid islands (*b*) and to the dome-shaped or the super-dome islands (*c*).

Again, as in the cases of the InAs/GaAs(001) QDs and of the InGaAs/GaAs(001) QRs (see Sec. 2 & 3 above), the differential conductivity spectra $\sigma_d(V_g)$ have been calculated from the measured $I_t(V_g)$ curves by numerical differentiation with non-linear smoothing. The spectra presented in Fig. 19, *a* were the results of the averaging of 25 $I_t(V_g)$ curves measured in the different areas on the sample surface 5 × 5 pixels by size. The peaks in the $\sigma_d(V_g)$ spectra were ascribed to the tunneling of the electrons from the valence band states in the p^+-Si buffer through the quantum confined hole states in the Ge/Si(001) pyramid islands into the free states above the Fermi level in the Pt AFM tip coating (Fig. 19, *b*).

We have failed to obtain the probe current images, which could be associated to the probability density patterns $|\chi(x, y)^2|$ of the quantum confined hole states in the pyramid Ge/Si(001) nanoislands. The most probable cause for this, in our opinion, were too small sizes (b = 50 ÷ 80 nm and h = 5 ÷ 8 nm) and too large surface density ($N_s \sim 10^{11}$ cm^{-2}) of the Ge nanoislands on the surface of this particular sample as compared to the AFM probe tip dimensions ($R_p \approx 35$ nm). As a result, the islands have been resolved poorly in the UHV AFM images due to the convolution artifact unlike the ambient air AFM (Fig. 17, *a*) measured with NT MDT® CSG-01 probes without metal coating ($R_p < 10$ nm, according to the vendor's specifications).

The differential conductivity spectra $\sigma_d(V_g) = dI_t/dV_g$ and the normalized differential conductivity spectra $(dI_t/dV_g)/(I_t/V_g)$ of the contact of the Pt coated AFM probe tip to the GeSi/Si(001) nanoislands grown at various temperatures and to the sample surface between the islands (WL) are presented in Fig. 20. The points on the sample surface where the initial $I_t(V_g)$ curves have been measured are marked in Fig. 21. The spectra presented in Fig. 20 were the results of the averaging of 25 $I_t(V_g)$ curves measured in the spots on the sample surface 5 × 5 pixels by size.

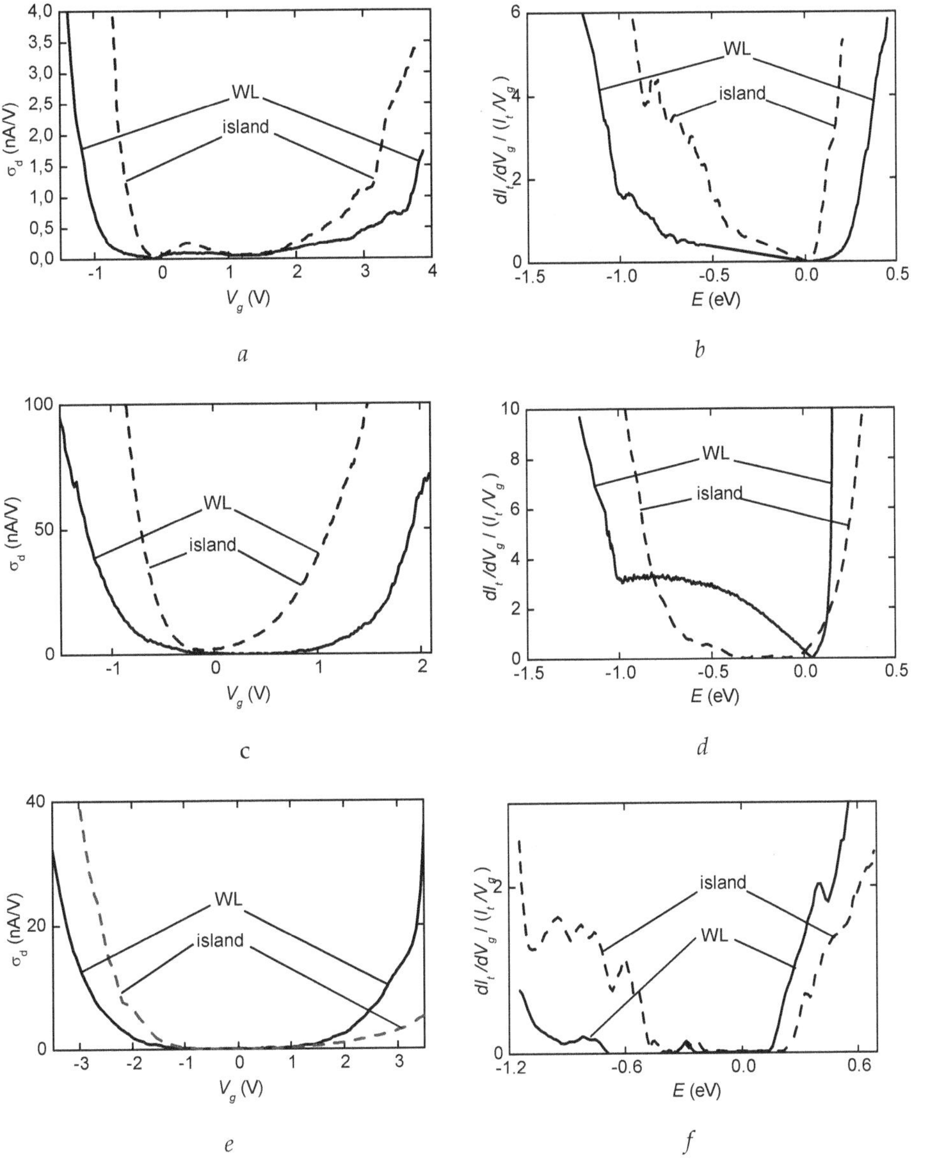

Fig. 20. The differential conductivity spectra $\sigma_d(V_g)$ (*a, c, e*) and the normalized differential conductivity spectra $(dI_t/dV_g)/(I_t/V_g)$ (*b, d, f*) of the contact of the Pt coated AFM probe tip to the GeSi/Si(001) nanoislands and to the sample surface between the islands (WL). The points on the sample surface where the initial $I_t(V_g)$ curves have been measured are denoted in Fig. 21. Reproduced partly from (Borodin et al., 2011) with permission from ©Pleiades Publishing, Ltd.

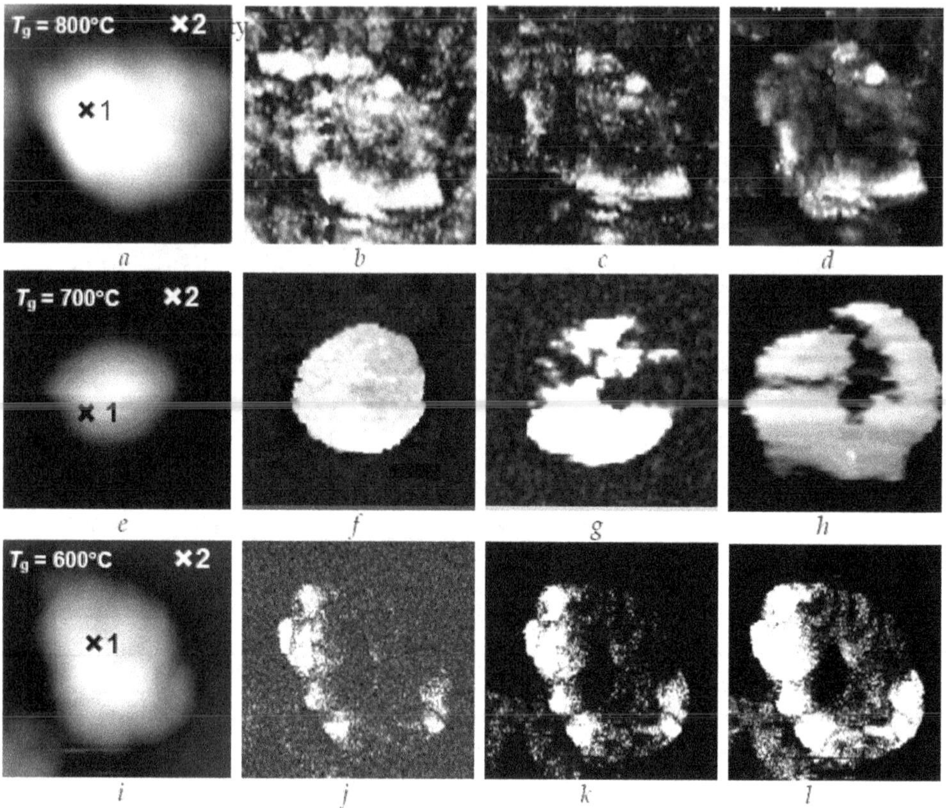

Fig. 21. The AFM (*a, e, i*) and the runnel current (*b – d, f – h, j – l*) images of the individual GeSi/Si(001) nanoislands grown by SMBE in GeH_4 ambient at various growth temperatures T_g. The points of measurements of the $I_t(V_g)$ curves the derivatives of which are presented in Fig. 20 are denoted in the AFM images (*a, e, i*). Reproduced partly from (Borodin et al., 2011) with permission from ©Pleiades Publishing, Ltd.

According to the theory of the tunnel spectroscopy, the normalized differential conductivity spectra $(dI_t/dV_g)/(I_t/V_g)$ of the contact of a metal STM tip to a sample is proportional to the LDOS at the sample surface. The relationship of the energies in the LDOS spectra to V_g, again, has been established taking into account the partial drop of V_g on the depletion layer of the tip-to-sample contact on the base of one-dimensional Poisson's equation (Feenstra & Stroscio, 1987). Unlike the case of the InAs/GaAs(001) QDs considered above in Sec. 2, the charge on the surface states on the boundary of the semiconductor with the native oxide has been neglected because the SiO_2/Si and GeO_2/Ge interfaces are known to be featured by low density of the surface states.

As it is evident from Fig. 20, tunnel spectra of the GeSi/Si(001) nanoislands grown at T_g = 700 and 800°C demonstrate the I-type conduction band alignment. Traditionally, the GeSi/Si(001) heterostructures had been being considered to be the II-type ones, i. e. the GeSi layer had been being treated as a potential barrier for the electrons with respect to the ones in Si (Berbezier & Ronda, 2009). However, there has been an increasing number of publications, both theoretical and experimental, where the $Ge_xSi_{1-x}/Si(001)$ heterostructures have been reported to be of the I type ones at low enough values of x, i. e. the GeSi layers were proven to be the potential wells for the electrons. More recent 30-band $k{\cdot}p$ calculations (El Kurdi et al., 2006) have demonstrated the pseudomorphic $Ge_xSi_{1-x}/Si(001)$ heterolayers to be the ones of the I type when $0.05 < x < 0.45$. In contrary, the GeSi/Si(001) nanoislands were reported to be always the II type heterostructures within the whole range of $0 < x < 1$ due to the nonuniform tensile strain of Si near the tops and the bottoms of the islands (El Kurdi et al., 2006). Anyway, the magnitude of the conduction band offset ΔE_c do not exceed several meV.

The $I_t(x, y)$ images of the GeSi/Si(001) islands were found to depend on V_g. At lower V_g corresponding to the extraction of the electrons From the electron states near the top of the valence band in the GeSi islands (see Fig. 19, *b*), the probe current images had more or less round shape (Fig. 21, *b, f, j*). At higher V_g the current image patterns transformed into the ones having a twofold-like symmetry Fig. 21, *c, g, k*). With further increasing of V_g, corresponding to the extraction of the electrons from the electron states near the Si valence band edge, the current images took a 4-fold-like symmetry Fig. 21, *d, h, l*) that was attributed to the elastic strain relaxation at the pyramidal island edges.

The values of ΔE_c at the $Ge_xSi_{1-x}/Si(001)$ nanoislands' heterointerface determined from the tunnel spectra of the individual islands presented in Fig. 20 are plotted *vs* the averaged Ge molar fraction in the nanoisland material $\langle x \rangle$ determined by PL spectroscopy and CRM (Filatov et al., 2008b, Mashin et al., 2010) in Fig. 22, *a*. Also, the theoretical dependence of ΔE_c on $\langle x \rangle$ calculated for a strained $Ge_xSi_{1-x}/Si(001)$ heterosturcture at 300K according to (Aleshkin & Bekin, 1997) is presented in Fig. 22, *a*. The theory predicts the strained $Ge_xSi_{1-x}/Si(001)$ heterostructures to be the ones of the I type when $0 < x < 0.44$, the conduction band minima in the strained GeSi are related to the Δ_4 valleys. The experimental data of the tunnel spectroscopy agree qualitatively with the theory by (Aleshkin & Bekin, 1997) as well as the one by (El Kurdi et al., 2006), although the magnitudes of ΔE_c at the respective values of $\langle x \rangle$ are larger than the ones predicted by the theory.

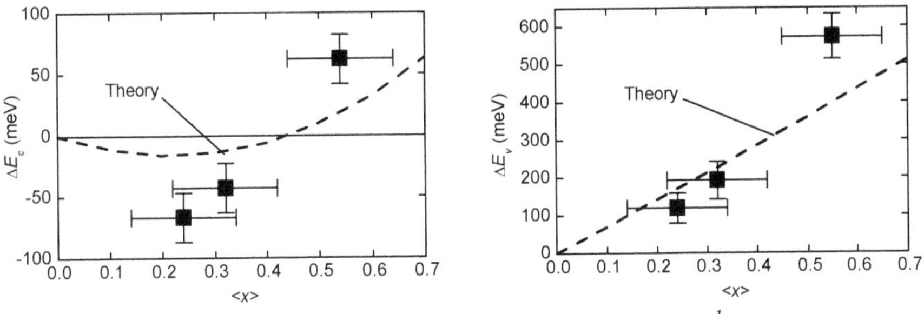

a *b*

Fig. 22. The dependencies of the conduction band offset ΔE_c (*a*) and of the valence band one ΔE_v (*b*) in the surface $Ge_xSi_{1-x}/Si(001)$ nanoislands on the averaged Ge molar fraction in the island material <>. The theoretical curves have been calculated for the strained $Ge_xSi_{1-x}/Si(001)$ layers according to (Aleshkin & Bekin, 1997).

This disagreement could be attributed probably to the partial strain relaxation in the surface GeSi/Si(001) nanoislands as compared to the pseudomorphic GeSi/Si(001) heterostructures or to the coherent GeSi/Si(001) nanoislands embedded into the single crystal Si.

Also, the values of the valance band offsets at the Ge_xSi_{1-x}/Si(001) nanoislands' heterointerface ΔE_v determined from the tunnel spectra presented in Fig. 20 are plotted *vs* <*x*> along with the ones calculated according to (Aleshkin & Bekin, 1997) in Fig. 22, *b*. A rather good agreement between the calculated and measured values of ΔE_v supports the conclusions about the type of the conduction band alignment in the self assembled GeSi/Si(001) nanoislands derived out from the tunnel spectroscopy data. Note that unlike the tunnel spectra of the Ge/Si(001) pyramid islands (Fig. 19, *a*), the tunnel spectra of the larger GeSi/Si(001) islands with lateral sizes $D > 100$ nm and height $h > 20$ nm (Fig. 20) do not exhibit any features, which could be attributed to the size quantization. This is not surprising taking into account the relatively large sizes of the islands grown at $T_g \geq 600$°C as compared to the de Broglie wavelength for the holes in GeSi alloy at 300K.

5. Conclusion

The results presented in this chapter demonstrate applicability of Tunneling AFM to the *ex situ* investigations of the LDOS of the quantum confined states in the InGaAs/GaAs and GeSi/Si nanostructures covered by the native oxide. The observed patterns of the probe current images agree with the results of the quantum confined states eigenfunction calculations reported in the literature. Besides, the application of Tunnelling AFM technique have brought some fundamental results on the basic properties of various nanostructures. Particularly, the angular patterning of the current images of the InGaAs/GaAs(001) quantum rings was related to the asymmetry of the confining potential. Also, the self assembled Ge_xSi_{1-x}/Si(001) nanoislands have been proven to be the I type heterostructures at $x < 0.45$.

6. Acknowledgement

The authors gratefully acknowledge the financial support from Russian Foundation of Basic Research (RFBR 10-02-90738-mob_st) and from Ministry of Science and Education, Russian Federation (NK-346P-25 and RNP 2.1.1.3615).

7. References

Baidus', N.V., Zvonkov, B.N., Filatov, D.O., Guschina, Yu.Yu., Karpovich, I.A. & Zdoroveischev, A.V. (2000). Investigaton of the overgrowth of the InAs nanoislands in the GaAs/InAs quantum dot heterostructures grown by Vapor Phase Epitaxy. *J. Surf. Investigations: Physics, Chemistry & Mechanics*, No. 7 (July 2000), pp. 71-75, ISSN 1027-4510.

Barticevic, Z., Pacheco, M. & Latge, A. (2000). Quantum rings under magnetic fields: Electronic and optical properties. *Phys. Rev. B*, Vol. 62, No. 11 (September, 2000), pp. 6963-6966, ISSN 1098-0121.

Berbezier, I. & Ronda, A. (2009). SiGe nanostructures. *Surf. Sci. Rep.*, Vol.64, No. 2 (February 2009), pp. 47-98, ISSN 0167-5729.

Borodin, P.A., Bukharaev, A.A., Filatov, D.O., Isakov, M.A., Shengurov, V.G., Chalkov, V.Yu. & Denisov, Yu.A. (2011). Investigation of the local density of states in self-assembled GeSi/Si(001) nanoislands by Combined Scanning Tunneling and Atomic Force Microscopy. *Semicond.*, Vol.45, No.3 (March 2011), pp.403–407, ISSN 1063-7826.

Bukharaev, A.A., Berdunov, N.V., Ovchinnikov, D.V. & Salikhov, K.M. (1998). Three-dimensional probe and surface reconstruction for Atomic Force Microscopy using a deconvolution algorithm. *Scanning Micros.*, Vol. 12, No. 1 (January 1998), pp. 225-234, ISSN 0891-7035.

El Kurdi, M., Sauvage, S., Fishman, G. & Boucaud, P. (2006). Band-edge alignment of SiGe/Si quantum wells and SiGe/Si self-assembled islands. *Phys. Rev. B*, Vol. 73, No. 11 (May 2006), p. 195327 (9 p.), ISSN 1098-0121.

Feenstra, R.M. & Stroscio, J.A. (1987). Tunneling spectroscopy of the GaAs (110) surface. *J. Vac. Sci. Technol. B* Vol. 5, No. 4 (April 1987), pp. 923-928, ISSN 0734-211X.

Filatov, D.O., Kruglova, M.V., Isakov, M.A., Siprova, S.V., Marychev, M.O., Shengurov, V.G., Chalkov, V.Yu. & Denisov, S.A. (2008a). Morphology and photoluminescence of self-assembled GeSi/Si nanoclusters grown by Sublimation Molecular-Beam Epitaxy in a GeH_4 ambient. *Bull. RAS: Physics*, Vol. 72, No. 2 (February 2008), pp. 249-252, ISSN 1062-8738.

Filatov, D.O., Kruglova, M.V., Isakov, M.A., Siprova, S.V., Marychev, M.O., Shengurov, V.G., Chalkov, V.Yu. & Denisov, S.A. (2008b). Photoluminescence of GeSi/Si nanoclusters formed by Sublimation Molecular-Beam Epitaxy in GeH_4 medium. *Semicond.*, Vol. 42, No. 9 (September 2008), pp. 1098-1103, ISSN 1063-7826.

Filatov, D.O., Lapshina, M.A., Isakov, M.A., Borodin, P.A. & Bukharaev, A.A. (2010). Tunnelling AFM study of the local density of states in the self assembled In(Ga)As/GaAs(001) quantum dots and rings. *J. Phys. Conf. Ser.* Vol. 245 (October, 2010), p. 012017 (4 p.), ISSN 1742-6596.

Filatov, D.O., Borodin, P.A. & Bukharaev, A.A. (2011). Study of local density of electron states in InGaAs/GaAs quantum rings by Combined STM/AFM. (2011). *J. Surf. Investigation. X-ray, Synchrotron, & Neutron Techniques,* Vol. 5, No. 3 (March 2011), pp. 547–553, ISSN 1027-4510.

Grandidier, B., Niquet, Y. M., Legrand, B., Nys, J. P., Prieste, C., Stievenard, D., Gerard, J.M. & Thierry-Mieg, V. (2000). Imaging the wave-function amplitudes in cleaved semiconductor quantum boxes. *Phys. Rev. Lett.*, Vol. 85, No. 5 (July 2000), pp. 1068-1073, ISSN 0031-9007.

Hasegawa, H. & Sawada, T. (1983). On the electrical properties of compound semiconductor interfaces in metaliinsulatorisemiconductor structures and the possible origin of interface states. *Thin Solid Films*, Vol. 103, No. 1-3 (January 1983), pp. 119-140, ISSN 0040-6090.

Kamins, T.I., Medeiros-Ribeiro, G., Ohlberg, D.A.A. & Stanley Williams, R. (1999). Evolution of Ge islands on Si(001) during annealing. *J. Appl. Phys.*, Vol. 85, No. 2 (February 1999), pp. 1159-1162, ISSN 1089-7550.

Karpovich, I.A., Zvonkov, B.N., Baidus', N.V., Tikhov, S.V. & Filatov, D.O. (2004a). Tuning the energy spectrum of the InAs/GaAs quantum dot structures by varying the thickness and composition of a thin double GaAs/InGaAs cladding layer. In:*Trends in Nanotechnology Research*, E.Dirote, pp.173-208, Nova Science, ISBN1-59454-091-8, New York.

Karpovich, I.A., Baidus', N.V., Zvonkov, B.N., Filatov, D.O., Levichev, S.B., Zdoroveishev, A.V. & Perevoshikov, V.A. (2004b). Investigation of the buried InAs/GaAs quantum dots by SPM combined with selective chemical etching *Phys.Low-Dim.Struct.*, No. 3/4 (March 2004), pp. 341-345, ISSN 0204-3467.

Lorke, A., Blossey, R., Garcia, J.M., Bichler, M. & Abstreiter, G. (2002). Morphological transformation of $In_yGa_{1-y}As$ islands, fabricated by Stranski–Krastanov growth. *Mater. Sci. Eng. B*, Vol. 88, No. 2-3 (January 2002), pp. 225-229, ISSN 0921-5107.

Maltezopoulos, T., Bolz, A., Meyer, C., Heyn, C., Hansen, W., Morgenstern, M. & Wiesendanger, R. (2003). Wave-function mapping of InAs quantum dots by Scanning Tunneling Spectroscopy. *Phys. Rev. Lett.*, Vol. 91, No. 19 (November 2003), p. 196804 (4 pp.), ISSN 0031-9007.

Mashin, A.I., Nezhdanov, A.V., Filatov, D.O., Isakov, M.A., Shengurov, V.G., Chalkov, V.Yu. & Denisov, S.A. (2010). Confocal Raman Microscopy of Self-assembled GeSi/Si(001) islands. *Semicond.*, Vol. 44, No. 11 (November 2010), pp. 1504–1510, ISSN 1063-7826.

Medeiros-Ribeiro, G., Bratkovski, A.M., Kamens, T.I., Ohlberg, D.A.A. & Stanley Williams, R. (1998). Shape transition of germanium nanocrystals on a silicon (001) surface from pyramids to domes. *Science*, Vol. 279, No. 5349 (January 1998), pp. 353-355, ISSN 1095-9203.

Stier, O., Grundmann, M. & Bimberg, D. (1999). Electronic and optical properties of strained quantum dots modeled by 8-band $k{\cdot}p$ theory. *Phys. Rev. B*, Vol. 59, No. 8 (February 1999), pp. 5688-5701, ISSN 1098-0121.

Suzuki, K., Kanisawa, K., Janer, C., Perraud, S., Takashina, K., Fujisawa, T. & Hirayama, Y. (2007). Spatial imaging of two-dimensional electronic states in semiconductor quantum wells. *Phys.Rev.Lett.*, Vol.98, No.13 (March 2007), p.136802 (4 p.), ISSN 0031-9007.

Svetlov, S.P., Shengurov, V.G., Tolomasov, V.A., Gorshenin, G.N. & Chalkov, V.Yu. (2001). A sublimation silicon Molecular Beam Epitaxy system. *Instrum. Exp. Tech.*, Vol. 44, No. 5 (May 2001), pp. 700-703, ISSN 1608-3180.

Yanev, V.; Rommel, M.; Lemberger, M.; Petersen, S.; Amon, B.; Erlbacher, T.; Bauer, A.J.; Ryssel, H.; Paskaleva, A.; Weinreich, W.; Fachmann, C.; Heitmann, J. & Schroeder, U. (2008). Tunneling atomic-force microscopy as a highly sensitive mapping tool for the characterization of film morphology in thin high-*k* dielectrics. *Appl. Phys. Lett.*, Vol. 92, No. 25 (June 2008), pp. 2910-2912, ISSN 0003-6951.

Zenkevich, A., Lebedinskii, Yu., Antonov, D., Gorshkov, O. & Filatov, D. (2011). Structure and electron transport in the ultrathin nanocomposite SiO_2/Si films with the metal nanoclusters. In: *Advances in Diverse Industrial Applications of Nanocompositess, B.* Reddy, pp. 317-340, InTech, ISBN 978-953-307-202-9, Vienna.

4

Quantum Mechanics of Semiconductor Quantum Dots and Rings

I. Filikhin, S.G. Matinyan and B. Vlahovic
North Carolina Central University
USA

1. Introduction

The progress of semiconductor physics in the decade 1970-1980 is connected with gradual deviation from the electronic band structure of ideal crystal of Bloch picture (Bloch, 1928) where, unlike atomic world with its discrete and precisely defined, in the limits of uncertainty relation, energy levels, energy of bound electron is a multivalued function of momentum in the energy band and density of states are continuous (For the earlier short but comprehensive survey see (Alferov, 1998)).

In principle, Bloch theory deals with infinite extension of lattice, with the understandable (and important) surface effects. The decreasing of the size of the object to a few micrometers principally does not change the picture of the extended crystal qualitatively. It takes a place until one reaches the scale where the size quantization essentially enters the game and we can speak about microscopic limit of matter. What generally divides macroscopic limit of the solid state from the microscopic one? It is defined by some correlation length (or, more generally, all such relevant lengths)): for carriers it is mean free path length l or Broglie length $l_B = h / p$ (p-momentum), which is smaller. One may say that the quantum mechanical properties of matter clearly reveal if $l / a \geq 1$, where a is the size of the lattice constant. In the opposite limit $l / a < 1$, matter is considered macroscopically.

In this light, it is worthy to remind that as long as 1962, L. V. Keldysh (Keldysh, 1962 as cited in Bimberg et al., 1999) considered electron motion in a crystal with periodic potential with the period that is much larger than the lattice constant. In this limit he discovered so called minizones and negative resistance. Just in this limit $l / a \geq 1$ we expect the size quantization with its discrete levels and coherence in the sense that electron can propagate across the whole system without scattering, its wave function maintains a definite phase. In this limit, mesoscopic (term coined by van Kampen (1981) relates to the intermediate scale dividing the macro and micro limits of matter) and nanoscopic objects (Quantum Wells (QW), Wires and Dots (QD)) shown very interesting quantum mechanical effects. In this limit many usual rules of macroscopic physics may not hold. For only one example, rules of addition of resistance both in series and parallel are quite different and more complicated (Landauer, 1970; Anderson et al., 1980; Gefen, et al., 1984).

Closing this brief introduction concerning some aspects of genuine quantum objects (QW, Q Wires, QD) we would like to emphasize the conditional sense of the notion of dimensions in this world: in the limit $l / a \geq 1$ dimensions are defined as difference between real spatial dimension (in our world D =3) and numbers of the confined directions: Quantum Well: D =2, Quantum Wire: D =1, Quantum Dot: D =0. However, for example, QD which will be one of our subject for study, has very rich structure with many discrete levels, their structure define the presence or absence of Chaos, as we will see below, inside QD. Minimal size of QD is defined by the condition to have at least one energy level of electron (hole) or both: $a_{\min} = \pi\hbar / \sqrt{2m^* \Delta E} \approx 4$ nm, where ΔE is average distance between neighboring energy levels. Maximal size of QD is defined by the conditions that all three dimensions are still confined. It depends, of course, on temperature: at room temperature it is 12 nm (GaAs), 20 nm (InAs) ($\Delta E \approx 3kT$). The lower temperature, the wider QD is left as quantum object with D =0 and the number of energy levels will be higher.

2. Effective model for semiconductor quantum dots

The effective potential method has been developed (Filikhin et al., 2006) to calculate the properties of realistic semiconductor quantum dot/ring (QD/QR) nanostructures with the explicit consideration of quantum dot size, shape, and material composition. The method is based on the single sub-band approach with the energy dependent electron effective mass. In this approach, the confined states of carriers are formed by the band gap offset potential. Additional effective potential is introduced to account for cumulative band gap deformations due to strain and piezoelectric effects inside the quantum dot nanostructure. The magnitude of the effective potential is selected in such a way as to reproduce experimental data for a given nanomaterial. Additionally, an analog of the Kane formula (Kane, 1957) is implemented in the model to take into account the non-parabolicity of the conduction/valence band. The resulting nonlinear eigenvalue problem for the Schrödinger equation is solved by means of the iterative procedure with the adjusted effective electron mass and non-parabolicity parameter, where in each iteration step the Schrödinger equation is numerically linearized and solved by the finite element method.

At present, simulations based on this approach are performed for the InGaAs/GaAs quantum dots and quantum rings of different sizes and configurations under different external conditions. The obtained results show that the residual strain and conduction band non-parabolicity effects greatly affect the device related properties of semiconductor quantum dots. The results are in good agreement with available experimental data, closely matching energy level and effective mass data extracted from capacitance-voltage experiments. The method also allows one to accurately simulate spin-orbital coupling effects for the electrons in excited states, as well as the presence of admixtures, such as Ga. Our calculations of the Coulomb shifts of the exciton complexes (positively and negatively charged trions, biexcitons) in the InGaAs/GaAs quantum dots with 22%-25% Ga fraction match very well both capacitance-voltage and photoluminescence measurements. To best reproduce the experimental data, Ga fraction in the InGaAs/GaAs quantum dots should not exceed 25%.

Commonly used numerical approaches, such as the 8-band ***kp***-theory, density functional theory, or atomistic pseudo-potential technique, take into account inter-band interactions, strain and piezoelectric effects in quantum dots in an *ab initio* manner. Such methods are very computationally intensive and time-consuming. The important advantage of the effective model is that the high accuracy of calculations is obtained at a very low computational cost - calculations can typically be completed using a 3 GHz PC with 1 GB of memory in less than 20 minutes. The effective potential method satisfactorily reproduces the results of the realistic simulations, thus offering an independent evaluation of the electronic confinement effects calculated within others models.

2.1 Formalism

2.1.1 Schrödinger equation and effective mass approximation

In the present review a semiconductor 3D heterostructure (QD or QR) is modeled utilizing a ***kp***-perturbation single sub-band approach with quasi-particle effective mass (Harrison, 2005; Manasreh, 2005; Yu & Cardona, 2005). The energies and wave functions of a single carrier in a semiconductor structure are solutions the Schrödinger equation:

$$(H_{kp} + V_c(\vec{r}))\Psi(\vec{r}) = E\Psi(\vec{r}) \tag{1}$$

Here H_{kp} is the single band ***kp***-Hamiltonian operator, $H_{kp} = -\nabla \frac{\hbar^2}{2m^*(\vec{r})}\nabla$, m^* is the electron/hole effective mass for the bulk, which may depend on coordinate, and $V_c(\vec{r})$ is the confinement potential. The confinement of the single carrier is formed by the energy misalignment of the conduction (valence) band edges of the QD material (index 1) and the substrate material (index 2) in the bulk. $V_c(\vec{r})$ is so called "band gap potential". The magnitude of the potential is proportional to the energy misalignment. The band structure of the single band approximation can be found in many textbooks (see, for example, (Harrison, 2005; Manasreh, 2005; Yu & Cadona, 2005). * (see the input below) $\Psi(\vec{r})$ and its derivative $1/m^*(\vec{n},\nabla)\Psi(\vec{r})$ on interface of QD and the substrate are continues.

2.1.2 The non-parabolicity of the conduction band. The Kane formula

Traditionally applied in the macroscopic scale studies parabolic electron spectrum needs to be replaced by the non-parabolic approach, which is more appropriate to nano-sized quantum objects (Wetzel et al., 1996; Fu et al., 1998). The Kane formula (Kane, 1957) is implemented in the model to take into account the non-parabolicity of the conduction band. The energy dependence of the electron effective mass is defined by the following formula:

$$\frac{m_0}{m^*} = \frac{2m_0 P^2}{3\hbar^2}\left(\frac{2}{E_g + E} + \frac{1}{E_g + \Delta + E}\right). \tag{2}$$

Here m_0 is free electron mass, P is Kane's momentum matrix element, E_g is the band gap, and Δ is the spin-orbit splitting of the valence band.

Taking into account the relation (2) the Schrödinger equation (1) is expressed as follows

$$(H_{kp}(E) + V_c(\vec{r}))\Psi(\vec{r}) = E\Psi(\vec{r}). \tag{3}$$

Here $H_{kp}(E)$ is the single band ***kp***-Hamiltonian operator $H_{kp}(E) = -\nabla \frac{\hbar^2}{2m^*(E,\vec{r})}\nabla$,

$m^*(E,\vec{r})$ is the electron/hole effective mass and $V_c(\vec{r})$ is the band gap potential. As a result, we obtain a non-linear eigenvalue problem.

Solution of the problem (3)-(2) results that the electron/hole effective mass in QD (or QR) varies between the bulk values for effective mass of the QD and substrate materials. The same it is given for the effective mass of carriers in the substrate. The energy of confinement states of carries is rearranged by the magnitude of the band gap potential V_c.

The Schrödinger equation (1) with the energy dependence of effective mass can be solved by the iteration procedure (Li et al., 2002; Voss, 2005; Filikhin et al., 2004, 2005).

$$\begin{aligned} &H_{kp}(m^{*k-1}_i)\Psi^k(\vec{r}) = E^k\Psi^k(\vec{r}), \\ &m^{*k}_i = f_i(E^k), \end{aligned} \tag{4}$$

where k is the iteration number, i refers to the subdomain of the system; $i=1$ for the QD, $i=2$ for the substrate. $H_{kp}(m^{*k}_i)$ is the Hamiltonian in which the effective mass does not depend on energy and is equal to the value of m^{*k}_i, f_i is the function defined by the relation (2). For each step of the iterations the equation (1) is reduced to Schrödinger equation with the effective mass of the current step which does not depend on energy. At the beginning of iterations the bulk value of the effective mass is employed. Obtained eigenvalue problem can be solved numerically (by the finite element method, for example). After that, a new value for effective mass is taken by using Eq. (2) and procedure is repeated. The convergence of the effective mass during the procedure has a place after 3-5 steps. As an example, the typical convergences for election effective mass and confinement energy of single electron are displayed in Fig. 1 for the InAs/GaAs QR (Filikhin et al., 2005). Description of other methods for the solution of the problem (3)-(2) can be found in (Betcke & Voss, 2011).

Remarks: at the first, in the present review the consideration was restricted by the electron and heavy hole carriers, and, the second, the Coulomb interaction was excluded. Often the linear approximation for the function $m^*_i/m_0 = f(E,r)$ is used. We also will apply the linear fit in the present chapter.

2.1.3 Effective approach for strained InAs/GaAs quantum structures: Effective potential

Here we propose the effective potential method to calculate the properties of realistic semiconductor quantum dot/ring nanostructures with the explicit consideration of quantum dot size, shape, and material composition. The method is based on the single sub-band approach with the energy dependent electron effective mass (Eq. (3)). In this approach, the confined states of carriers are formed by the band gap offset potential. Additional effective potential is introduced to simulate the cumulative band gap deformations due to strain and piezoelectric effects inside the quantum dot nanostructure. The magnitude of the effective potential is selected in such a way that it reproduces experimental data for a given nanomaterial.

We rewrite the Schrödinger equation (3) in the following form:

$$(H_{kp}(E) + V_c(\vec{r}) + V_s(\vec{r}))\Psi(\vec{r}) = E\Psi(\vec{r})\,. \tag{5}$$

Here $H_{kp}(E)$, as before, is the single band ***kp***-Hamiltonian operator $H_{kp}(E) = -\nabla \frac{\hbar^2}{2m^*(E,\vec{r})}\nabla$. As previously, $m^*(E,\vec{r})$ is the electron (or hole) effective mass, and $V_c(\vec{r})$ is the band gap potential, $V_s(\vec{r})$ is the effective potential. $V_c(\vec{r})$ is equal zero inside the QD and is equal to V_c outside the QD, where V_c is defined by the conduction band offset for the bulk (see Section 1.22). The effective potential $V_c(\vec{r})$ has an attractive character and acts inside the volume of the QD. This definition for the effective potential is schematically illustrated by Fig. 2 for the conduction band structure of InAs/GaAs QD. In the figure, the confinement potential of the simulation model with effective potential V_s is denoted as "strained". The band gap potential for the conduction band (valence band) can be determinate as V_c =0.594 eV (V_c =0.506 eV). The magnitude of the effective potential can be chosen to reproduce experimental data. For example, the magnitude of V_s for the conduction (valence) band chosen in (Filikhin et al., 2009) is 0.21 eV (0.28 eV). This value was obtained to reproduce results of the 8-th band ***kp***-calculations of (Schliwa et al., 2007) for InAs/GaAs QD. To reproduce the experimental data from (Lorke et al., 2000), the V_s value of 0.31 eV was used in (Filikhin, et al. 2006) for the conduction band.

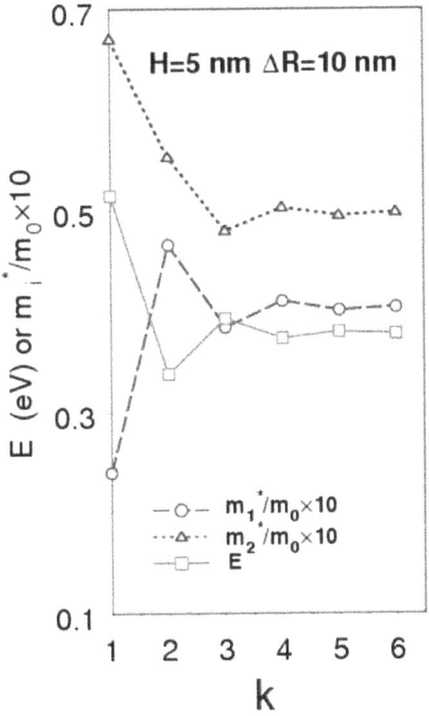

Fig. 1. Convergence of the iterative procedure (4) for the confinement energy E (solid line) and electron effective mass m^*_i/m_0 calculated for InAs/GaAs QR (dashed line) and GaAs substrate (dotted line). Here the height of QR is H, radial width is ΔR and inner radius is R_1 (R_1 =17 nm), V_c =0.77 eV.

Possibility for the substitution of the function describing the strain distribution in QD and the substrate was firstly proposed in (Califano & Harrison, 2000). Recent works (Zhao & Mei, 2011; Li, Bin & Peeters, 2011) in which the strain effect taken into account rigorously applying the analytical method of continuum mechanics allow us to say that the approximation of the effective potential is appropriate.

In the next sub-section of the section 2 we will review the results obtained in both these approximations as the non-parabolic one as well as the effective potential method.

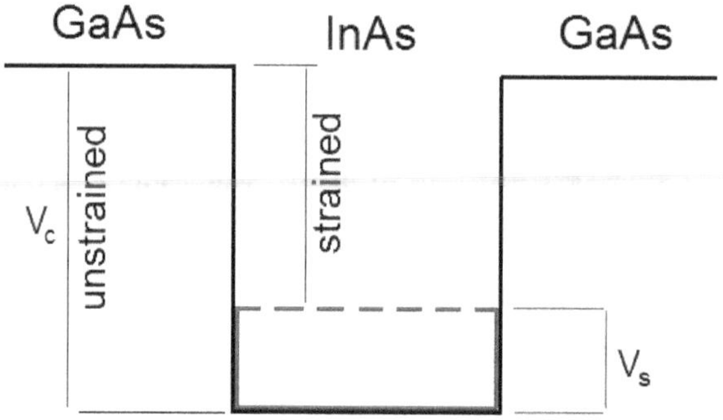

Fig. 2. Effective potential V_s and band gap structure of the conductive band of InAs/GaAs QD.

2.2 Electron energy in quantum rings with varieties of geometry: Effect of non-parabolicity

In this section a model of the InAs/GaAs quantum ring with the energy dispersion defined by the Kane formula (2) (non-parabolic approximation) based on single sub-band approach is considered. This model leads to the confinement energy problem with three-dimensional Schrödinger equation in which electron effective mass depend on the electron energy. This problem can be solved using the iterative procedure (4). The ground state energy of confined electron was calculated in (Filikhin et al., 2004, 2005, 2007a) where the effect of geometry on the electron confinement states of QR was studied and the non-parabolic contribution to the electron energy was estimated. The size dependence of the electron energy of QR and QD was subject of several theoretical studies (Li & Xia, 2001; Li et al., 2002). We present here, unlike the previous papers, a general relation for the size dependence of the QR energy.

Consider is semiconductor quantum ring located on the substrate. Geometrical parameters of the semi-ellipsoidal shaped QR are the height H, radial width ΔR and inner radius R_1. It is assumed that $H / \Delta R \ll 1$ which is appropriate technologically. QR cross section is schematically shown in Fig. 3. The discontinuity of conduction band edge of the QR and the substrate forms a band gap potential, which leads to the confinement of electron.

The band gap potential $V_c(\vec{r})$ is equal to zero inside the QR ($V_c(\vec{r})$=0) and it is equal to the confinement potential E_c outside of the QR: The spatial dependence of the electron effective mass is given as $m^*(E,\vec{r}) = m_i^*(E)$, i=1,2,3, where m_1^* is the effective mass in the material of QR ($\vec{r} \in$ E1), and $m_2^*(E)$, m_3^* are the effective mass of the substrate material ($\vec{r} \in$ E2 and

E3). Within each of the regions E1, E2 and E3 m_i^* does not depend on the coordinates. The effective mass m_3^* is equal to a constant bulk value. The energy dependence of the electron effective mass from the E1 and E2 subdomains is defined by the formula (2) (Kane, 1957). The equation (1) satisfies the asymptotical boundary conditions: $\Psi(\vec{r})|_{|\vec{r}|\to\infty} \to 0$, $\vec{r} \in$ substrate and $\Psi(\vec{r})|_{\vec{r}\in S} = 0$, where S is free surface of QR. On the surface of boundaries with different materials the wave function and the first order derivative $(\vec{n}, \vec{\nabla}\Psi) / m_i^*$ are continuous (the surface normal $\vec{n}$).

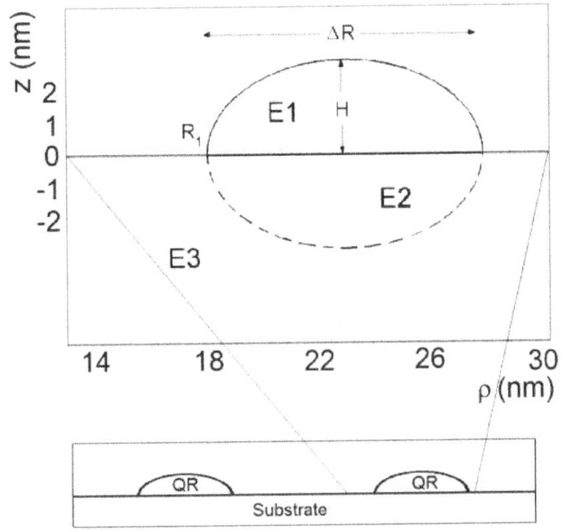

Fig. 3. Profile of cross section of quantum ring (E1) and substrate (E2 and E3). Cylindrical coordinates ρ and z shown on axis.

The Schrödinger equation (3) was numerically solved by the finite element method and iterative procedure (4). The following typical QR/substrate structures with experimental parameters were chosen: InAs/GaAs and CdTe/CdS. The parameters of the model are given in Tabl. 1 for the each hetero-structure.

QR/Substrate	m^*_1 / m^*_2	m^*_1 / m^*_2 (eV)	$\frac{2m_0P_1^2}{\hbar^2} / \frac{2m_0P_2^2}{\hbar^2}$	Δ_1 / Δ_2
InAs/GaAs	0.024/0.067	0.77	22.4/24.6	0.34/0.49
CdTe/CdS	0.11/0.20	0.66	15.8/12.0	0.80/0.07

Table 1. Parameters of the QR and substrate materials

It has to be noted that the effective mass substrate calculated for the InAs/GaAs and CdTe/CdS QRs is slightly differ from the bulk values within area E2. One can consider a simpler model when the properties of the area E2 and E3 are similar. It means that the wave function of electron does not penetrated by surface of QR (area E1) essentially. The simple model does not change qualitative results of these calculations.

Analysis of the results of numerical calculations shows that the ground state energy of QR can be best approximated as a power function of the inverse values of the height and the radial width:

$$E \approx a(\Delta R)^{-\gamma} + bH^{-\beta}, \qquad (6)$$

where the coefficients γ=3/2 and β=1 were obtained numerically by the least square method. An example of this relation is illustrated in Fig. 4 for InAs/GaAs QR. Parameters a and b remain constant except for extremely low values of H and ΔR. Our analysis also reveals a significant numerical difference between the energy of QR electron ground states, calculated in non-parabolic and parabolic approximations. The results of the calculation with parabolic approximation are represented in the Fig. 4 by the dashed lines. Computation of the electron confinement energy of QRs for different materials shows that the non-parabolic contribution is quite significant when chosen QR geometrical parameters are close to those of the QRs produced experimentally: H < 7 nm, R < 30 nm for InAs/GaAs, H < 5 nm, R < 20 nm for CdTe/CdS. Magnitude of this effect for InAs/GaAs can be greater than 30%. According with this fact the coefficients a and b in Eq. (6) also depend on the approximation used: a/b =3.4/1.9 for the non-parabolic and a/b =6.2/3.0 for parabolic approximation.

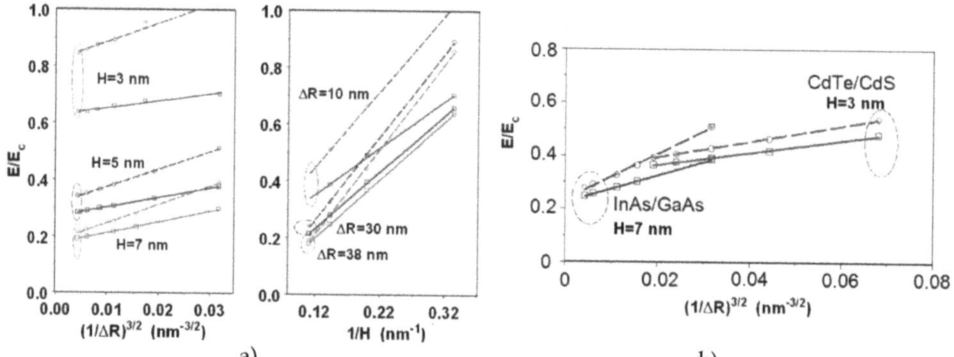

a) b)

Fig. 4. a) Normalized electron ground state energy of semi-ellipsoidal shape InAs/GaAs QR with parabolic (dashed line) and non-parabolic (solid lines) approximation as function of the QR size (R_1 =17 nm). b) Normalized electron confinement energy of QRs of various materials in the parabolic (dashed line) and non-parabolic (solid lines) approximation.

As it can be seen from the Fig. 4b), coefficients γ and β in the relation (6) do not depend on QR/substrate materials. Their values are defined by geometry and by the boundary conditions of the applied model. The model described above corresponds to the boundary condition as "hard wall at one side" (top side of the QR). For the model without the walls when the QR embedded into the substrate one can obtain γ =1, and β =1/3. In contrast with it, the coefficients a and b depend on the QR/substrate material set essentially.

Concluding, we have shown that for wide QR sizes the non-parabolicity effect does considerably alter the energy of the electron states, especially when the height or width of QR is relatively small.

2.3 The C-V measurements and the effective model: Choosing the parameters

The well-established process of QDs formation by epitaxial growth and consecutive transformation of QDs into InAs/GaAs quantum rings (QR) (Lorke et al., 2000) allows the production of 3D structures with a lateral size of about 40-60 nm and a height of 2-8 nm. In produced QDs and QRs it is possible directly to observe discrete energy spectra by applying capacitance-gate-voltage (CV) and far-infrared spectroscopy (FIR). In this section we will show how the effective model works using as an example the CV data. We use results of the CV experiment from (Lorke et al., 2000; Emperador et al., 2000; Lei et al., 2010) for QD and QR.

The effective mass of an electron in QD and QR is changing from the initial bulk value to the value corresponding to the energy given by the Kane formula (2). Results of the effective model calculations for the InAs/GaAs QR are shown in Fig. 5. The effective mass of an electron in the InAs QR is close to that of the bulk value for the GaAs substrate. Since the effective mass in the QD is relatively smaller, as it is clear from Fig. 5, for QD the electron confinement is stronger; the s-shell peak of the CV trace is lower relative upper edge of conduction band of GaAs. The lower s-shell peak corresponds to the tunneling single electron into the QD. The pictures is a starting point for the choosing the parameters of the effective potential model. In this section we follow the paper (Filikhin et al., 2006a) where the semi-ellipsoidal InAs/GaAs QD has been considered. The average sizes of InAs/GaAs QD reported in (Lorke et al., 2000) were: H =7 nm (the height) and R=10 nm (the radius). A cross section of the quantum dot is shown in Fig. 6a). The quantum dot has rotation symmetry. Thus the cylindrical coordinate was chosen in Eq. (5) which defines the effective model. For each step of iterative procedure (4) the problem (3-2) is reduced to a solution of the linear eigenvalue problem for the Schrödinger equation.

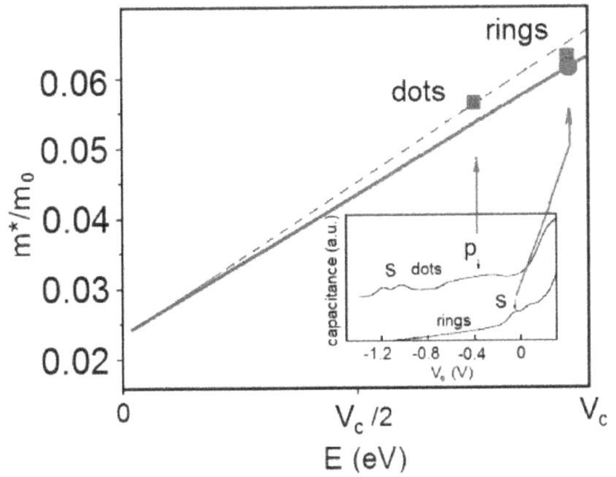

Fig. 5. Calculated (circle) and experimentally obtained by (Lorke et al., 2000; Emperador et al., 2000) (squares) values for the electron effective mass and the confinement energies of the electron s- and p-levels of QD and QR. The solid line is obtained by the Kane formula (2), and the dashed line connects the bulk values of the effective mass. The insert: the capacitance-gate voltage traces (Lorke et al., 2000).

Taking into account the axial symmetry of the quantum dot (ring) considered, this equation may be written in the cylindrical coordinates (ρ, z, ϕ) as follows:

$$(-\frac{\hbar^2}{2m^*}(\frac{\partial^2}{\partial\rho^2}+\frac{1}{\rho}\frac{\partial}{\partial\rho}-\frac{l^2}{\rho^2}+\frac{\partial^2}{\partial z^2})+V_c(\rho,z)+V_s(\rho,z)-E)\Phi(\rho,z)=0. \quad (7)$$

The wave function is of the form: $\Psi(r)=\Phi(\rho,z)exp(il\phi)$, where l=0, ± 1, $\pm 2...$ is the electron orbital quantum number. For each value of the orbital quantum number l, the radial quantum numbers $n=0,1,2,...$ are defined corresponding to the numbers of the eigenvalues of (4) which are ordered in increasing. The effective mass m^* must be the mass of electron for QD or for the substrate depending on the domain of the Eq. (3) is considered.

The wave function $\Phi(\rho,z)$, and its first derivative in the form $\frac{\hbar^2}{2m^*}(\vec{n},\nabla)\Phi$, have to be continuous throughout the QD/substrate interface, where $\vec{n}$ is the normal vector to the interface curve. The Neumann boundary condition $\frac{\partial}{\partial\rho}\Phi(\rho,z)=0$ is established for $\rho=0$.

The asymptotical boundary conditions is $\Phi(\rho,z)\to 0$, when $\rho\to\infty$, $|z|\to\infty$ (QD is located near the origin of z-axes).

When quantum dots are in an external perpendicular magnetic field, as it will be considered below, the magnetic potential term must be added to the potentials of Eq. (7) (Voskoboinikov et al., 2000) in the form $V_m(\rho)=\frac{1}{2m^*}(\beta\hbar l+\frac{\beta^2}{4}\rho^2)$, where $\beta=eB$, B is the magnetic field strength, and e is the electron charge. We consider the case of a magnetic field normal to the plane of the QD and do not take into account the spin of electron because the observed Zeeman spin-splitting is small.The confinement potential in Eq. (7) was defined as follows: $V_c=0.7(E_g^S-E_g^{QD})$; $V_c=0.77$ eV. The parameters of the QD and substrate materials were $m^*_{bulk,1}/m^*_{bulk,2}$ =0.024/0.067, E_g^{QD}/E_g^S =0.42/1.52, $\frac{2m_0P_1^2}{\hbar^2}/\frac{2m_0P_2^2}{\hbar^2}$ =20.5/24.6, Δ_1/Δ_2 =0.34/0.49. The magnitude of the effective potential V_s was chosen as 0.482 eV. There are three electron confinement states: the s, p, and d, as shown in the Fig. 6b). The energy of the s single electron level measured from the top of the GaAs conduction band can be obtained from CV experimental data. To explain it, in Fig. 6c) the capacitance-gate-voltage trace from (Miller et. al., 1997) is shown. The peaks correspond to the occupation of the s and p energy shells by tunneled electrons. The Coulomb interaction between electrons results to the s-shell splits into two levels and the p-shell splits into four levels taking into account the spin of electron and the Pauli blocking for fermions. The gate voltage-to-energy conversion coefficient $f=7$ ($\Delta E=e\Delta V_g/f$) was applied to recalculate the gate voltage to the electron energy. The value of the effective potential V_s was chosen in order to accurately reproduce the observed s-wave level localization with respect to the bottom of GaAs conduction band. The approximate size of this energy region is 180 meV.

The non-parabolic effect causes a change in the electron effective mass of QD with respect to the bulk value. According to the relation Eq. (2), the effective electron mass for InAs is sufficiently increased from the initial value of 0.024 m_0 to 0.054 m_0, whereas for GaAs substrate it is slightly decreased from $0.067\,m_0$ to $0.065\,m_0$ within the region where the wave function is out of the quantum dot. The obtained value of the electron effective mass of InAs in QD is close to the one ($0.057\,m_0 \pm 0.007$) extracted in (Miller et. al., 1997) from the CV measurements of orbital Zeeman splitting of the p level.

Appling the obtained effective model, one can take into account the effect the Coulomb interaction between electrons (the Coulomb blockade). The goal is to reproduce the C-V data presented in Fig. 6 for the InAs QD. The calculations (Filikhin et al., 2006a) have been carried out using the perturbation procedure, proposed in (Warburton et al., 1998). The Coulomb energy matrix elements were calculated by applying single electron wave functions obtained from the numerical solution of Eq. (7). Both the direct terms of E_{ij}^{c} and the exchange terms E_{ij}^{x} of the Coulomb energy between electron orbitals with angular momentum projection of $\pm i$ and $\pm j$ were calculated (notation is given in (Warburton et al., 1998)). The results of calculations of the electron energies of the s, p and d levels are shown in Fig. 7 (*Cal.* 2). The s shell Coulomb energy was found to be close to the experimental value which is about 20 meV.

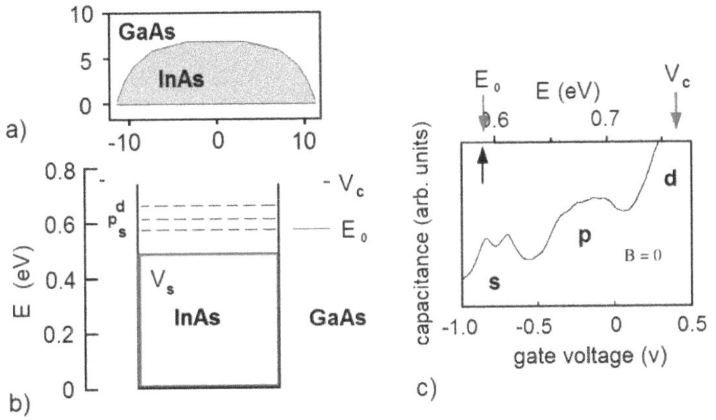

Fig. 6. a) A cross section of the quantum dot. The dimensions are given in nm. b) Localization of the s, p and d single electron levels relatively to the bottom of the GaAs conduction band. V_c is the band-gap potential, V_s is the effective potential simulating the sum of the band-gap deformation potential, the strain-induced potential and the piezoelectric potential. c) The capacitance-gate-voltage trace (Miller et. al., 1997). The peaks correspond to the occupation of the s and p energy shells by tunneled electrons. The arrows denote the s level (E_0) and the bottom of the GaAs conduction band.

Returning to the Fig. 5 we have to note that the effective potential obtained for InAs/GaAs QD has to be corrected for the case of the InAs/GaAs quantum rings. The reason is the topological, geometrical dependence of the depth of the effective potential. This dependence is weak for the considered QD and QR. The corresponding V_s potentials have the magnitude of 0.482 eV and 0.55 eV for QD and QR, respectively. Accordingly to the experimental data the electron effective mass in quantum dots and rings is changing from $0.024\,m_0$ to $(0.057 \pm 0.007)\,m_0$ (Miller et. al., 1997) and $0.063\,m_0$ (Lorke et al., 2000), respectively. The Kane's formula describes these variations well as it is shown in Fig. 5. The calculated values for the effective masses for quantum dots and rings are 0.0543 m_0 and 0.0615 m_0, respectively (Filikhin et al., 2006).

Correct choice of the average QD profile is important for an analysis of the C-V data. It was shown in (Filikhin et al., 2008), where the calculation of the energy shifts due to the Coulomb interaction between electrons tunneling into the QD was performed for comparison with the C-V experiments.

One can see in Fig. 7 that the agreement between our results and the experimental data is satisfactory well. Slight disagreement can be explained by uncertainty in the QD geometry which has not been excluded by available experimental data. In (Filikhin et al., 2008) it was shown that small variations of the QD cross section lead to significant changes in the levels presented in Fig. 7. The variations of the QD profile we considered are shown in Fig. 8a, and the results of calculations for the electron energies are presented in Fig. 8b) for s, p and d -shell levels. The results of the calculations shown in Fig. 8 reveal rather high sensitivity to these variations of the QD profile. In particular, the spectral levels shift is noticeable due to a small deformation of the QD profile. Thus, we have seen that the average QD profile is important when we are comparing the result of the calculations and the experimental data.

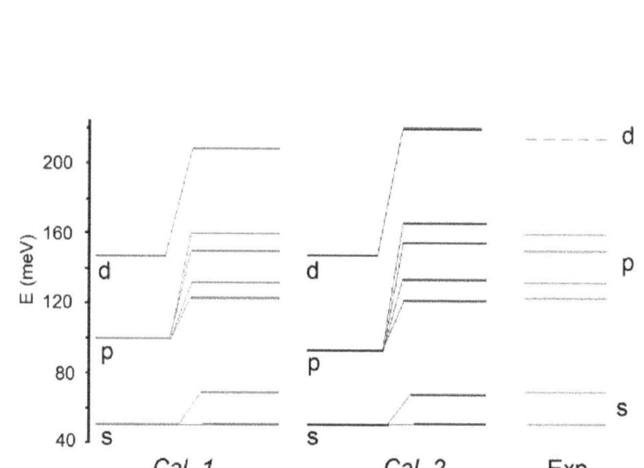

Fig. 7. Energies of the electrons occupying a few first levels of the quantum dot at zero magnetic field. The calculations *Cal.* 1 are that of parabolic model (Warburton et al., 1998). Our calculations are denoted by *Cal.* 2. The splitting of the single electron levels of a corresponding energy shell is presented. CV experimental data are taken from (Warburton et al., 1998).

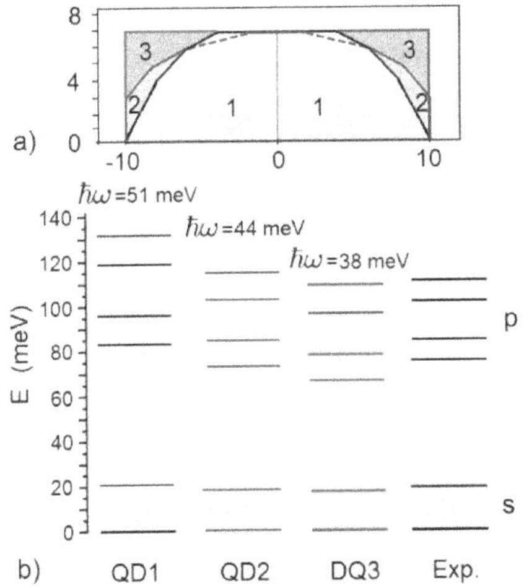

Fig. 8. a) Cross sections of the QD. The dimensions are given in nm. b) Excitation energies of the electrons occupying s and p -energy shells of the InAs/GaAs quantum dot for various QD profiles are shown in Figure 7a). CV experimental data are taken from (Warburton et al., 1998). Here $\hbar\omega$ is the excitation energy $\hbar\omega = E_{(0,0)} - E_{(0,1)}$, where $E_{(n,l)}$ is a single electron energy of the (n,l) state.

Finally, we may conclude that the effective model of QD/substrate semiconductor structure with the energy dependent effective mass and realistic 3D geometry taken into account, can quantitatively well interpret the CV spectroscopy measurements.

2.4 Electron effective mass in the InAs/GaAs QD

In this section we present the effective model based on another version of the band structure model for InAs/GaAs QDs proposed in (Filikhin et al., 2008). The cross section of the semi-ellipsoidal shaped InAs QD embedded in a GaAs substrate is shown in Fig. 6a). Band gap structure model was defined by choosing for the conduction band κ^{CB} =0.54, and for the valence band κ^{VB} =0.46 (Duque et al., 2005). Using experimental values $E_{g,1}$ =0.42 eV, $E_{g,2}$ =1.52 eV we obtain V_c =0.594 eV for the conduction band and. V_c .=0.506 eV for the valence band. The band structure model for InAs/GaAs QDs is shown in Fig. 9.

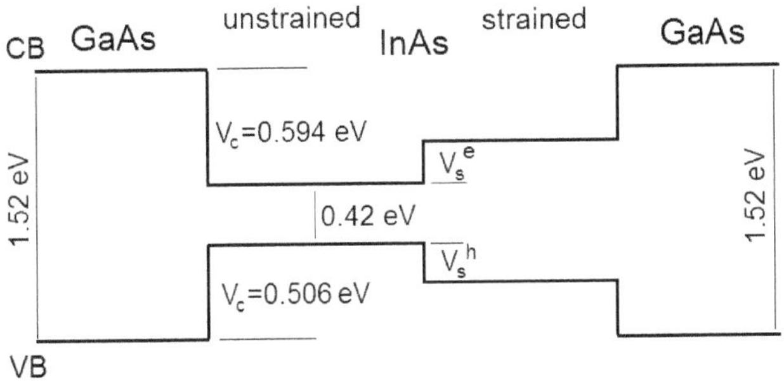

Fig. 9. Band structure model for InAs/GaAs QDs. CB (VB) is conduction (valence) band.

Bulk effective masses of InAs and GaAs are $m^*_{0,1}$=0.024 m_0 and $m^*_{0,2}$ =0.067 m_0, respectively. For the effective mass of the heavy hole, a value of m^* =0.4 m_0 for both the QD and the substrate was used. The band gap model just described is for "unstrained" InAs/GaAs structures. Realistic models for QDs must take into account the band-gap deformation potential, the strain-induced potential, and the piezoelectric potential, in addition to the band-gap potential. These effects can be included by introducing an effective potential V_s. The magnitude of the potential has been chosen (Filikhin et al., 2006) to reproduce experimental data and the value of 0.31 eV was used for V_s. The effect of non-parabolicity, taken into account in the effective model, leads to a change of the effective electron mass in the QD relative to its bulk value. For the QD under study, the effective mass for InAs increases from the initial bulk value of 0.024 m_0 to 0.057 m_0 which coincides with the experimental value $0.057m_0 \pm 0.007m_0$, obtained in CV measurements through the Zeeman splitting of p-shell levels. This result is shown in Fig. 10. In accordance with Eqs. (2)-(3), the effective electron masses in the s, p and d states are different. The value of the effective mass, mentioned

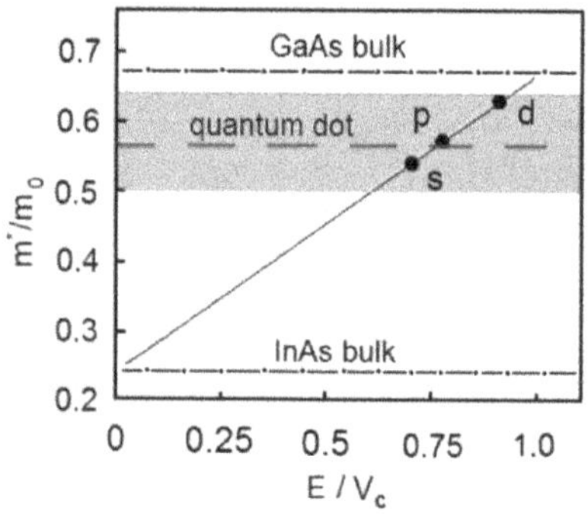

Fig. 10. Effective mass of electron and single electron energy of s, p, d-levels in InAs/GaAs QD. Dashed line corresponds to the experimental value. The grey color stripe shows the experimental uncertainty.

above, corresponds to the one for the p-state. The effective mass for s-shell is slightly less and is equal to 0.054 m_0. The differences of the effective masses are small and cannot be extracted from this experiment due to the large experimental uncertainties (Miller et al., 1998).

2.5 Experimental data for InAs/GaAs QR and the effective model

In this section we continue the description of the effective model use on the example of InAs/GaAs quantum ring. The geometry of the self-assembled QRs, reported in (Lorke et al., 2000), is shown in Fig. 11 (Geometry 1). The InGaAs QRs have a height of about 2 nm, an outer diameter of about 49 nm, and an inner diameter of about 20 nm. Also, three-dimensional QR geometry (Geometry 2), which follows from the oscillator model (Lei et al., 2010) is used. The confinement of this model is given by the parabolic potential:

$U(r) = \frac{1}{2} m^* \omega (r - r_0)^2$, where ω, r_0 are parameters (Chakraborty & Pietiläinen, 1994). The QR geometry is dictated by the relation between the adopted oscillator energy and a length l as follows (Szafran & Peeters, 2005):

$$l = \sqrt{2\hbar / m^* \omega} . \tag{8}$$

Here the width d for the considered rings is defined by $d = 2l$. The obtained geometry with the parameters m^* and ω from (Lei et al., 2010) is shown in Fig. 11 (Geometry 2); m^*=0.067 m_0 and ω=15 meV. The center radius of QR is 20 nm.

Results of the effective model calculations for the ground state energy of electron in a magnetic field are shown in Fig. 12. (Filikhin et al. 2011a) The picture of the change of the orbital quantum number of the ground state is similar to that obtained in (Lei et al., 2010) with the oscillator model. The change occurred at 2.2 T and 6.7 T. The obtained energy fits the experimental data rather well.

It has to be noted here that one cannot reproduce this result using the geometry proposed in (Lei et al., 2010) (Geometry 1) for this QR. The correspondence between the confinement potential parameters of the oscillator model and the real sizes of quantum objects has to be established by Eq. (8). Only using the geometry followed from Eq. (8) we reproduce result of (Lei et al., 2010), as is shown in Fig. 12. The strength parameter of the effective potential, in the case of the Geometry 2, was chosen to be 0.382 eV, which is close to that for QD from (Filikhin et al., 2008), where V_s =0.31 eV. The difference is explained by the topology dependence of the effective potentials (see section above and also (Filikhin et al., 2006)).

Note that the considered QRs are the plane quantum rings with the condition H <<D (for height and diameter of QR), which enhances the role of the lateral size confinement effect. To qualitatively represent the situation shown in Fig. 12, one can used an approximation for the 3D QR based on the formalism of one dimensional ideal quantum ring. Additional electron energy, due to the magnetic field, can be calculated by the relation: $E = \hbar^2 / (2m^* R^2)(l + \Phi / \Phi_0)^2$ (see for instance (Emperador et al., 2000)), where fluxes are $\Phi = \pi R^2 B$, $\Phi_0 = h / e$.. ($\Phi_0 = 4135.7$ T nm²); R is radius of the ideal ring. The Aharonov-Bohm (AB) (Aharonov & Bohm, 1959) period ΔB (Aronov & Sharvin, 1987) is given by the relation: $\Delta B = \Phi_0 / \pi / R^2$. Using the root mean square (rms) radius for R (R=20.5 nm), one can obtain $\Delta B / 2$ =1.56 T and $\Delta B / 2 + \Delta B$ =4.68 T for the ideal ring. This result is far from the result of 3D calculations shown in Fig. 12 where $\Delta B / 2 \approx 2.2$ T and $\Delta B / 2 + \Delta B \approx 6.7$ T are determined. Note here that the electron root mean square radius $R_{n,l}$ is defined by the relation $R^2{}_{n,l} = \int |\Phi^N{}_{n,l}(\rho,z)|^2 \rho^3 d\rho dz$, where $\Phi^N{}_{n,l}(\rho,z)$ is the normalized wave function of electron state described by the quantum numbers (n,l).

One can obtain better agreement by using the radius for the most probable localization of the electron $R_{loc.}$, defined at the maximum of the square of the wave function. The electron is mostly localized near 17.1 nm, for B=0. With this value, the ideal ring estimation leads to the values for $\Delta B / 2$ and $\Delta B / 2 + \Delta B$ as 2.25 T and 6.75 T, respectively. That agrees with the result of the 3D calculations (see Fig. 12). Obviously, the reason for this agreement is the condition $H << D$, for the considered QR geometry as it was mentioned above. The mostly localized position of the electron in QR depends weakly on the magnetic field. We present .. as a function of the magnetic field B in Fig. 13. $R_{loc.}(B)$ is changed in an interval of ±1 nm around the mean value $R_{loc.}(0)$ of 17 nm. It is interesting to note that the magnetization of a single electron QR demonstrates the same behavior as it does for $R_{loc.}(B)$ if the one dimensional ring is used (see (Voskoboynikov et al., 2002) for details).

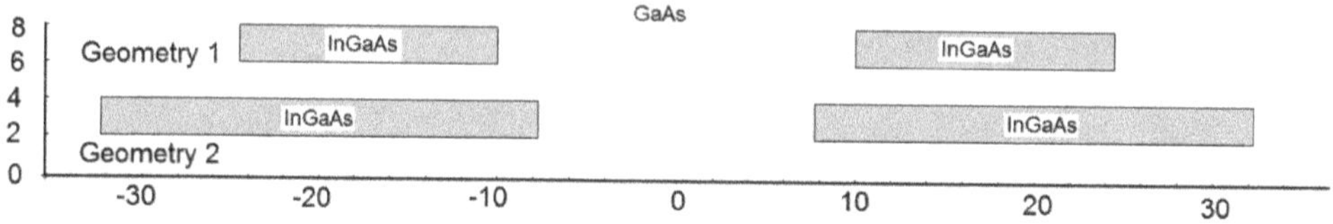

Fig. 11. QR cross section profile corresponding to Geometry 1 and Geometry 2; sizes are in nm.

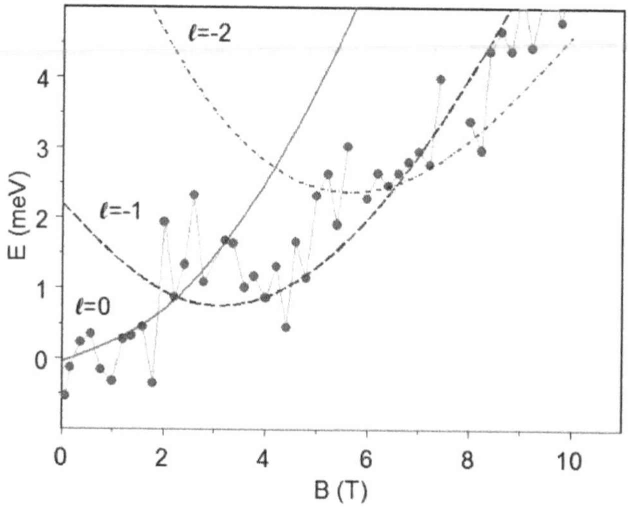

Fig. 12. Additional energy of an electron in QR in a magnetic field B. The C-V experimental energies (circles) were obtained in (Lei, et al. 2010) by using a linear approximation $\Delta E = e\Delta V_g / f$, with the lever arm $f = 7.84$. The curves $l = 0,-1,-2$ are the results of our calculations multiplied by a factor of 1.18 (Lei, et al. 2010).

Additionally we compare the results of calculations for the QR geometry parameters corresponding to Geometry 1 and Geometry 2 in Fig. 11 with the far-infrared (FIR) data, reported in (Emperador, et al. 2000). The results are presented in Fig. 14. One can see that the QR geometry proposed in (Lei et al., 2010) leads to a significant difference between the FIR data and the effective model calculations (see Fig. 14a), whereas the results obtained with Geometry 2 are in satisfactory agreement with the data (Fig. 14b). Again we conclude that the QR geometry of (Lei et al., 2010) does not provide an adequate description of electron properties of the InAs/GaAs QRs measured in (Lorke et al., 2000; Lei et al., 2010).

To summarize, we wish to point out that the problem of reliable theoretical interpretation of the C-V (and FIR) data for InAs/GaAs quantum rings is far from resolved. Obtained geometry can be considered as a possible version of geometry for experimentally fabricated QR.

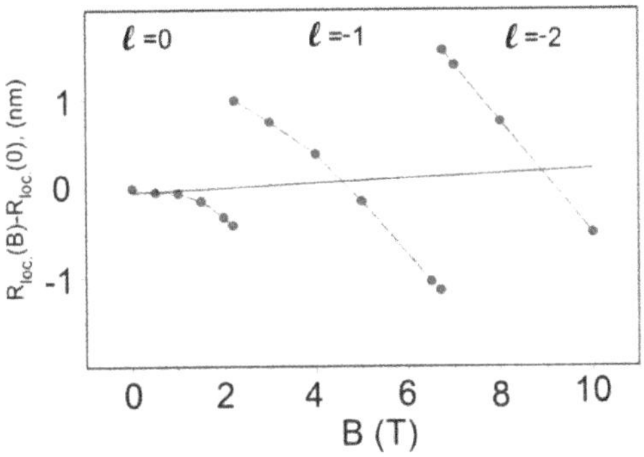

Fig. 13. The radius ($R_{loc.}$) of the most localized position of an electron as a function of a magnetic field B. The electron of the ground state is considered. The circles indicate the calculated values and the solid line indicates the result of the least squares fitting of the calculated values. The orbital quantum number of the ground state is shown.

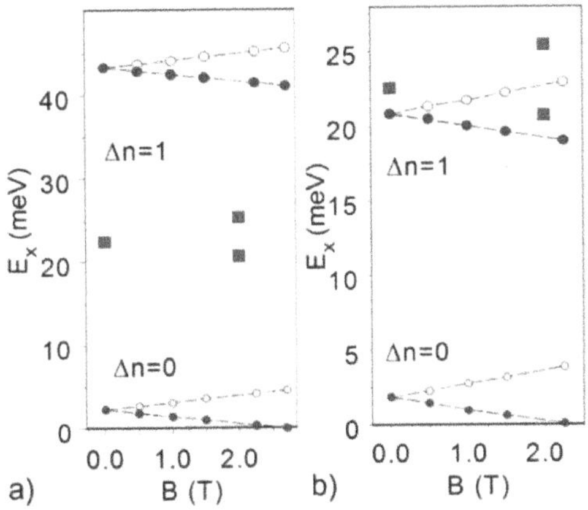

Fig. 14. Solid squares represent the observed resonance positions (Emperador, et al. 2000) of the FIR transmissions at various magnetic field B. Calculated energies of the excited states with $|\Delta l|=1$ are marked by the circles. a) QR with shape given by Geometry 1, b) QR with shape given by Geometry 2. The orbital quantum number of the ground state is $l=0$. The quantum number n is changed as shown.

2.6 Material mixing in InGaAs/GaAs quantum dots

The fabrication process of nano-sized self-assembled InAs/GaAs quantum dots and quantum rings may be accompanied by the material mixing in the initially pure InAs QDs due to interdiffusion of the QD/substrate materials. This mixing cannot be precisely controlled, resulting in QDs with spatially inhomogeneous Ga fractions that are not well specified. In this section we show an application of the effective model to study $In_xGa_{1-x}As$ QDs with significant Ga fractions.

InAs QDs having a semi-ellipsoidal shape embedded into the GaAs substrate are considered (see Fig. 6a)). The effective potential V_s =0.31 eV, which was found in (Filikhin et al., 2008) for pure InAs QDs, reproduces the capacitance-gate-voltage experiments satisfactory well (in Table 2 these results described in the column "0% Ga (V_s =0.31eV)"). It was assumed that a realistic approach must therefore take into account material mixing. The results of the effective model calculations for Ga fractions of 10%, 20% and 25% are listed in Table 2 (Filikhin et al., 2009). The calculations was performed varying the Ga fraction in QDs for strength parameters V_s^e =0.21 eV and V_s^h =0.28 eV of the potential. The effective electron mass, the band gap and the effective potential for $In_x Ga_{1-x} As$ changed linearly with respect to the value of the Ga fraction, assuming a homogeneous distribution of Ga in the QD volume. The experimental value of the transition energy for recombination of an exciton pair (E_{ex}) in the ground state is matched by calculations corresponding to a Ga fraction of approximately 22% in the QDs. Thus we conclude that the data obtained in CV and PLexperiments to this QD may be related with mixing in QD of 22%. It has to note that calculations with the 22% in the QDs (V_s =0.21eV) and pure InAs QDs (used V_s =0.31eV) demonstrate some uncertainties in the QD geometry and the Ga fraction and may lead to non-unique descriptions of the same experimental data.

Ga fraction	10%	20%	25%	0% (V_s =0.31eV)	Exp.
m^*/m_0	0.050	0.056	0.057	0.057	0.057 ± 0.007
$\Delta E(e)$ $\Delta E(h)$	238 245	205 217	188 151	185 206	204
$e_1 - e_0$ $e_2 - e_1$	50 55	48 53	46 52	46 52	44 49
$h_0 - h_1$ $h_1 - h_2$	10 12	10 11	9 11	10 11	
E^c_{e0e0}	21.0	20.9	20.8	20.8	21.5 (or 18.9)
E^c_{e0e1}	18.1	18.0	17.9	18.0	24 (or 13.0)
E^c_{e1e1}	17.0	17.0	16.9	17.0	~18.0
E^c_{h0h0}	25.1	24.9	24.7	25.1	24
E^c_{e0h0}	22.8	22.6	22.5	22.7	33.3
E_{ex}	1014	1075	1160	1106	1098
d_{00}	0.08	0.08	0.08	0.08	0.4 ± 0.1

Table 2. Calculated single electron (hole) energy-level spacing $e(h)$, electron (hole) binding energy $\Delta E(e)$ ($\Delta E(h)$), electron-electron, electron-hole and hole-hole Coulomb energies $E^c_{\alpha\beta}$ ($\alpha,\beta = e,h$), excitonic band gap E_{ex} (in meV), exciton dipole moment d_{00} (in nm) and effective mass of the QD material for semi-ellipsoidally shaped InGaAs QDs (Ga fraction in %) embedded in GaAs. Electron (hole) energy of the ground state is measured from the GaAs conduction (valence) band. The value of the effective mass is given for the p -wave electron level.

In (Filikhin et al. 2009) it was brought argument for existence of the essential mixing of the Ga-fraction in QD. The effective model with the material mixing was tested by comparison with available experimental data for the Coulomb shifts of the transition energies for positive (X^+) and negative (X^-) charged trions and biexcitons (XX) as a function of the neutral exciton (X) recombination energy. Results of these calculations for various base size parameters of QDs are depicted in Fig. 15, along with experimental data. The root mean square fit of the experimental data from (Rodt et al., 2005) shown by the dashed lines in Fig. 15. The vertical line shows the transition energy that corresponds to the limit of the QD sizes for which there are only two electron and two heavy hole levels. In this case the Coulomb shifts are calculated by combinations of the Coulomb energies of electron-electron, electron-hole and hole-hole pairs:

$$E(XX)-(X)=E_{ee}^{c}-E_{eh}^{c}+E_{hh}^{c}-E_{eh}^{c},\ E(X^{-})-(X)=E_{ee}^{c}-E_{eh}^{c},\ E(X^{+})-(X)=E_{hh}^{c}-E_{eh}^{c}.$$

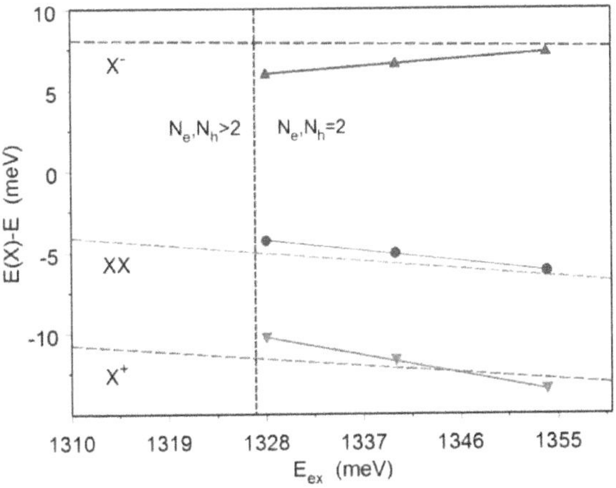

Fig. 15. Coulomb shifts of transition energies for positively (X^+) and negatively (X^-) charged trions and biexcitons (XX) as a function of neutral exciton (X) recombination energy. Results of the calculations for various base size parameters of QDs are marked by solid triangles (X^- , X^+) and dots (XX). The dashed lines correspond to root mean square fits to experimental data from (Rodt et al., 2005). The solid lines correspond to root mean square fits to the calculated results. The vertical line shows the transition energy, which corresponds to the limit of the QD sizes for which there are only two electrons and two heavy holes levels. The amount of the Ga fraction in our calculations is equal to 25%.

When there are several interacting carrier pairs, the calculations must be performed with more intricate scheme using perturbation theory. The value of the Ga fraction in our calculations was 25%. Calculations were performed for three QD geometries. A lens-shaped geometry with a height of 3.5 nm and base sizes of 9 nm, 10 nm and 11 nm were used. The effective model results in Fig. 15 demonstrate qualitative agreement with the experimental data for the aforementioned confinement region. The calculated results are very sensitive to the value of the Ga fraction. In particular, increasing the fraction shifts the X^+ and X^- energies to the region of large exciton energies (X). At the same time, the Coulomb shifts decrease in absolute value within the region of the X -energies with N_h =2. Decreasing the Ga fraction gives the opposite results.

We can conclude that in the framework of an effective model one can reproduce the CV and PL experimental data for InGaAs/GaAs QDs. In these calculations the amount of the Ga-fraction was taken to be about 22%. Taking into account this value for Ga-fraction we also reproduce the measured transition energies and Coulomb shifts for excitonic complexes (X^- , X^+ , XX) in the limit of two interacting pairs of carriers in the QDs.

3. Quantum chaos in single quantum dots

3.1 Quantum chaos

Quantum Chaos concerns with the behavior of quantum systems whose classical counterpart displays chaos. It is quantum manifestation of chaos of classical mechanics.

The problem of quantum chaos in meso - and nano-structures has a relatively long history just since these structures entered science and technology. The importance of this problem is related to wide spectrum of the transport phenomena and it was actively studied in the last two decades (Beenakker & van Houten, 1991; Baranger & Stone, 1989; Baranger et al., 1991). One of the main results of these studies, based mainly on the classical and semi-classical approaches, is that these phenomena sensitively depend on the geometry of these quantum objects and, first of all, on their symmetry: Right - Left (RL) mirror symmetry, up-down symmetry and preserving the loop orientation inversion symmetry important in the presence of the magnetic field (Whitney et al., 2009; Whitney et al., 2009a).

These results are well -known and discussed widely. There is another, actively studied in numerous fields of physics, aspect which ,in essence, is complimentary to the above mentioned semi classical investigations: Quantum Chaos with its inalienable quantum character , including, first of all, Nearest Neighbor level Statistics (NNS) which is one of the standard quantum-chaos test.

Mathematical basis of the Quantum Chaos is a Random Matrix Theory (RMT) developed by Wigner, Dyson, Mehta and Goudin (for comprehensive review see book (Beenakker & van Houten, 1991)). RMT shows that the level repulsion of quantum systems (expressed by one of the Wigner-Dyson -like distributions of RMT) corresponds to the chaotic behavior and, contrary, level attraction described by Poisson distribution tells about the absence of chaos in the classical counterpart of the quantum system. This theorem-like statement checked by numerous studies in many fields of science. For the completeness, we add that there are other tests of Quantum Chaos based on the properties of the level statistics: Δ_3 statistics (spectral rigidity $\Delta_3(L)$), Number variance $\Sigma_2(L)$), spectral form-factor, two- and multipoint correlation functions, two level cluster function $Y_2(E)$ etc. They play an important subsidiary role to enhance and refine the conclusions emerging from the NNS.

The present review surveys the study of the NNS of nanosize quantum objects - quantum dots (QD) which demonstrate atom-like electronic structure under the regime of the size confinement. To use effectively NNS, we have to consider so called weak confinement regime where the number of levels can be of the order of several hundred. QD of various shape embedded into substrate are considered here under the effective model (Filikhin et al., 2010). We use the sets of QD/substrate materials (Si/SiO_2, $GaAs/Al_{0.7}Ga_{0.25}As$, GaAs/InAs).

3.2 The nearest neighbor spacing statistics

For the weak confinement regime (for the Si/SiO$_2$ QD, the diameter $D \geq 10$ nm), when the number of confinement levels is of the order of several hundred (Filikhin et al. 2010), we studied NNS statistics of the electron spectrum. The low-lying single electron levels are marked by E_i, $i = 0,1,2,...N$. One can obtain the set $\Delta E_i = E_i - E_{i-1}$, $i = 1,2,3,...N$ of energy differences between neighboring levels. An example of the energy spectrum and set of the neighbor spacings for Si/SiO$_2$ QD are in Fig. 16. We need to evaluate the distribution function $R(\Delta E)$, distribution of the differences of the neighboring levels. The function is normalized by $\int R(\Delta E) d\Delta E = 1$. For numerical calculation, a finite-difference analog of the distribution function is defined by following relation:

$$R_j = N_j / H_{\Delta E} / N , \; j = 1,...M ,$$

where $\sum N_j = N$ represents total number of levels considered, $H_{\Delta E} = ((\Delta E)_1 - (\Delta E)_N) / M$ is the energy interval which we obtained by dividing the total region of energy differences by M bins. N_j ($j = 1,2,...M$) is the number of energy differences which are located in the j -th bin.

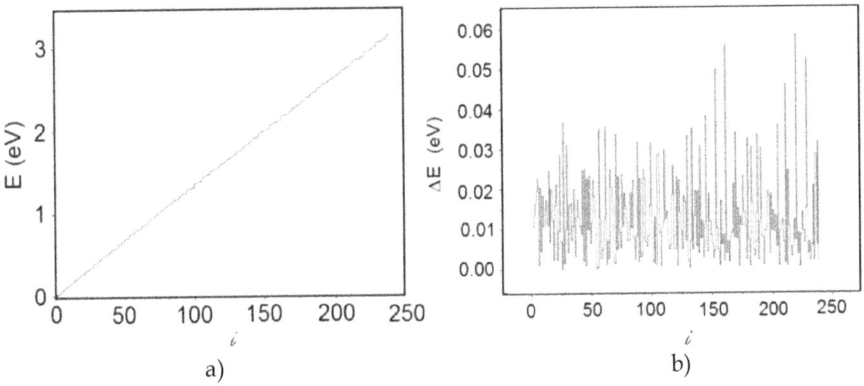

Fig. 16. a) The energy levels and b) the neighbor spacings $\Delta E_i = E_i - E_{i-1}$, $i = 1,2,..N$,of the spherical Si/SiO$_2$ QD with diameter D=17 nm.

The distribution functions $R(\Delta E)$ is constracted using the smoothing spline method. If R_j, $j = 1,2,..M$, are calculated values of the distribution functions corresponding to ΔE_j, the smoothing spline is constructed by giving the minimum of the form $\sum_{j=1}^{M} (R_j - R(\Delta E_j))^2 + \int R''(\Delta E)^2 d(\Delta E) / \lambda$. The parameter $\lambda > 0$ is controlling the concurrence between fidelity to the data and roughness of the function sought for. For $\lambda \to \infty$ one obtains an interpolating spline. For $\lambda \to 0$ one has a linear least squares approximation.

We studied neighbouring level statistics of the electron/hole spectrum treated by way considered above. The Si quantum dots having strong difference of electron effective mass in two directions is considered as appropriate example for the study of role of the effective mass asymmetry. In this study we do not include the Coulomb potential between electrons and holes. The shape geometry role is studied for two and three dimensions.

3.3 Violation of symmetry of the QD shape and nearest neighbor spacing statistics

Distribution functions for the nearest neighboring levels are calculated for various QD shapes (Filikhin et al., 2010). Our goal here to investigate the role of violation of the symmetries of QD shape on the chaos. The two and three dimensional models are considered. Existing of any above mentioned discrete symmetry of QD shape leads to the Poisson distribution of the electron levels.

In Fig. 17 the numerical results for the distribution functions of Si/SiO_2 QD are presented. The QD has three dimensional spherically shape. We considered the two versions of the shape. The first is fully symmetrical sphere, and the second shape is a sphere with the cavity damaged the QD shape. The cavity is represented by semispherical form; the axis of symmetry for this form does not coincide with the axis of symmetry of the QD. In the first case, the distribution function is the Poisson-like distribution. The violation symmetry in the second case leads to non-Poisson distribution.

We fit the non-Poissonian distribution function $R(\Delta E)$ using the Brody distribution (Brody et al., 1999):

$$R(s) = (1+\beta)bs^{\beta}\exp(-bs^{1+\beta}), \tag{9}$$

with the parameter β=1.0 and $b = (\Gamma[(2+\beta)/(1+\beta)]/D)^{1+\beta}$, D is the average level spacing. Note that for the Poisson distribution the Brody parameter is equal zero.

If the QD shape represents a figure of rotation (cylindrical, ellipsoidal and others) then the 3D Schrödinger equation is separable. In cylindrical coordinates the wave function is written by the following form $\psi(\vec{r}) = \Phi(\rho,z)\exp(il\varphi)$, where $l = 0,\pm1,\pm3,...$ is the electron orbital quantum number. The function $\Phi(\rho,z)$ is a solution of the two dimensional equation for cylindrical coordinates ρ and z.

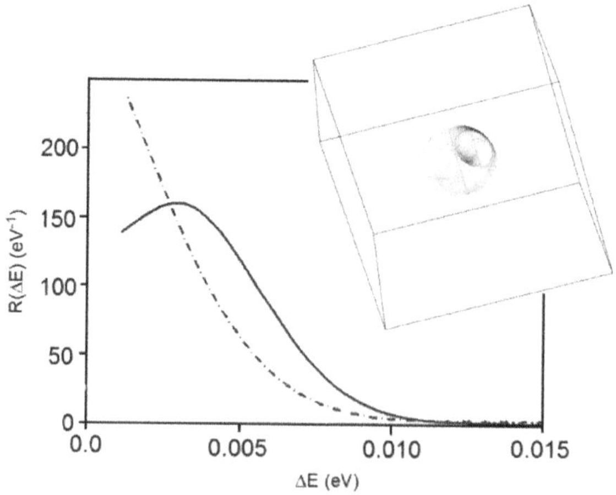

Fig. 17. Distribution functions for electron neighboring levels in Si/SiO_2 QD for spherical-like shape with cut. The Brody parameter β=1.0. The geometry of this QD is shown in 3D. The QD diameter is 17 nm (inset).

Our results for the distribution function for the ellipsoidal shaped Si/SiO_2 QD are presented in Fig. 18a) (left). In the inset we show the cross section of the QD. The fitting of the calculated values for $R(\Delta E)$ gives the Poisson-like distribution. For the case of QD shape with the break of the ellipsoidal symmetry (Fig. 18b) (left)) by the cut below the major axis we obtained a non-Poisson distribution.

Fig. 18 (right) shows the that slightly deformed rhombus-like shape leads to the NNS with Brody parameter $\beta = 1$ (10). It is obvious why systems with different discrete symmetries reveal Poisson statistics: the different levels of the mixed symmetry classes of the spectrum of the quantum system are uncorrelated.

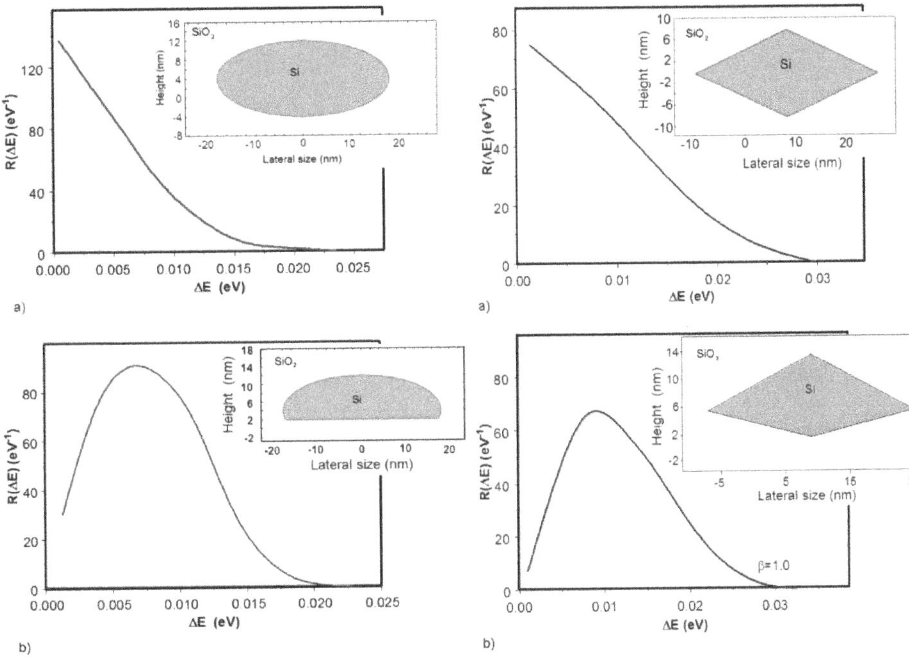

Fig. 18. (Left) Distribution functions for electron neighboring levels in Si/SiO_2 QD for different shapes: a) ellipsoidal shape, b) ellipsoidal like shape with cut. Brody parameter β is defined to be equal 1.02 for the fitting of this distribution. The 3D QD shape has rotation symmetry. Cross section of the shapes is shown in the inset.
(Right). Violation of the shape Up-Down symmetry for Si/SiO_2 QD. Distribution functions for electron neighboring levels in Si/SiO_2 QD for different shapes: a) with rhombus cross section, b) with slightly deformed rhombus cross section. The 3D QD shape has rotation symmetry. The Brody parameter β for the curve fitting this distribution is shown. Cross section of the shapes is shown in the inset.

In Schrödinger equation (7) in the asymptotical region of ρ one can neglect the two terms $\frac{1}{\rho}\frac{\partial}{\partial\rho}$ and $-\frac{l^2}{\rho^2}$ of this equation. The solution of Eq. (7) can demonstrate the same properties of the solution of the Schrödinger equation for 2D planar problem in Cartesian coordinates with the same geometry of QD shape in the asymptotical region. We illustrate this fact by Fig. 19. In this figure the violation of the shape Up-Down symmetry for 2D Si/SiO_2 QD is clarified. We compare the distribution functions for QD with "regular" semi-ellipsoidal shape (dashed curve in Fig. 19 a) and for QD with the semi-ellipsoidal shape having the cut (solid curve) as it are shown in Fig. 19 b). In the first case there is Up-Down symmetry of the QD shape. Corresponding distribution functions is Poissonian type. In second case the symmetry is broken by cut. The level statistics become non Poissonian. We have qualitative the same situation as for QD having rotation symmetry in 3D, presented in Fig. 18 (left) for the QD shape with rotation symmetry in cylindrical coordinates. The relation between the symmetry of QD shape and NNS is presented by Fig. 20 where we show the results of calculation of NNS for the 2D InAs/GaAs quantum well (QW). The two types of the statistics are presented in Fig. 20(left). The Poisonian distribution corresponds to shapes shown in Fig. 20 (b)-(d)(left) with different type of symmetry. The non-Poissonian distribution has been obtained for the QW shape with cut (a) which violated symmetry of initial shape (b), which is square having left-right symmetry, up-down symmetry, and diagonal reflection symmetry. The shape of the Fig. 20c) has only diagonal reflection symmetry. In Fig. 20d) the left-right symmetry of the shape exists only. The electron wave function of the high excited state, which contour plot is shown with the shape contour in Fig. 20(left), reflects the symmetry properties of the shapes.

Concluding, we can note that, obviously, the topological equivalent transformations of QD shape (keeping at least one discrete symmetry) do not lead to the non Poissonian distribution of the levels.

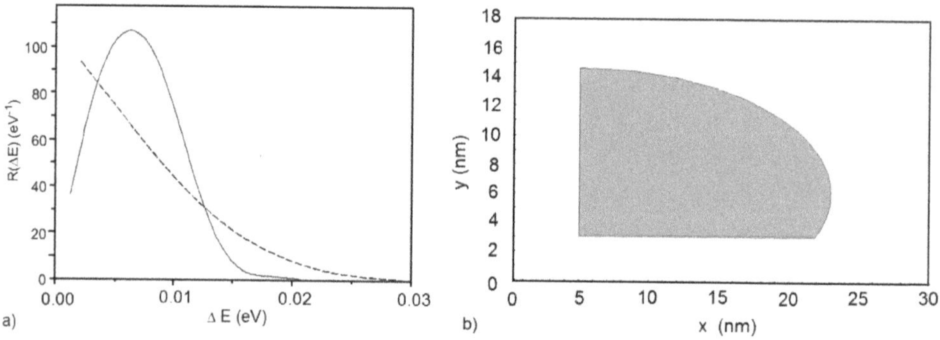

Fig. 19. Violation of the shape Up-Down symmetry for two dimensional Si/SiO_2 QD. a) Distribution functions for electron neighboring levels for the "regular" semi-ellipsoidal shape (dashed curve), for the semi-ellipsoidal shape with the cut (solid curve). b) The shape of the QD with cut (in Cartesian coordinates).

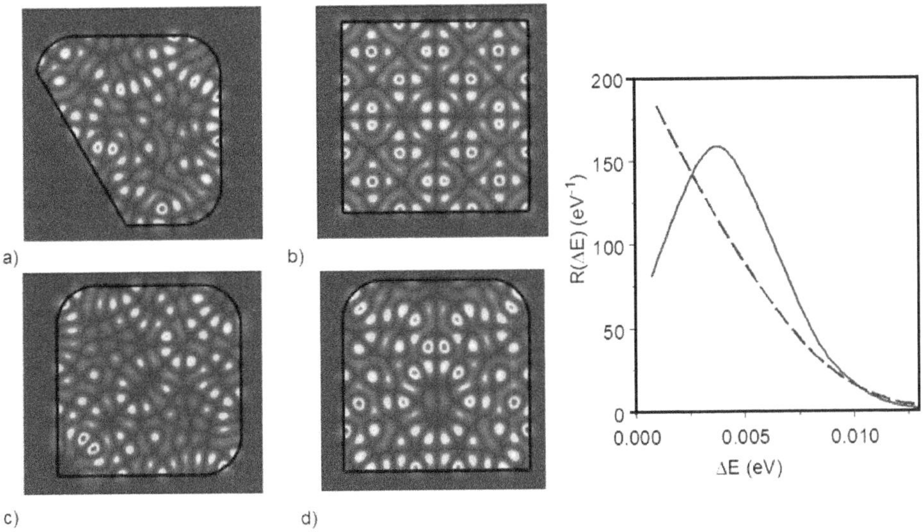

Fig. 20. Shape of the 2D InAs/GaAs quantum dots (Left). The black curves mean the perimeters. The electron wave function contour plots of the excited state (with energy about 0.5 eV are shown). The corresponding types of the level statistics are shown (Right). The shape a) leads to non-Poissonian statistics (solid curve). The shapes b)-d) result to the Poissonian statistics (dashed curve).

4. Double quantum dots and rings: New features

4.1 Disappearance of quantum chaos in coupled chaotic quantum dots

In the previous section, we investigated the NNS for various shape of the single quantum dots (SQD) in the regime of the weak confinement when the number of the levels allows to use quite sufficient statistics. Referring for details to (Filikhin et al., 2010), we briefly sum up the main conclusions of previous section: SQDs with at least one mirror (or rotation) symmetry have a Poisson type NNS whereas a violation of this symmetry leads to the Quantum Chaos type NNS.

In this section we study quantum chaotic properties of the double QD (DQD). By QD here we mean the three dimensional (3D) confined quantum object, as well its 2D analogue - quantum well (QW). In three dimensional case we use an assumption of the rotational symmetry of QD shape. The presented effective approach is in good agreement with the experimental data and previous calculations in the strong confinement regime (Filikhin et al., 2010). Here, in the regime of weak confinement, as in (Filikhin et al., 2010), we also do not consider Coulomb interaction between electron and hole: Coulomb effects are weak when the barrier between dots is thin leading to the strong interdot tunneling and dot sizes are large enough. In these circumstances, studied in detail in (Bryant, 1993) (see also for short review a monograph (Bimberg et al., 1999), one may justify disregard of the Coulomb effects. The physical effect, we are looking for, has place just for thin barriers; to have sufficient level statistics, we need large enough QDs (≥100 nm for InAs/GaAs QW).

Thus, we consider tunnel coupled two QDs with substrate between, which serves as barrier with electronic properties distinct from QD. Boundary conditions for the single electron Schrödinger equation are standard. We take into account the mass asymmetry inside as well outside of QDs (Filikhin et al., 2010). To avoid the complications connected with spin-orbit coupling, *s*-levels of electron are only considered in the following. We would like to remind that the selection of levels with the same quantum numbers is requisite for study of NNS and other types of level statistics.

Whereas at the large distances between dots each dot is independent and electron levels are twofold degenerate, expressing the fact that electron can be found either in one or in the other isolated dot, at the smaller inter-dot distances the single electron wave function begins to delocalize and extends to the whole DQD system. Each twofold degenerated level of the SQD splits by two, difference of energies is determined by the overlap, shift and transfer integrals (Bastard, 1990). Actually, due to the electron spin, there is fourfold degeneracy, however that does not change our results and below we consider electron as spinless. Note that the distance of removing degeneracy is different for different electron levels. This distance is larger for levels with higher energy measured relative to the bottom quantum well (see Fig. 23 below). By the proper choice of materials of dots and substrate one can amplify the "penetration" effects of the wave function.

Below we display some of our results for semiconductor DQDs. The band gap models are given in (Filikhin et al., 2010). Fig. 21 shows distribution function for two Si/SiO_2 QDs of the shape of the 3D ellipsoids with a cut below the major axis. Isolated QD of this shape, as we saw in the previous section, is strongly chaotic. It means that distribution function of this QD can be well fitted by Brody formula with the parameter which is close to unity (Filikhin et al, 2010). We see that the corresponding up-down mirror symmetric DQD shows Poisson-like NNS. Note that these statistics data involved 300 confined electron levels, which filled the quantum well from bottom to upper edges. We considered the electron levels with the orbital momentum l=0, as was mentioned above. The orbital momentum of electron can be defined due to rotational symmetry of the QD shape.

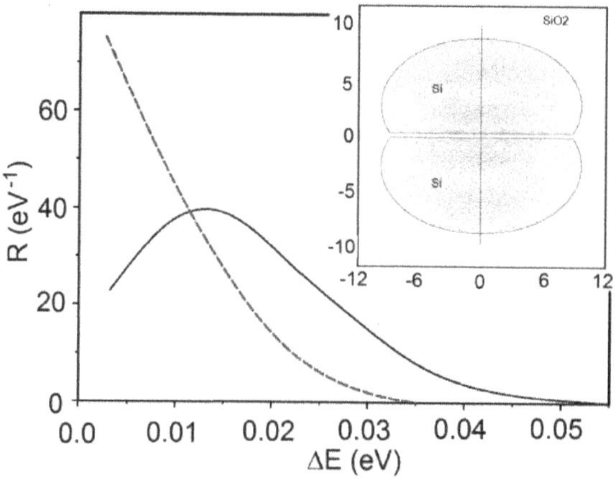

Fig. 21. The electron wave function of the ground state is shown by the contour plot. (The lower figure) Distribution functions for energy differences of the electron neighbori between QDs in InAs/GaAs DQD.

In Fig. 22, SQD (2D quantum well) without both type of symmetry reveals level repulsion, two tunnel coupled dots show the level attraction. From the mirror symmetry point of view, the chaotic character of such single object is due to the lack of the R-L and up-down mirror symmetries. The symmetry requirements in this case, for the coupled dots are less restrictive: presence of one of the mirror symmetry types is sufficient for the absence of quantum chaos.

Dependence NNS on the interdot distance shows a gradual transition to the regular behaviour with intermediate situation when Poisson-like behavior coexists with chaotic one: they combine but the level attraction is not precisely Poisson-like. Further decreasing distance restores usual Poisson character (see Fig. 22). Fig. 23 shows how the degeneracy gradually disappears with the distance b between QDs in InAs/GaAs DQD.

Finally, we would like to show the disappearance of the Quantum Chaos when chaotic QW is involved in the "butterfly double dot" (Whitney, 2009) giving huge conductance peak in the semi-classical approach. Fig. 24 shows the NNS for chaotic single QW of (Whitney, 2009) by dashed line. Mirror (up-down and L-R) symmetry is violated. The NNS for an L-R mirror symmetric DQW is displayed by solid line in Fig. 24. It is clear that Quantum Chaos disappears.

We conjecture that the above mentioned peak in conductance of (Whitney, 2009) and observed here a disappearance of Quantum Chaos in the same array are the expression of the two faces of the Quantum Mechanics with its semi-classics and genuine quantum problem of the energy levels of the confined objects, despite the different scales (what seems quite natural) in these two phenomena (several micrometers and 10–100 nm, wide barrier in the first case and narrow one in the second). We have to emphasize here that the transport properties are mainly the problem of the wave function whereas the NNS is mainly the problem of eigenvalues. Similar phenomena are expected for the several properly arranged coupled multiple QDs and QD superlattices. In the last case, having in mind, for simplicity, a linear array, arranging the tunnel coupling between QDs strong enough, we will have wide mini-bands containing sufficient amount of energy levels and the gap between successive mini-bands will be narrow. Since the levels in the different mini-bands are uncorrelated, the overall NNS will be Poissonian independently of the chaotic properties of single QD. We would like to remark also that our results have place for 3D as well as for 2D quantum objects. It is important to notice that the effect of reduction of the chaos in a system of DQD could appear for interdot distances larger than considered, for instance in figure 22, if an external electrical field is applied. By properly designed bias the electric field will amplify wave function "penetration" effectively reducing a barrier between QDs.

Thus, we have shown that the tunnel coupled chaotic QDs in the mirror symmetric arrangement have no quantum chaotic properties, NNS shows energy level attraction as should to be for regular, non-chaotic systems. These results are confronted with the huge conductance peak found by the semi-classical method in (Whitney, 2009). We think that our results have more general applicability for other confined quantum objects, not only for the quantum nanostructures, and may be technologically interesting. Concerning the last issue, problem is what easer: try to achieve regular, symmetric shape of SQDs, or, not paying attention to their irregular, chaotic shape arrange more or less symmetric mutual location (Ponomarenko et al., 2008).

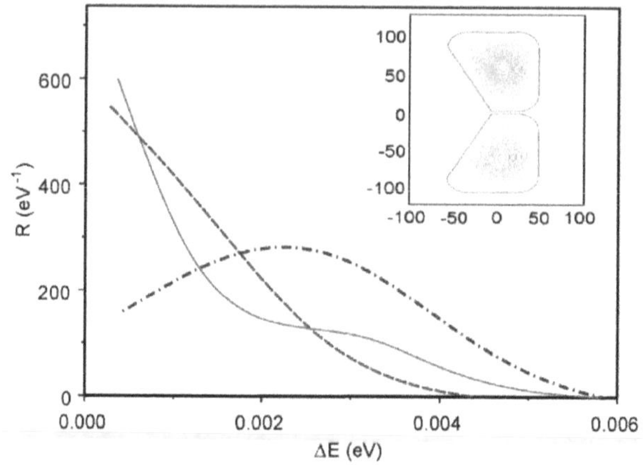

Fig. 22. Distribution functions for energy differences of the electron neighboring levels in the 2D InAs/GaAs DQW calculated for various distances b between QWs. Dashed (solid) line corresponds to b =4 nm (b =2 nm). Distribution functions of single QW is also shown by the dot-dashed line. The DQW shape is shown in inset (sizes are in nm).

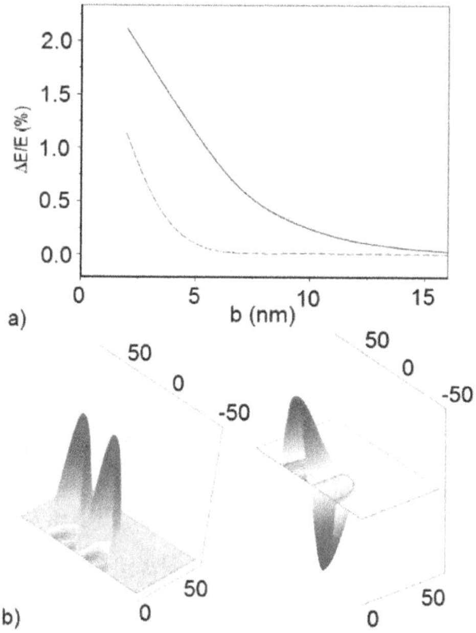

Fig. 23. (The upper figure) Doublet splitting ΔE of single electron levels dependence on the distance b between QDs in InAs/GaAs DQD. The ground state (E =0.23 eV) level splitting is ΔE expressed by dashed line. The solid line corresponds to doublet splitting of a level which is close to upper edge of the quantum well (E =0.56 eV). The shape of DQD is the same as in Fig. 21 (The lower figure). The electron wave functions of the doublet state: the ground state (left) and first excited state (right), are shown.

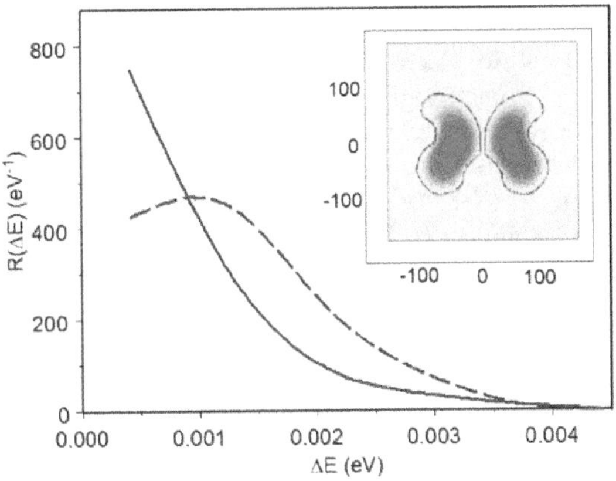

Fig. 24. Distribution functions for electron neighboring levels in InAs/GaAs single QW (dashed line) and DQW (solid line). Shape of DQW is shown in the inset. The electron wave function of the ground state is shown by the contour plot in the inset. Data of the statistics include 200 first electron levels.

4.2 Electron transfer between pair of concentric quantum rings in magnetic field

Quantum rings are remarkable meso- and nanostructures due to their non-simply connected topology and attracted much attention last decade. This interest supported essentially by the progress in the fabrication of the structures with wide range of geometries including single and double rings. This interest rose tremendously in the connection to the problem of the persistent current in mesoscopic rings (Buttiker et al., 1983) Transition from meso - to nano - scale makes more favorable the coherence conditions and permits to reduce the problem to the few or even to single electron.

Application of the transverse magnetic field B leads to the novel effects: Whereas the quantum dots (QDs) of the corresponding shape (circular for two dimensional (2D), cylindrical or spherical for 3D) has degeneracy in the radial n and orbital l quantum numbers, QR due to the double connectedness in the absence of the magnetic field B has degeneracy only in l, and the nonzero B lifts the degeneracy in l, thus making possible the energy level crossing at some value of B, potentially providing the single electron transition from one state to the another.

Use the configurations with double concentric QR (DCQR) reveals a new pattern: one can observe the transition between different rings in the analogy with atomic phenomena. For the DCQR, the 3D treatment is especially important when one includes the inter-ring coupling due to the tunneling. The dependence on the geometries of the rings (size, shape and etc.) becomes essential.

We investigate the electron wave function localization in double concentric quantum rings (DCQRs) when a perpendicular magnetic field is applied. In weakly coupled DCQRs can be arisen the situation, when the single electron energy levels associated with different rings may be crossed. To avoid degeneracy, the anti-crossing of these levels has a place. In this DCQR the electron spatial transition between the rings occurs due to the electron level anti-crossing. The anti-crossing of the levels with different radial quantum numbers (and equal orbital quantum numbers) provides the conditions when the electron tunneling between rings becomes possible. To study electronic structure of the semiconductor DCQR, the single sub-band effective mass approach with energy dependence was used (see section 2 of this Chapter). Realistic 3D geometry relevant to the experimental DCQR fabrication was employed taken from (Kuroda et al., 2005; Mano et al., 2005). The GaAs QRs and DQRs rings, embedded into the $Al_{0.3}Ga_{0.7}As$ substrate, are considered (Filikhin, et al., 2011). The strain effect between the QR and the substrate materials was ignored here because the lattice mismatch between the rings and the substrate is small. Due to the non-parabolic effect taken into account by energy dependence effective mass of electron in QR, the effective mass of the electron ground state is calculated to be the value of 0.074 m_0 that is larger than the bulk value of 0.067 m_0 . For the excited states, the effective mass will increase to the bulk value of the $Al_{0.3}Ga_{0.7}As$ substrate. Details of this calculation one can find in (Filikhin, et al. 2011).

Electron transfer in the DCQR considered is induced by external factor like a magnetic or electric fields. Probability of this transfer strongly depends on the geometry of DCQR. The geometry has to allow the existing the weakly coupled electron states. To explain it, we note that DCQR can be described as a system of double quantum well. It means that there is duplication of two sub-bands of energy spectrum (see (Manasreh, 2005) for instance) relative the one for single quantum object. In the case of non-interacting wells (no electron tunneling between wells) the each sub-band is related with left or right quantum well. The wave function of the electron is localized in the left or right quantum well. When the tunneling is possible (strong coupling state of the system), the wave function is spread out over whole volume of the system. In a magnetic field, it is allowed an intermediate situation (weak coupled states) when the tunneling is possible due to anti-crossing of the levels. Anti-crossing, of course, is consequence of the impossibility to cross of levels with the same space symmetry (von Neumann & Wigner, 1929; Landau & Lifshitz, 1977).

There is a problem of notation for states for DCQR. If we consider single QR (SQR) then for each value of the orbital quantum number $|l|=0,1,2...$ in Eq. (7) we can definite radial quantum number $n=1,2,3,...$ corresponding to the numbers of the eigenvalues of the problem (7) in order of increasing. One can organize the spectrum by sub-bands defined by different n . When we consider the weakly coupled DCQR, in contrast of SQR, the number of these sub-bands is doubled due to the splitting the spectrum of double quantum object (Bastard, 1990). Electron in the weakly coupled DCQR can be localized in the inner or outer ring. In principle, in this two ring problem one should introduce a pair of separate sets of quantum numbers (n_i,l) where index $i=1,2$ denoted the rings where electron is localized. However, it is more convenient, due to the symmetry of the problem, to have one pair (n,l) numbers ascribed to both rings (inner or outer), in other words, we use a set of quantum numbers $(n,l),p$ where p is dichotomic parameter attributed to the electron localization ("inner" or "outer").

Since we are interested here in the electron transition between rings and, as we will see below, this transition can occur due to the electron levels anti-crossing followed a tunneling, we concentrate on the changing of the quantum numbers n . The orbital quantum numbers must be equal providing the anti-crossing of the levels with the same symmetry (see Landau & Lifshitz, 1977). Thus, the anti-crossing is accompanied by changing the quantum numbers n and p of the $(n,l),p$ set.

Strongly localized states exist in the DCQR with the geometry motivated by the fabricated DCQR in (Kuroda et al., 2005; Mano et al., 2005). The wave functions of the two s-states of the single electron with $n=1,2$ are shown in Fig. 25, where the electron state $n=1$ is localized in outer ring, and the electron state $n=2$ is localized in inner ring. Moreover all states of the sub-bands with $n=1,2$, and $|l|=1,2,3...$ are well localized in the DCQR. The electron localization is outer ring for $n=1$, $|l|=0,1,2,...$, and inner ring for $n=2$, $|l|=0,1,2....$

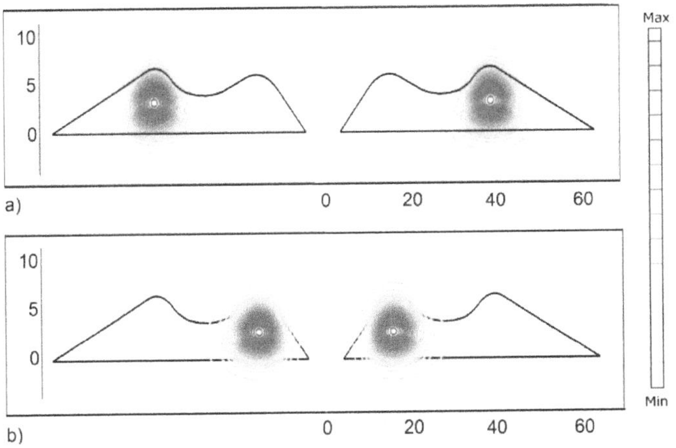

Fig. 25. The squares of wave functions for the a) (1,0) ,outer ($E=0.072$ eV) and b) (2,0) ,inner ($E=0.080$ eV) states are shown by contour plots. The contour of the DCQR cross-section is given. The sizes are in nm.

The difference of properties of the two sub-bands can be explained by competition of two terms of the Hamiltonian of Eq. (7) and geometry factor. The first term includes first derivative of wave function over ρ in kinetic energy; the second is the centrifugal term. For $|l|\neq 0$ the centrifugal force pushes the electron into outer ring. One can see that the density of the levels is higher in the outer ring. Obviously, the geometry plays a role also. In particular, one can regulate density of levels of the rings by changing a ratio of the lateral sizes of the rings.

Summarizing, one can say that for $B=0$ the well separated states are only the states $(1,l),p$ and $(2,l),p$. Thus, used notation is proper only for these states. The wave functions of the rest states $(n>2,l)$ are distributed between inner and outer rings. These states are strongly coupled states.

Crossing of electron levels in the magnetic field B are presented in Fig. 26 There are crossings of the levels without electron transfer between the rings. This situation is like when we have crossing levels of two independent rings. There are two crossings when the orbital quantum number of the lower state is changed due to the Aharonov-Bohm effect. It occurs at about 0.42 T and 2.5 T. There are two anti-crossings: the first is at 4.8 T, another is at 5.2 T. These anti-crossings are for the states with different n; the first are states (1,0) and (2,0) and the second are states (1,-1) and (2,-1). In these anti-crossings the possibility for electron tunneling between rings are realized. In Fig. 27 we show how the root mean square (rms) of the electron radius is changed due to the tunneling at anti-crossing. One can see from Fig. 26 that the electron transition between rings is only possible when the anti-crossed levels have different radial quantum numbers and equal orbital quantum numbers, in accordance of (von Neuman & Vigner, 1929).

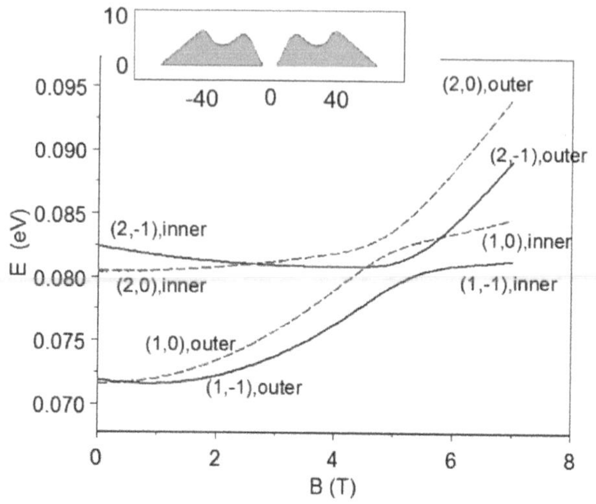

Fig. 26. Single electron energies of DCQR as a function of magnetic field magnitude B. Notation for the curves: the double dashed (solid) lines mean states with l=0 (l=-1) with n =1,2. The quantum numbers of the states and positions of the electron in DCQR are shown. The cross section of the DCQR is given in the inset.

Transformation of the profile of the electron wave function during the process of anti-crossing with increasing B is given in Fig. 28. The electron state (1,-1), outer is considered as "initial" state of an electron (B=0). The electron is localized in outer ring. Rms radius is calculated to be R=39.6 nm. For B=5.2 T the second state is the tunneling state corresponding to the anti-crossing with the state (0,-1). The wave function is spreaded out in both rings with R=32.7 nm. The parameter p has no definite value for this state. The "final" state is considered at B=7 T. In this state the electron was localized in inner ring with R=17.6 nm. Consequently connecting these three states of the electron, we come to an electron trapping, when the electron of outer ring ("initial" state) is transferred to the inner ring ("final" state). The transfer process is governed by the magnetic field.

Note that the energy gap between anti-crossed levels which one can see in Fig. 26 can be explained by the general theory for double interacting quantum well (Bastard, 1990). The value of the gap depends on separation distance between the rings, governed by the overlapping wave functions corresponding to the each ring, and their spatial spread which mainly depends on radial quantum number of the states (Filikhin et al., 2011).

Other interesting quantum system is one representing QR with QD located in center of QR. The cross section of such heterostructure ($GaAs/Al_{0.3}Ga_{0.7}As$) is shown in Fig. 29a. In Fig. 29b we present the results of calculations for electron energies of the (1,0) and (3,0) states in the magnetic field B (Filikhin et al., 2011). Once more we can the level anti-crossing (for about of 12.5 T). This anti-crossing is accompanied by exchange of electron localization between the QD and the QR. In other words if initial state (for B<12.5 T) of electron was the state (1,0),outer, then the "final" state (for B >12.5 T) will be (1,0),inner. It can be considered as one of possibilities for trapping of electron in QD.

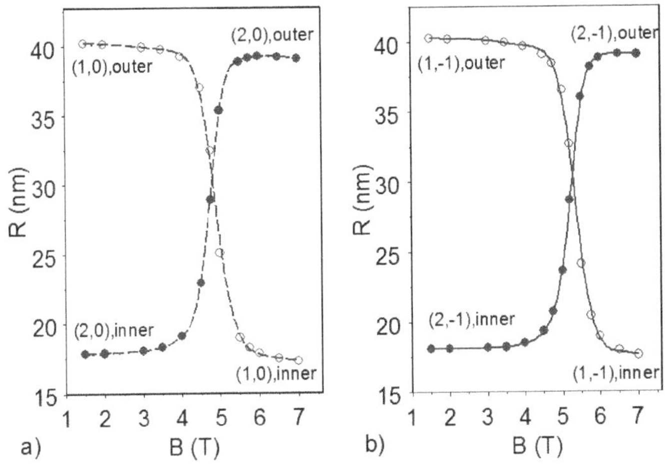

Fig. 27. Rms radius of an electron in DCQR as a function of magnetic field for the states a) $((n=1,2),l=0)$ and b) $((n=1,2),l=-1)$ near point of the anti-crossing. The calculated values are shown by solid and open circles. The dashed (solid) line, associated with states of $l=0$ ($l=-1$), fits the calculated points.

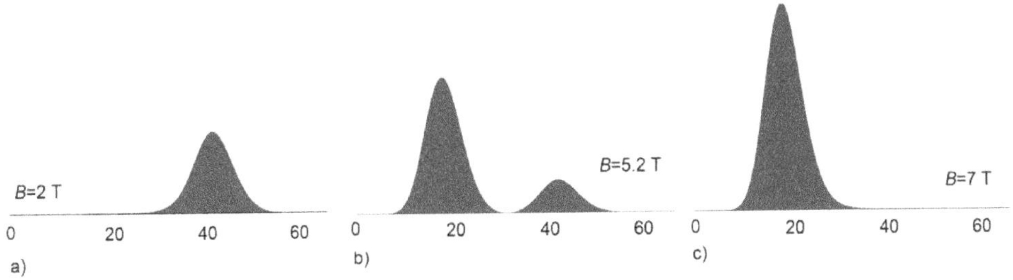

Fig. 28. Profiles of the normalized square wave function of electron in the states a) $(1,-1)$,outer; b) $(1,-1)$,n/a and c) $(1,-1)$,inner for different magnetic field B. The a) is the "initial" state (B=0) with R=39.6 nm, the b) is the state of electron transfer (B=5.2 T) with R=32.7 nm, the c) is the "final" state (B=7 T) with R=17.6 nm. The radial coordinate ρ is given in nm (see Fig. 26 for the DCQR cross section).

One can see from Fig. 29b that the energy of the dot-localized state grows more slowly than the envelope ring-localized state. At the enough large B the dot-localized state becomes the ground state (Szafran et al., 2004). In other words, when the Landau orbit of electron becomes smaller then dot size, electron can enter the dot without an extra increase of kinetic energy.

Concluding, we made visible main properties of this weakly coupled DCQD established by several level anti-crossings that occurred for the states with different radial quantum number n (n =1,2) and equal orbital quantum number l. One may conclude that the fate of the single electron in DCQRs is governed by the structure of the energy levels with their crossing and anti-crossing and is changing with magnetic field. The above described behavior is the result of the nontrivial excitation characteristic of the DCQRs. Effect of the trapping of electron in inner QR (or QD) of DCQR may be interesting from the point of view of quantum computing.

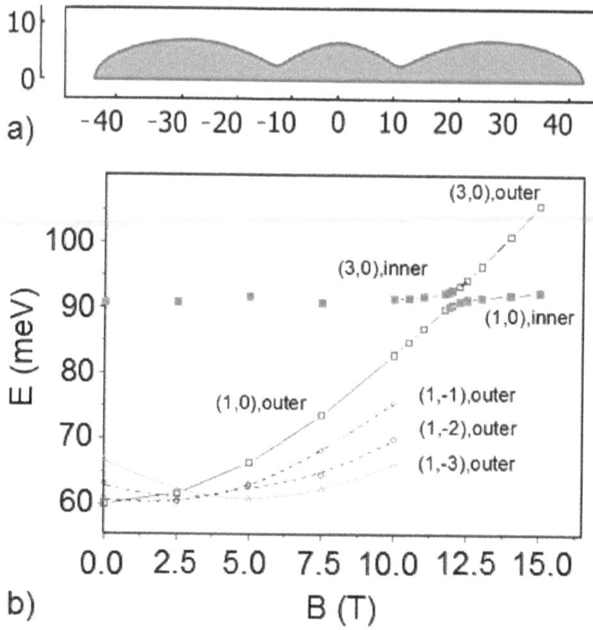

Fig. 29. a) Cross section of the QR with QD system. Sizes are given in nm. b) Energies of the (1,0) and (3,0) states in the magnetic field B for the QR with QD system. The open symbols show that the electron is localized in the ring. The solid squares show that the electron localized in QD.

5. Acknowledgment

This work is supported by NSF CREST award; HRD-0833184 and NASA award NNX09AV07A

6 References

Aharonov Y. & Bohm, D. (1959). Significance of Electromagnetic Potentials in the Quantum Theory, *Physical Review*, Vol. 115, Issue 3, (1959), pp. 485–491

Alferov, Zh.I. (1998).The history and future of semiconductor heterostructures, *Semiconductors*, Vol. 32, No. 1 (January 1998), pp. 1-14, ISSN 1063-7826 (Print); 1090-6479 (Online)

Anderson P.W. ; Thouless D.J.; Abrahams, E.; Fisher, D. S. (1980). New method for a scaling theory of localization. *Physical Review B*, Vol. 22, Issue 8, (1980), pp. 3519–3526, ISSN 1550-235x (online), 1098-0121 (print)

Aronov A.G. & Sharvin, Yu.V. (1987). Magnetic flux effects in disordered conductors, *Review Modern Physics*. Vol. 59, (1987), pp. 755–779, ISSN 1539-0756 (online), 0034-6861 (print)

Baranger, H.U. & Stone, A.D. (1989). Quenching of the Hall resistance in ballistic microstructures: A collimation effect, *Physical Review Letters*, Vol. 63, (1989), pp. 414–417, ISSN 1079-7114 (online) 0031-9007 (print)

Baranger, H.U.; DiVincenzo, D.P.; Jalabert R.A. & Stone, A.D. (1991). Classical and quantum ballistic-transport anomalies in micro-junctions, *Physical Review B*, Vol. 44, Issue 19, (1991), pp. 10637–10675, ISSN 1550-235x (online), 1098-0121 (print)

Bastard, G. (1990).*Wave Mechanics applied to Semiconductor Heterostructures*, John Wiley and Sons Inc. (Jan 01, 1990), p. 357, ISBN 0-471-21708-1, New York, NY (USA)

Beenakker, C.W.J. & van Houten, H. (1991). *Solid State Physics: Advances in Research and Applications*, Ed. H. Ehrenreich and D. Turnbull, Academic Press (April 2, 1991), Vol. 44, (1991), p.1-454, ISBN-10: 0126060444, N-Y, USA

BenDaniel D. J. & Duke C. B. (1966). Space-Charge Effects on Electron Tunneling, *Physical Review*, Vol. 152, Issue 2, (1966), pp. 683–692,

Betcke, M.M. & Voss, H. (2011). Analysis and efficient solution of stationary Schrödinger equation governing electronic states of quantum dots and rings in magnetic field, https://www.mat.tu-harburg.de/ins/forschung/rep/rep143.pdf; to appear in CiCT (2011)

Bimberg, D.; Grundmann, M.; Ledentsov, N.N. (1999). *Quantum Dot Heterostructure*, John Wiley and Sons Inc. (1999), ISBN 0-471-97388-2, Chichester, England

Bloch, F. (1928). Über die Quantenmechanik der Elektronen in Kristallgittern, *Zeitschrift fur Physik*, Vol. 52, Issue 7-8, (1928), pp. 555–600, ISSN 0939-7922 (Print), 1431-5831 (Online)

Brody, T. A.; Flores, J.; French, J. B.; Mello, P. A.; Pandey, A.; Wong, S. S. M. (1981). Random-matrix physics: spectrum and strength fluctuations, *Reviews of Modern Physics*, Vol. 53, Issue 3, (1981) pp. 385–479, ISSN 1539-0756 (online), 0034-6861 (print).

Bryant, G.W (1993). Electrons in coupled vertical quantum dots: Interdot tunneling and Coulomb correlation, *Physical Review B*, Vol. 48, Issue 11, (1993), pp. 8024–8034, ISSN 1550-235x (online), 1098-0121 (print)

Bryant, G.W. (1992). Indirect-to-direct crossover of laterally confined excitons in coupled quantum wells. *Physical Review B*, Vol. 46, Issue 3, (1992), pp. 1893–1896, ISSN 1550-235x (online), 1098-0121 (print)

Buttiker, M. , Imry Y. & Landauer, R. (1983). Josephson behavior in small normal one-dimensional rings, *Physics Letters A*, Vol. 96, Issue 7, 1(8 July 1983), pp. 365-367, ISSN 0375-9601

Califano, M. & Harrison, P. (2000). Presentation and experimental validation of a single-band, constant-potential model for self-assembled InAs/GaAs quantum dots, *Physical Review B*, Vol. 61, (2000), pp. 10959–10965, ISSN 1550-235x (online), 1098-0121 (print)

Chakraborty, T. & Pietiläinen, P. (1994). Electron-electron interaction and the persistent current in a quantum ring. *Physical Review B*, Vol. 50, pp. 8460–8468, ISSN 1550-235x (online), 1098-0121 (print)

Duque, C.A.; Porras-Montenegro, N.; Barticevic, Z.; Pachecoc, M.; Oliveira, L.E. (2005). Electron-hole transitions in self-assembled InAs/GaAs quantum dots: Effects of applied magnetic fields and hydrostatic pressure, *Microelectronics Journal* 36, (2005), pp. 231-233, ISSN 0026-2692

Emperador, A.; Pi, M.; Barranco, M.; Lorke, A. (2000). Far-infrared spectroscopy of nanoscopic InAs rings, *Physical Review B,* Vol. 62, (2000), pp. 4573. 4573–4577, ISSN 1550-235x (online), 1098-0121 (print)

Filikhin I, E. Deyneka & Vlahovic, B (2007a) Numerical modeling of experimentally fabricated InAs/GaAs quantum rings, *Molecular Simulation,* Vol. 33, Issue 7, (2007), pp. 589-592, ISSN 0892-7022 (Print), 1029-0435 (Online)

Filikhin, I.; Deyneka, E. & Vlahovic, B. (2006a). Non parabolic model for InAs/GaAs quantum dot capacitance spectroscopy, *Solid State Communications* 140, (2006), pp. 483–486, ISSN 0038-1098

Filikhin, I.; Deyneka, E.; Melikyan, H. & Vlahovic, B. (2005). Electron States of Semiconductor Quantum Ring Electron States of Semiconductor Quantum Ring with Geometry and Size Variations, *Journal Molecular Simulation,* Vol. 31, (2005), pp. 779-782, Print ISSN 0892-7022, Online ISSN 1029-0435

Filikhin, I.; Matinyan, S.; Nimmo, J. & Vlahovic, B. (2011). Electron Transfer between Weakly Coupled Concentric Quantum Rings, *Physica E: Low-dimensional Systems and Nanostructures,* Vol. 43, No. 9, (July 2011), pp. 1669-1676, ISSN: 1386-9477

Filikhin, I.; S.G. Matinyan, S.G.; Schmid, B.K. & Vlahovic, B. (2010). Electronic and level statistics properties of Si/SiO2 quantum dots, *Physica E: Low-dimensional Systems and Nanostructures,* Vol. 42, Issue 7, (May 2010), pp. 1979-1983, ISSN: 1386-9477.

Filikhin, I.; Suslov, V. M.; Wu, M. & Vlahovic, B. (2008). Effective approach for strained InAs/GaAs quantum structures. *Physica E: Low-dimensional Systems and Nanostructures,* Vol. 40, (2008), pp. 715–723, ISSN: 1386-9477

Filikhin, I.; Suslov, V.M. & Vlahovic, B. (2006). Modeling of InAs/GaAs quantum ring capacitance spectroscopy in the nonparabolic approximation, *Physical Review B,* Vol. 73, (2006), pp. 205332-205336, ISSN 1550-235x (online), 1098-0121 (print)

Filikhin, I.; Suslov, V.M. & Vlahovic, B. (2009) InGaAs/GaAs quantum dots within an effective approach, *Physica E: Low-dimensional Systems and Nanostructures,* Vol. 41, (2009), pp. 1358–1363, ISSN: 1386-9477

Filikhin, I.N., Deyneka, E. & Vlahovic, B. (2004). Energy dependent effective mass model of InAs/GaAs quantum ring, *Modelling and Simulation in Materials Science and Engineering,* Vol. 12, (2004), pp. 1121-1126, ISSN 0965-0393 (Print); ISSN 1361-651X (Online)

Filikhin, I; Suslov, V.M. & Vlahovic, B. (2011a). C-V Data and Geometry Parameters of Self-Assembled InAs/GaAs Quantum Rings, accepted for publication in *Journal of Computational and Theoretical Nanoscience* (2011), ISSN:1546-1955

Fu, H.; Wang, L.-W.; Zunger, A. (1998). Applicability of the k.p method to the electronic structure of quantum dots, *Physical Review B,* Vol. 57, (1998), pp. 9971, ISSN 1550-235x (online), 1098-0121 (print)

Gefen Y.; Imry, Y.; Azbel, M. Ya. (1984). Quantum Oscillations and the Aharonov-Bohm Effect for Parallel Resistors. *Physical Review Letters,* Vol. 52, Issue 2, (1984), pp. 129–132, ISSN 1079-7114 (online); 0031-9007 (print)

Harrison, P. (2005). *Quantum Wells, Wires And Dots: Theoretical And Computational Physics of Semiconductor Nanostructures,* Wiley-Interscience, (2005), pp. 1-497, ISBN:9780470010808

Kane, E.O. (1957), Band Structure of Indium Antimonide, *Journal of Physics and Chemistry of Solids*, Pergamon Press, Vol. 1, (1957), pp. 249-261, ISSN 0022-3697

Keldysh, L. V. (1962). *Soviet Physics Solid State*, Vol. 4, (1962), pp. 2265, ISSN: 0038-5654

Landau, L.D. & Lifschitz, E. M. *Quantum Mechanics (Non-Relativistic Theory)*, 3rd ed. Pergamon Press, (1977), ISBN 0080209408, Oxford, England

Landauer, R. (1970). Electrical resistance of disordered one-dimensional lattices, *Philosophical Magazine*, Vol. 21, Issue 172, (1970) pp. 863-867, ISSN 1478-6443

Lei, W.; Notthoff, C.; Lorke, A.; Reuter, D. & Wieck, A. D. (2010). Electronic structure of self-assembled InGaAs/GaAs quantum rings studied by capacitance-voltage spectroscopy, Applied Physics Letters, Vol. 96, (2010), pp. 033111-033114, Print: ISSN 0003-6951, Online: ISSN 1077-3118

Li, Bin & Peeters F. M. (2011). Tunable optical Aharonov-Bohm effect in a semiconductor quantum ring. *Physical Review B*, Vol. 83, (28 March 2011) pp. 115448-13, ISSN 1550-235x (online), 1098-0121 (print)

Li, S.S. & Xia, J.B. (2001). Electronic states of InAs/GaAs quantum ring, *Journal Applied Physics*, Vol. 89, (2001), pp. 3434-3438, Print: ISSN 0021-8979, Online: ISSN 1089-7550

Li, Y.; Voskoboynikov, O.; Lee, C.P. (2002). Computer simulation of electron energy states for three-dimensional InAs/GaAs semiconductor quantum ring, *Proceedings of the International Conference on Modeling and Simulating of Microsystems* (San Juan, Puerto Rico, USA), p. 543, Computational Publication, (2002), Cambridge

Lorke, A.; Luyken, R.J.; Govorov, A.O. & Kotthaus, J.P.; Garcia, J.M. & Petroff, P. M. (2000). Spectroscopy of Nanoscopic Semiconductor Rings, *Physical Review Letters*, Vol. 84, Issue 10, (March, 2000), pp. 2223–2226, ISSN 1079-7114 (online); 0031-9007 (print)

Manasreh, O. (2005). *Semiconductor Heterojunctions and Nanostructures (Nanoscience & Technology)*, McGraw-Hill, (2005), p. 554, ISBN 0-07-145228-1, New York

Miller, B.T.; Hansen, W.;Manus, S.; Luyken, R.J.; Lorke, A; Kotthaus, J.P.; Huant, S.; Medeiros-Ribeiro, G; Petroff, P.M. (1997). Few-electron ground states of charge-tunable self-assembled quantum dots, *Physical Review B*, Vol. 56, (1997), pp. 6764–6769, ISSN 1550-235x (online), 1098-0121 (print)

Ponomarenko, L.A.; Schedin, F.; Katsnelson, M.I.; Yang, R.; Hill, E. W.; Novoselov K. S.; Geim A. K. (2008). Chaotic Dirac Billiard in Graphene Quantum Dots, *Science*, Vol. 320, No. 5874, (18 April 2008), pp. 356-358, ISSN 0036-8075 (print), 1095-9203 (online).

Rodt, S.; Schliwa, A.; Pötschke, K.; Guffarth, F.& Bimberg, D. (2005). Correlation of structural and few-particle properties of self-organized InAs⁄GaAs quantum dots, *Physical Review B*, Vol. 71, Issue 15, (2005), pp. 155325-155232, ISSN 1550-235x (online), 1098-0121 (print)

Schliwa, A.; Winkelnkemper, M. & Bimberg, D. Impact of size, shape, and composition on piezoelectric effects and electronic properties of In(Ga)As⁄GaAs quantum dots, *Physical Review B*, Vol. 76, (2007) pp. 205324-205341, ISSN 1550-235x (online), 1098-0121 (print)

Szafran, B. & Peeters, F.M. (2005). Few-electron eigenstates of concentric double quantum rings, *Physical Review B*, Vol. 72, (2005), pp. 155316-1555325, ISSN 1550-235x (online), 1098-0121 (print)

Szafran, B.; Peeters, F.M. & Bednarek, S. (2004). Exchange energy tuned by asymmetry in artificial molecules, *Physical Review B*, Vol. 70, Issue 20, (2004), pp. 205318-205323, ISSN 1550-235x (online), 1098-0121 (print).

T. Kuroda, T. Mano, T. Ochiai, S. Sanguinetti, K. Sakoda, G. Kido and N. Koguchi, (2005). Optical transitions in quantum ring complexes, *Physical Review B*, Vol. 72, Issue 20, (2005), pp. 205301–205309, ISSN 1550-235x (online), 1098-0121 (print)

T. Mano, T. Kuroda, S. Sanguinetti, T. Ochiai, T. Tateno, J. Kim, T. Noda, M. Kawabe, K. Sakoda, G. Kido, and N. Koguchi, (2005). Self-Assembly of Concentric Quantum Double Rings, *Nano Letters* , Vol. 5, No. 3, (2005), pp. 425–428, Print Edition ISSN 1530-6984, Web Edition ISSN 1530-6992

V. Neumann, J. & Wigner, E. (1929). Über das Verhalten von Eigenwerten bei adiabatischen Prozessen, *Physikalische Zeitschrift*, Vol. 30, (1929), pp. 467-468, OCLC Number 1762351

Voskoboynikov, O.; Li, Yiming; Lu, Hsiao-Mei; Shih, Cheng-Feng & Lee C.P. (2002). Energy states and magnetization in nanoscale quantum rings, *Physical Review B*, Vol. 66, (2002), pp. 155306-155312, ISSN 1550-235x (online), 1098-0121 (print).

Voss, H. (2005), Electron energy level calculation for a three dimensional quantum dot, Advances in Computational Methods in Sciences and Engineering 2005, selected papers from the *International Conference of Computational Methods in Sciences and Engineering 2005* (ICCMSE 2005) pp. 586 - 589, Leiden, The Netherlands; Editors: Theodore Simos and George Maroulis. ISBN 10 9067644412, 9067644439, 9067644447

Warburton R. J.; Miller, B. T.; Durr, C. S.; Bodefeld, C.; Karrai, K.; Kotthaus, J.P.; Medeiros-Ribeiro, G.; Petroff, P. M.; Huant, S. (1998). Coulomb interactions in small charge-tunable quantum dots: A simple model, *Physical Review B*, Vol. 58, Issue 24, (1998), pp. 16221–16231, ISSN 1550-235x (online), 1098-0121 (print)

Wetzel, C.; Winkler, R.; Drechsler M. & Meyer, B. K.;Rössler, U.; Scriba J. & Kotthaus, J. P.; Härle, V. & Scholz, F. (1996). Electron effective mass and nonparabolicity in Ga0.47In0.53As/InP quantum wells, *Physical Review B*, Vol. 53, (1996), pp. 1038–1041, ISSN 1550-235x (online), 1098-0121 (print)

Whitney, R.S.; Marconcini P. & Macucci, M. (2009a). Huge Conductance Peak Caused by Symmetry in Double Quantum Dots, *Physical Review Letters*, Vol. 102, (2009), pp. 186802-186806, ISSN 1079-7114 (online) 0031-9007 (print)

Whitney, R.S.; Schomerus, H. & Kopp, M. (2009). Semiclassical transport in nearly symmetric quantum dots. I. Symmetry breaking in the dot, *Physical Review E*, Vol. 80, Issue 5, (2009), 056209-056225, ISSN 1550-2376 (online), 1539-3755 (print)

Yu, P. & Cardona, M. (2005). *Fundamentals of Semiconductors: Physics and Materials Properties* (3rd ed.). Springer. Section 2.6, (2005), pp. 68, ISBN 3-540-25470-6

Zhao Q. & Mei, T. (2011). Analysis of electronic structures of quantum dots using meshless Fourier transform k·p method, *Journal of Applied Physics*, Vol. 109, Issue 6, (2011), pp. 063101-13, Print: ISSN 0021-8979, Online: ISSN 1089-7550

5

Electron Scattering Through a Quantum Dot

Leonardo Kleber Castelano[1], Guo-Qiang Hai[2] and Mu-Tao Lee[3]
[1]*Departamento de Física, Universidade Federal de São Carlos*
[2]*Instituto de Física de São Carlos, Universidade de São Paulo*
[3]*Departamento de Química, Universidade Federal de São Carlos*
Brazil

1. Introduction

Electron scattering and transport through quantum dots (QDs) in a semiconductor nanostructure have been intensively studied (Engel & Loss, 2002; Fransson et al., 2003; Konig & Martinek, 2003; Koppens et al., 2006; Qu & Vasilopoulos, 2006; Zhang et al., 2002). The spin-dependent transport properties are of particular interest for its possible applications, *e.g.*, the QD spin valves (Konig & Martinek, 2003), the quantum logic gates using coupled QDs, as well as the spin-dependent transport in single-electron devices (Seneor et al., 2007), etc.. In such systems, the electron-electron exchange potential and the electron spin states have been utilized and manipulated (Burkard et al., 2000; Gundogdu et al., 2004; Sarma et al., 2001; Wolf et al., 2001). A thorough quantitative understanding of spin-dependent transport properties due to electron-electron interaction is therefore important for a successful construction of these devices. Theoretically the transport through QDs has been studied by different approaches such as transfer matrix, nonequilibrium Green's functions, random matrix theory, as well as those methods built on the Lippmann-Schwinger (L-S) equation (Castelano et al., 2007a;b).

In this chapter, we develop a theoretical method to study electron scattering through a quantum dot (QD) of N-electrons embedded in a semiconductor nanostructure. We construct the scattering equations including electron-electron interaction to represent the process of a free electron scattered by the QD confined in a two-dimensional (2D) or in a quasi-one-dimensional (Q1D) semiconductor system. The generalized multichannel Lippmann-Schwinger equations(Bransden & McDowell, 1977; Joachain, 1975) are solved for these systems by using the method of continued fractions (MCF). As an example, we apply this method to a one-electron QD case and obtain scattering cross-sections in 2D and conductances in Q1D systems resulting from both the singlet- and triplet-coupled continuum states of two electrons (incident and QD electron) during the electron transport.

This chapter is organized as follows. In Sec. 2 we present our general theoretical approach and numerical method. In Sec. 3, we describe the electron scattering through a quantum dot in a 2D system. The scattering for a quantum dot confined in a Q1D system is presented in Sec. 4. In Sections 5 and 6, we show our numerical results for the scattering through a one-electron QD within both the one-channel and the multichannel models. We conclude in Sec. 7.

2. Theoretical approach

2.1 The system: Incident electron + quantum dot

The system under investigation consists of an incident free electron and a quantum dot of N electrons as shown schematically in Fig. 1. The incident electron is scattered by both the QD potential and by the confined electrons inside the QD. The Schrödinger equation of the system is given by

$$(H - \mathcal{E}_i)\Psi_i(\tau; \mathbf{r}_{N+1}, \sigma_{N+1}) = 0\,, \qquad (1)$$

where τ represents collectively the spatial and spin coordinates of the N electrons localized in the QD and $\mathbf{r}_{N+1} = (x_{N+1}, y_{N+1})$ and σ_{N+1} denote the spatial and spin coordinates of the incident electron. The total energy of the system is $\mathcal{E}_i$, where the subscript i represents a set of quantum numbers required to uniquely specify the initial quantum state of the system. Explicitly, the total Hamiltonian of the system can be written as

$$H = H_0(\mathbf{r}_{N+1}) + H_{\mathrm{QD}}(\tau) + V_{\mathrm{int}}(\mathbf{r}_1, \mathbf{r}_2, ..., \mathbf{r}_N, \mathbf{r}_{N+1})\,, \qquad (2)$$

where $H_0(\mathbf{r}_{N+1}) = -\hbar^2\nabla^2_{N+1}/2m^* + V_{\mathrm{QD}}(\mathbf{r}_{N+1})$, $H_{\mathrm{QD}}(\tau)$ is the Hamiltonian of the QD of N electrons, and V_{int} is the interaction potential between the incident electron at $\mathbf{r}_{N+1}$ and the N electrons in the QD

$$V_{\mathrm{int}}(\mathbf{r}_1, \mathbf{r}_2, ..., \mathbf{r}_N, \mathbf{r}_{N+1}) = \frac{e^2}{\epsilon_0^*}\sum_{i=1}^{N}\frac{1}{|\mathbf{r}_{N+1} - \mathbf{r}_i|}, \qquad (3)$$

where ϵ_0^* is the dielectric constant of the semiconductor material and m^* is the electron effective mass. The Hamiltonian for an unperturbed QD is given by

$$H_{\mathrm{QD}}(\tau) = \sum_{i=1}^{N}\left(-\frac{\hbar^2}{2m^*}\nabla_i^2 + V_{\mathrm{QD}}(\mathbf{r_i})\right) + \frac{e^2}{\epsilon_0^*}\sum_{i\neq j}^{N}\frac{1}{|\mathbf{r}_i - \mathbf{r}_j|}, \qquad (4)$$

where the first term in the *rhs* of Eq. (4) describes N independent electrons in the QD of confinement potential $V_{\mathrm{QD}}(\mathbf{r})$ and the second term gives the Coulomb interactions among these electrons. The eigenenergy and eigenfunction of this N-electron QD are denoted by ε_n and Φ^n, respectively. They are determined by the following Schrödinger equation

$$H_{\mathrm{QD}}(\tau)\Phi^n = \varepsilon_n\Phi^n, \qquad (5)$$

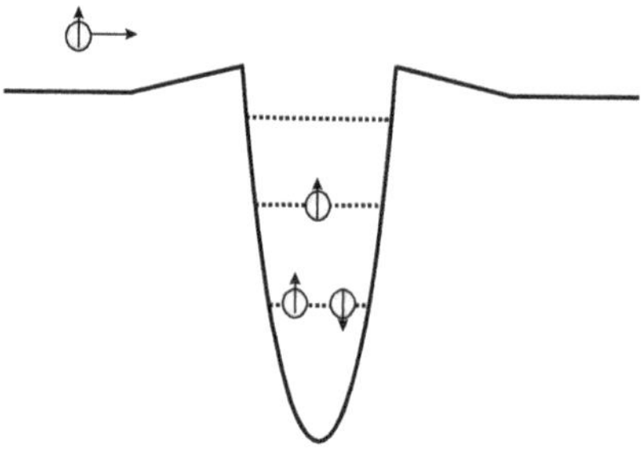

Fig. 1. Representation of the incident electron and the target, which in this case is a quantum dot containing 3 electrons.

with $n = 0, 1, 2, 3...$. The ground state of the QD is labeled by $n = 0$ and the excited states by $n \geq 1$. The eigenstates of the QD can be obtained using, *e.g.*, the restricted or unrestricted Hartree-Fock (HF) methods (Szabo & Ostlund, 1982).

2.2 Scattering equations including electron-electron interaction

In order to extract scattering properties of the system (QD + incident electron), we can write the total wave-function Ψ_i of the system as a superposition of the QD wave-function Φ^n and the incident electron wave-function,

$$|\Psi_i\rangle = \sum_{n=0}^{\infty} |\mathcal{A}(\Phi^n \psi_{ni})\rangle, \tag{6}$$

where ψ_{ni} describes the wave-functions of the incident (scattered) electron in the continuum states corresponding to a quantum transition from an initial state i to a final state n. The operator $\mathcal{A}$ warrants the antisymmetrization property between the QD electrons and the incident electron, defined by,

$$\mathcal{A} = \frac{1}{\sqrt{N+1}} \sum_{p=1}^{N+1} (-1)^{N+1-p} \mathcal{P}_{N+1,p} \tag{7}$$

where $\mathcal{P}_{N+1,p}$ is the permutation operator which exchanges the electrons at $\mathbf{r}_{N+1}$ and $\mathbf{r}_p$. From Eqs. (1), (2) and (6), we obtain

$$\sum_{n=0}^{\infty} \left(-\frac{\hbar^2}{2m^*} \nabla^2_{N+1} + V_{\text{QD}} + H_{\text{QD}} + V_{\text{int}} \right) |\mathcal{A}(\Phi^n \psi_{ni})\rangle = \mathcal{E}_i \sum_{n=0}^{\infty} |\mathcal{A}(\Phi^n \psi_{ni})\rangle. \tag{8}$$

The total energy of the system $\mathcal{E}_i$ is composed of two parts. The first part is the kinetic energy of the incident (scattering) electron and the second is the energy of the N-electron QD in a particular configuration, *i.e.*, $\mathcal{E}_i = \frac{\hbar^2 k_i^2}{2m^*} + \varepsilon_i = \frac{\hbar^2 k_n^2}{2m^*} + \varepsilon_n$, for different eigenstates of the QD $(i, n = 0, 1, 2, ...)$ or different scattering channels. These different channels appear because the incident electron can probably be scattered inelastically, leaving the QD in a different state from its initial. A projection of Eq. (8) onto a particular QD state $|\Phi^m\rangle$ leads to the following scattering equation for the incident electron,

$$\frac{\hbar^2}{2m^*} \left(\nabla^2 + k_m^2 \right) \psi_{mi}(\mathbf{r}) = \sum_{n=0}^{\infty} V_{mn}(\mathbf{r}) \psi_{ni}(\mathbf{r}) \tag{9}$$

for $i, m = 0, 1, 2, ...,$ where $\mathbf{r} = \mathbf{r}_{N+1}$ and $V_{mn} = V_{mn}^{\text{st}} + V_{mn}^{\text{ex}}$ with V_{mn}^{st} the static potential and V_{mn}^{ex} the exchange potential due the nonlocal interaction, giving by

$$V_{mn}^{\text{st}}(\mathbf{r}) = V_{\text{QD}}(\mathbf{r}) \delta_{mn} + \frac{e^2}{\epsilon_0^*} \sum_{j=1}^{N} \langle \Phi^m | \frac{e^{-\lambda |\mathbf{r} - \mathbf{r_j}|}}{|\mathbf{r} - \mathbf{r_j}|} | \Phi^n \rangle, \tag{10}$$

and

$$V^{\text{ex}}_{mn}(\mathbf{r})\psi_{ni}(\mathbf{r}) = (H_0(\mathbf{r}) - \frac{\hbar^2 k_m^2}{2m^*})\langle\Phi^m|\mathcal{A}'(\Phi^n\psi_{ni})\rangle + \frac{e^2}{\epsilon_0^*}\sum_{j=1}^{N}\langle\Phi^m|\frac{1}{|\mathbf{r}-\mathbf{r_j}|}|\mathcal{A}'(\Phi^n\psi_{ni})\rangle, \quad (11)$$

respectively, where $\mathcal{A}' = \sum_{p=1}^{N}(-1)^{N+1-p}\mathcal{P}_{N+1,p}$. In Eq. (10) we have introduced a screening $e^{-\lambda|\mathbf{r}-\mathbf{r}'|}$ on the direct Coulomb potential for two reasons: (i) the ionized impurities in the semiconductor nanostructure and/or the external electrodes screen the direct Coulomb potential and (ii) at $|\mathbf{r}| \rightarrow \infty$ limit the scattering potential should decay faster than $1/|\mathbf{r}|$. The screening length is given by λ^{-1}. Notice that we do not consider the screening on the exchange potential because this potential is non-zero inside the QD only. Inclusion of the screening on the exchange potential in Eq. (11) is possible but it will not affect much our results and considerably complicates the numerical calculation. The scattering equation is a system of coupled integro-differential equations. The corresponding generalized L-S equation for such a multichannel scattering problem is given by

$$\psi_{mi}(\mathbf{r}) = \varphi_i(\mathbf{r})\delta_{mi} + \sum_{n=0}^{\infty}\int d\mathbf{r}' G^{(0)}(\mathbf{k}_m, \mathbf{r}, \mathbf{r}')V_{mn}(\mathbf{r}')\psi_{ni}(\mathbf{r}'), \quad \text{for } i, m = 0, 1, 2\ldots \quad (12)$$

with an incident plane wave $\varphi_i(\mathbf{r}) = e^{i\mathbf{k}_i.\mathbf{r}} = e^{ik_ix}$ in the x-direction.

2.3 Method of continued fractions

The method of continued fractions (MCF) (Horacek & Sasakawa, 1984) is an iterative method to solve the L-S equation. To apply this method for a multi-channel scattering we have firstly to rewrite Eq. (12) in a matrix form:

$$\widetilde{\Psi} = \widetilde{\varphi} + \widetilde{G}^{(0)}\widetilde{V}\widetilde{\Psi}. \quad (13)$$

In the first step to start the MCF, we use the scattering potential $\widetilde{V} = V^{(0)}$ and the free electron wave-function $\widetilde{\varphi} = |\varphi^{(0)}\rangle$ in Eq. (13). Afterwards, we define the nth-order weakened potential as

$$V^{(n)} = V^{(n-1)} - \frac{V^{(n-1)}|\varphi^{(n-1)}\rangle\langle\varphi^{(n-1)}|V^{(n-1)}}{\langle\varphi^{(n-1)}|V^{(n-1)}|\varphi^{(n-1)}\rangle}, \quad (14)$$

where

$$|\varphi^{(n)}\rangle = \widetilde{G}^{(0)}V^{(n-1)}|\varphi^{(n-1)}\rangle. \quad (15)$$

The nth-order correction of the $\mathbf{T}$ matrix can be obtained through

$$T^{(n)} = \langle\varphi^{(n-1)}|V^{(n-1)}|\varphi^{(n)}\rangle + \langle\varphi^{(n)}|V^{(n)}|\varphi^{(n)}\rangle$$
$$\times\left[\langle\varphi^{(n)}|V^{(n)}|\varphi^{(n)}\rangle - T^{(n+1)}\right]^{-1}\langle\varphi^{(n)}|V^{(n)}|\varphi^{(n)}\rangle. \quad (16)$$

Hence, we can stop the iteration when the potential $V^{(N)}$ becomes weaker enough. In the numerical calculation, we start with $T^{(N+1)} = 0$ and evaluate $T^{(N)}, T^{(N-1)}, \ldots,$ and $T^{(1)}$.

Therefore the $\mathbf{T}$ matrix is given by

$$\mathbf{T} = \langle \varphi^{(0)} | V^{(0)} | \varphi^{(0)} \rangle + T^{(1)} \frac{\langle \varphi^{(0)} | V^{(0)} | \varphi^{(0)} \rangle}{\langle \varphi^{(0)} | V^{(0)} | \varphi^{(0)} \rangle - T^{(1)}}. \tag{17}$$

3. Quantum dot embedded in a two-dimensional system

The Green's function $G^{(0)}(\mathbf{k}, \mathbf{r}, \mathbf{r}')$ in 2D is given by

$$G^{(0)}(\mathbf{k}, \mathbf{r}, \mathbf{r}') = -\frac{2m^*}{\hbar^2}(i/4)H_0^{(1)}(k|\mathbf{r} - \mathbf{r}'|), \tag{18}$$

where $H_0^{(1)}$ is the usual zero order Hankel's function (Morse & Feshbach, 1953).

At $|\mathbf{r}| \to \infty$ limit, the asymptotic form of Eq. (12) for the scattered wave-function in a 2D system is given by

$$\psi_{mi}(\mathbf{r}) \underset{|\mathbf{r}| \to \infty}{\longrightarrow} e^{ik_i x}\delta_{mi} + \frac{2m^*}{\hbar^2}\sqrt{\frac{i}{k_m}}\frac{e^{+ik_m r}}{\sqrt{r}} f_{k_m, k_i}(\theta), \tag{19}$$

where $f_{k_m, k_i}(\theta)$ is the scattering amplitude

$$f_{k_m, k_i}(\theta) = -\frac{1}{4}\sqrt{\frac{2}{\pi}} < \mathbf{k}_m | T(E) | \mathbf{k}_i > \tag{20}$$

with

$$< \mathbf{k}_m | T(E) | \mathbf{k}_i > = \sum_{n=0}^{\infty} \int d\mathbf{r}' e^{-i\mathbf{k}_m \cdot \mathbf{r}'} V_{mn}(\mathbf{r}') \psi_{ni}(\mathbf{r}').$$

The momenta of the initial and final states of the incident (scattered) electron are $\mathbf{k}_i$ and $\mathbf{k}_m$, respectively, and θ is the scattering angle between them. It is evident from Eq. (12) and its boundary condition Eq. (19) that the different scattering channels are coupled to each other through the interaction potential V_{mn}.

In the above procedure in dealing with the electron scattering through a QD, both the electron-electron exchange and correlation interactions are present in this system. However, a complete correlation effect is difficult to include in a practical calculation. In order to do so, besides an exact solution for the N-electron QD, a full sum over all the intermediate states n in the scattering equation [Eq. (9)] is needed, which is a formidable task in a self-consistent calculation. In an alternative way, the correlation effects can be considered by adding an effective correlation potential in the scattering equation (Joachain, 1975). In the present work, we focus on the exchange effects on the scattering process and limit the sum over n to a few lowest energy levels of the QD. For this reason, we prefer to call the nonlocal interaction potential V_{mn}^{ex} in Eq. (11) as exchange potential, though the correlation can be partially included.

The differential cross-section (DCS) for a scattering from initial state i (*i.e.* the incident electron of kinetic energy $E_i = \frac{\hbar^2 k_i^2}{2m^*}$ and the QD in the state ε_i) to final state m (*i.e.* $E_m = \frac{\hbar^2 k_m^2}{2m^*}$ and the QD in the state m) is given by

$$\sigma_{mi}(\theta) = \frac{k_m}{k_i^2}|f_{k_m, k_i}(\theta)|^2. \tag{21}$$

The integral cross-section (ICS) which is an energy dependent quantity can be found by

$$\Gamma_{mi}(E_i) = \int_0^{2\pi} \sigma_{mi}(\theta) d\theta. \tag{22}$$

When the incident electron is scattered to a state of the same energy and the QD keeps in the same state ($m = i$), the scattering is called elastic. Otherwise, the scattering is inelastic. A possible scattering is the so-called super-elastic scattering ($E_m > E_i$) where the incident electron is scattered out with a higher energy by an QD initially in an excited state. Because the different scattering channels are coupled to each other, we have to solve the multichannel L-S equation to obtain the scattering probabilities through different channels simultaneously for the same total energy of the system.

3.1 Partial wave expansion

In two dimensions the angular momentum basis is given by (Adhikari, 1986),

$$\Theta_l(\phi) = \sqrt{\frac{\kappa_l}{2\pi}} \cos(l\phi) \tag{23}$$

where $l = 0, 1, 2, ...,$ $\kappa_l = 2$ for $l \neq 0$ and $\kappa_l = 1$ for $l = 0$. In applying the partial wave expansion in the multi-channel scattering problem Eq. (12), we expand all functions, *i.e.*, the incident free electron wavefunction $\varphi_i(\mathbf{r})$, the Green's function $G^{(0)}(\mathbf{k}_m, \mathbf{r}, \mathbf{r}')$, and the scattered electron wavefunction $\psi_{mi}(\mathbf{r})$, in the angular momentum basis as follows,

$$\varphi_i(\mathbf{r}) = \sum_{l,l'=0}^{\infty} \sqrt{\frac{\kappa_l}{2\pi}} i^l J_l(kr) \delta_{ll'} \Theta_l(\phi_r) \Theta_{l'}(\phi_k), \tag{24}$$

and

$$\psi_{mi}(\mathbf{r}) = \sum_{l,l'=0}^{\infty} \psi_{mi}^{l,l'}(k,r) \Theta_l(\phi_r) \Theta_{l'}(\phi_k), \tag{25}$$

where ϕ_r and ϕ_k are the variables due to expansion on the position $\mathbf{r}$ and momentum $\mathbf{k}$, respectively. The expansion on the Green's function yields the following expression,

$$G^{(0)}(\mathbf{k}_m, \mathbf{r}, \mathbf{r}') = -\frac{i\pi}{2} \sum_{l=0}^{\infty} \sqrt{\frac{\kappa_l}{2\pi}} J_l(k_m r_<) H_l^{(1)}(k_m r_>) \Theta_l(\phi_r) \Theta_l(\phi_{r'}), \tag{26}$$

where $k = k_i$, $r_< = \min(r, r')$, $r_> = \max(r, r')$, $J_l(k_m r)$ $(Y_l(k_m r))$ is the Bessel (Neumann) function and $H_l^{(1)}(k_m r) = J_l(k_m r) + iY_l(k_m r)$ is the Hankel function (Morse & Feshbach, 1953). Using the partial wave expansion the Lippmann-Schwinger equation can be reduced to a set of radial equations. The radial Lippmann-Schwinger equation corresponding to Eq. (12) is given by,

$$\psi_{mi}^{l,l'}(k,r) = \sqrt{\frac{\kappa_l}{2\pi}} i^l J_l(kr) \delta_{ll'} \delta_{mi} + \sum_{l''=0}^{\infty} \sum_{n=0}^{\infty} \int_0^{\infty} r' dr' g_0^l(k_m, r, r') V_{mn}^{l,l''}(r') \psi_{ni}^{l'',l'}(r'), \tag{27}$$

where

$$g_0^l(k_m, r, r') = \frac{-i\pi}{2} \sqrt{\frac{\kappa_l}{2\pi}} J_l(k_m r_<) H_l^{(1)}(k_m r_>) \tag{28}$$

and

$$V_{mn}^{l,l''}(r') = \int_0^{2\pi} d\phi_{r'} \Theta_l(\phi_{r'}) V_{mn}(\mathbf{r}') \Theta_{l''}(\phi_{r'}). \tag{29}$$

We see that, when the partial wave method is used, there is a change in the continuum variable ϕ to a partial wave l. Consequently, the wave function $\psi_{mi}(\mathbf{r})$ becomes a matrix function with elements $\psi_{mi}^{l,l'}(k,r)$.

The partial wave expansion for the exchange potential is a little subtle due to its non-locality. Here, we show some details about how the partial wave expansion is applied in this case. We take as an example the exchange potential which couples the channels n and m for a single electron spin-orbital α [see Eq. (11)],

$$V_{mn}^{\text{ex}}(\mathbf{r})\psi_{ni}(\mathbf{r}) = -\frac{e^2}{\epsilon_0^*} \zeta_\alpha^n(\mathbf{r}) \int d\mathbf{r}_1 \zeta_\alpha^{m*}(\mathbf{r}_1) \frac{1}{|\mathbf{r}-\mathbf{r}_1|} \psi_{ni}(\mathbf{r}_1). \tag{30}$$

The partial wave expansion of the spin-orbital function is given by

$$\zeta_\alpha^n(\mathbf{r}) = \sum_{l=0}^{\infty} \zeta_{n\alpha}^l(r) \Theta_l(\phi_r). \tag{31}$$

The product of two different functions can also be expanded in the angular momentum basis as follows,

$$\psi_{ni}(\mathbf{r})\zeta_\alpha^{m*}(\mathbf{r}) = \sum_{l,l'} \Pi_{ni;m\alpha}^{l,l'}(r) \Theta_l(\phi_r) \Theta_{l'}(\phi_k), \tag{32}$$

where

$$\Pi_{ni;m\alpha}^{l,l'}(r) = \sum_{\lambda,\lambda'} \frac{\psi_{ni}^{\lambda,l'}(k,r)\zeta_{m\alpha}^{\lambda'*}(r)}{2\sqrt{2\pi}} \sqrt{\frac{\kappa_\lambda \kappa_{\lambda'}}{\kappa_l}} \left(\delta_{l,\lambda+\lambda'} + \delta_{l,|\lambda-\lambda'|}\right). \tag{33}$$

Using the above relation, we obtain Eq. (30) in the partial wave expansion form,

$$V_{mn}^{\text{ex}}(\mathbf{r})\psi_{ni}(\mathbf{r}) = -\frac{e^2}{\epsilon_0^*} \zeta_\alpha^n(\mathbf{r}) \sum_{l,l'} \Theta_l(\phi_r)\Theta_{l'}(\phi_k) \times \int_0^\infty r_1 dr_1 \Pi_{ni;m\alpha}^{l,l'}(r_1) \int_0^{2\pi} \frac{\Theta_l(\theta)d\theta}{\sqrt{r^2+r_1^2-2rr_1\cos(\theta)}}, \tag{34}$$

where $\theta = \phi_r - \phi_{r_1}$. To solve the angular integral we use the generating function of the Legendre Polynomials (Morse & Feshbach, 1953),

$$\frac{1}{\sqrt{r^2+r_1^2-2rr_1\cos(\theta)}} = \sum_{j=0}^{\infty} \frac{r_<^j}{r_>^{j+1}} P_j(\cos\theta), \tag{35}$$

where $r_< = \min(r, r_1)$, $r_> = \max(r, r_1)$ and $P_j(\cos\theta)$ are the Legendre Polynomials. Thus the angular integral that we need to solve is

$$c_{l,j} = \int_0^{2\pi} d\theta \Theta_l(\theta) P_j(\cos\theta). \tag{36}$$

Substituting the Eqs. (35) and (36) into Eq. (34) we finally obtain the exchange potential

$$V_{mn}^{\text{ex}}(\mathbf{r})\psi_{ni}(\mathbf{r}) = -\frac{e^2}{\epsilon_0^*}\zeta_\alpha^n(\mathbf{r})\sum_{l,l'}\Theta_l(\phi_r)\Theta_{l'}(\phi_k)\sum_{j=0}^{\infty}\int_0^{\infty} r_1 dr_1 \Pi_{ni;m\alpha}^{l,l'}(r_1)c_{l,j}\frac{r_<^j}{r_>^{j+1}}. \tag{37}$$

In the numerical calculations, we firstly evaluate the coefficients $c_{l,j}$ given by Eq. (36). Then the integration on r_1 in Eq. (37) is performed for each iteration in the MCF. Finally we multiply the result by $-\frac{e^2}{\epsilon_0^*}\zeta_\alpha^n(\mathbf{r})$.

Within the one-channel approximation ($i = m = n = 0$), the calculations can be further simplified by using the concept of phase shift. Considering a central potential $V(r)$ ($l = l' = l''$), Eq. (27) becomes

$$\psi^l(k,r) = \sqrt{\frac{\kappa_l}{2\pi}} i^l J_l(kr) + \int_0^{\infty} r' dr' g_0^l(k,r,r')V(r')\psi^l(k,r') \tag{38}$$

where $\psi^l(k,r) = \psi_{00}^{l,l}(k,r)$. To define the phase-shift we write the asymptotic form of the above equation as

$$\psi^l(k,r) \underset{r\to\infty}{\longrightarrow} A_l\sqrt{\frac{1}{kr}}\cos(kr - \frac{l\pi}{2} - \frac{\pi}{4} - \Delta_l), \tag{39}$$

where Δ_l is the phase-shift. Comparing the coefficients of e^{ikr} and e^{-ikr} of Eq. (39) with the asymptotic form of Eq. (38) one can obtain the following relations

$$A_l = 2\sqrt{\frac{\kappa_l}{\pi}} i^l e^{i\Delta_l}, \tag{40}$$

and

$$e^{i\Delta_l}\sin\Delta_l = \frac{-\pi}{2i^l}\int_0^{\infty} r' dr' J_l(kr')V(r')\psi^l(r'). \tag{41}$$

On the other hand, from the definition of the scattering amplitude in Eq. (19), we can express the scattering amplitude f_{k_0,k_0} in terms of the phase-shift (Adhikari, 1986) Δ_l,

$$f_{k_0,k_0}(\theta) = 2\sum_{l=0}^{\infty}\sqrt{\frac{\kappa_l}{\pi}} e^{i\Delta_l}\sin\Delta_l\Theta_l(\theta). \tag{42}$$

The corresponding DCS is $\sigma_{00}(\theta) = \left|f_{k_0,k_0}(\theta)\right|^2/k$ and the ICS is given by

$$\Gamma_{00} = \frac{4}{k}\sum_{l=0}^{\infty}\kappa_l\sin^2\Delta_l. \tag{43}$$

4. Quantum dot confined in a quasi-one-dimensional structure

In this section, we study the electron scattering through a QD confined in a quasi-one-dimensional structure. The quantum dot is considered to be confined in the y-direction and the incident (scattered) electron moves in the x-direction. Far from the QD, the electron is free to propagate in the x-direction. In this limit, the Schrödinger equation in the y-direction is given by

$$\left[-\frac{\hbar^2}{2m^*}\frac{d^2}{dy^2} + V_c(y)\right]\chi_n(y) = \varepsilon_n\chi_n(y). \tag{44}$$

We choose the confining potential as being parabolic $V_c(y) = \frac{1}{2}m^*\omega_y^2y^2$. The solution of equation (44) for this potential is given by (Merzbacher, 1970):

$$\chi_n(y) = \frac{e^{-\frac{y^2}{2l_y^2}}}{(\pi l_y^2)^{1/4}} \frac{H_n(\frac{y}{l_y})}{\sqrt{2^n!}} \tag{45}$$

where H_n are the Hermite's polinomials (Morse & Feshbach, 1953), $\varepsilon_n = \hbar\omega_y(n+1/2)$, and $l_y = \sqrt{\hbar/m^*\omega_y}$. The eigenfunctions $\chi_n(y)$ are also called transversal modes.

As the basis composed of $\chi_n(y)$ is complete, we are able to expand the wave-function in such a system on this basis,

$$\Psi_i(\mathbf{r}) = \sum_{n=0}^{\infty} \chi_n(y)\psi_{ni}(x), \tag{46}$$

where i refers to the incident wave-vector. By introducing this result into the Schrödinger equation, multiplying it by $\chi_m^*(y)$, and integrating in the y-direction, we find the following coupled equations:

$$\left(\frac{\hbar^2}{2m^*}\frac{d^2}{dx^2} + \frac{\hbar^2k_n^2}{2m^*}\right)\psi_{ni}(x) = \sum_{m=0}^{\infty} V_{m,n}(x)\psi_{mi}(x), \tag{47}$$

where $\frac{\hbar^2k^2}{2m^*} = \frac{\hbar^2k_n^2}{2m^*} + \varepsilon_n$ and

$$V_{m,n}(x) = \int dy\chi_m^*(y)V(\mathbf{r})\chi_n(y). \tag{48}$$

The Green's function is defined as being the solution of the equation:

$$\left(\frac{\hbar^2}{2m^*}\frac{d^2}{dx^2} + \frac{\hbar^2k_n^2}{2m^*}\right)G_n(x,x') = \delta(x-x'), \tag{49}$$

which allows to rewrite the solution of Eq. (47) as a Lippmann-Schwinger equation in one-dimension,

$$\psi_{ni}(x) = \varphi_n(x) + \frac{2m^*}{\hbar^2}\sum_{m=0}^{\infty}\int dx'G_n(x,x')V_{m,n}(x')\psi_{mi}(x'), \tag{50}$$

where $\varphi_n(x) = exp(ik_nx)\delta_{n,i}/\sqrt{k_n}$. The Green's function to each sub-band in a Q1D system is equal to:

$$G_n(x,x') = \frac{-i}{2k_n}e^{ik_n|x-x'|}. \tag{51}$$

Because the energy of the incident electron is $\frac{\hbar^2k^2}{2m^*} = \frac{\hbar^2k_n^2}{2m^*} + \varepsilon_n$, there is the possibility of $\varepsilon_n > \frac{\hbar^2k^2}{2m^*}$ and of k_n being a pure imaginary number. In such a situation, we must replace k_n by $i|k_n|$ in Eq. (50) and the eigenfunctions $\psi_{ni}(x)$ are not localized anymore.

Taking the limit $x \to \infty$ in Eq. (50), we obtain:

$$\psi_{ni}(x) \underset{x\to\infty}{\longrightarrow} \frac{e^{ik_nx}}{\sqrt{k_n}}\left[\delta_{ni} + \frac{m^*}{i\hbar^2}\sum_{m=0}^{\infty}\int dx'\frac{e^{-ik_nx'}}{\sqrt{k_n}}V_{m,n}(x')\psi_{mi}(x')\right]. \tag{52}$$

The scattering matrix T can be found through the following result

$$T_{ni} = \sum_{m=0}^{\infty} \int_{-\infty}^{\infty} dx' \frac{e^{-ik_n x'}}{\sqrt{k_n}} V_{m,n}(x')\psi_{mi}(x'). \tag{53}$$

Another quantity that we can obtain is transmission probability t_{ni}, which by definition satisfies the following equation (Vargiamidis et al., 2003):

$$\Psi_i(\mathbf{r}) \underset{x\to\infty}{\longrightarrow} \sum_{n=0}^{\infty} t_{ni} \frac{e^{ik_n x}}{\sqrt{k_n}} \chi_n(y). \tag{54}$$

Multiplying Eq. (52) by $\chi_n(y)$, then adding $n = 0$ to ∞, and comparing the resulting equation to Eq. (54) we obtain the following expression for t_{ni}:

$$t_{ni} = \delta_{ni} + \frac{m^*}{i\hbar^2} \sum_{m=0}^{\infty} \int_{-\infty}^{\infty} dx' \frac{e^{-ik_n x'}}{\sqrt{k_n}} V_{m,n}(x')\psi_m(x'). \tag{55}$$

So we can relate the matrix T with the scattering transmission probability t_{ni} by:

$$t_{ni} = \delta_{ni} + \frac{m^*}{i\hbar^2} T_{ni}. \tag{56}$$

These quantities are useful to determine the conductance (Fisher & Lee, 1981; Imry & Landauer, 1999). In the Q1D system with multiple scattering channels, the conductance can be obtained by using the Landauer-Büttiker equation (Buttiker et al., 1985; Landauer, 1957; 1970; 1975),

$$G = \frac{e^2}{\pi\hbar} Tr(tt^{\dagger}), \tag{57}$$

where t is the matrix whose elements are exactly given by Eq. (55).

5. Applications and numerical results

In the previous sections, we presented a theoretical model that describes the quantum scattering through a quantum dot with N-electrons confined. However, we apply this model to the case where only one electron is confined in the quantum dot. Although this is the simplest case, it reveals basic information for a more complicated system. In this section, we describe the details of this particular case considering the elastic and inelastic scattering in the 2D system in sub-sections 5.1 and 5.2, respectively. The scattering through a confined QD in the Q1D system will be discussed in sub-section 5.3.

5.1 Elastic scattering

Here we describe in details how the elastic scattering can be accounted for. To do so, we start by considering the electron in the ground state of energy ε_1. The total Hamiltonian for this system (incident electron + confined electron) is given by:

$$H(\mathbf{r}_1, \mathbf{r}_2) = \frac{-\hbar^2\nabla_2^2}{2m^*} + V_{QD}(\mathbf{r}_2) + H_{QD}(\mathbf{r}_1) + V(\mathbf{r}_1, \mathbf{r}_2) \tag{58}$$

where H_{QD} is the QD Hamiltonian and $V(\mathbf{r}_1, \mathbf{r}_2)$ is the Coulomb interaction potential between the pair of electrons. The total wave function should be written as linear combination of

Slater determinants, as shown in sub-section 2.2. There are four possible combinations for two electrons,

$$|\overline{\Phi^1}, \overline{\psi_{11}} >= \frac{1}{\sqrt{2}} \begin{vmatrix} \Phi^1(\mathbf{r}_1)\beta(1) & \psi_{11}(\mathbf{r}_1)\beta(1) \\ \Phi^1(\mathbf{r}_2)\beta(2) & \psi_{11}(\mathbf{r}_2)\beta(2) \end{vmatrix}, \tag{59}$$

$$|\Phi^1, \psi_{11} >= \frac{1}{\sqrt{2}} \begin{vmatrix} \Phi^1(\mathbf{r}_1)\alpha(1) & \psi_{11}(\mathbf{r}_1)\alpha(1) \\ \Phi^1(\mathbf{r}_2)\alpha(2) & \psi_{11}(\mathbf{r}_2)\alpha(2) \end{vmatrix}, \tag{60}$$

$$|\Phi^1, \overline{\psi_{11}} >= \frac{1}{\sqrt{2}} \begin{vmatrix} \Phi^1(\mathbf{r}_1)\alpha(1) & \psi_{11}(\mathbf{r}_1)\beta(1) \\ \Phi^1(\mathbf{r}_2)\alpha(2) & \psi_{11}(\mathbf{r}_2)\beta(2) \end{vmatrix}, \tag{61}$$

$$|\psi_{11}, \overline{\Phi^1} >= \frac{1}{\sqrt{2}} \begin{vmatrix} \psi_{11}(\mathbf{r}_1)\alpha(1) & \Phi^1(\mathbf{r}_1)\beta(1) \\ \psi_{11}(\mathbf{r}_2)\alpha(1) & \Phi^1(\mathbf{r}_2)\beta(2) \end{vmatrix}, \tag{62}$$

where $\Phi^1(\mathbf{r})$ is the wave function of the confined electron in the QD, $\alpha(i)$ and $\beta(i)$ correspond to spin-up ($\uparrow$) and spin-down ($\downarrow$), respectively. The index (i) denotes to which electron the spin refers to.

Because the total Hamiltonian (Eq. (58)) commutes with the total spin operator (S^2) and its component in the z-direction (S_z), the Hamiltonian eigenfunctions must be eigenfunctions of both S_z and S^2. The first two determinants of Slater in equations (59 and 60) are eigenfunctions of S_z and S^2, but the equations (61 and 62) are not eigenfunctions of S^2. Thus, we have to construct linear combinations between these Slater determinants (Eqs. (61 and 62)) in order to obtain eigenfunctions of S_z and S^2. These combinations can be written as follows:

$$\begin{aligned} |\Psi^s >&= \frac{1}{\sqrt{2}} \left[|\Phi^1, \overline{\psi_{11}} > + |\psi_{11}, \overline{\Phi^1} > \right] = \\ &= \frac{1}{\sqrt{2}} \left[\psi_{11}(\mathbf{r}_1)\Phi^1(\mathbf{r}_2) + \psi_{11}(\mathbf{r}_2)\Phi^1(\mathbf{r}_1) \right] \left(\frac{|\downarrow,\uparrow> - |\uparrow,\downarrow>}{\sqrt{2}} \right) \end{aligned} \tag{63}$$

and

$$\begin{aligned} |\Psi^t >&= \frac{1}{\sqrt{2}} \left[|\Phi^1, \overline{\psi_{11}} > - |\psi_{11}, \overline{\Phi^1} > \right] = \\ &= \frac{1}{\sqrt{2}} \left[\psi_{11}(\mathbf{r}_1)\Phi^1(\mathbf{r}_2) - \psi_{11}(\mathbf{r}_2)\Phi^1(\mathbf{r}_1) \right] \left(\frac{|\downarrow,\uparrow> + |\uparrow,\downarrow>}{\sqrt{2}} \right). \end{aligned} \tag{64}$$

Equation (63) corresponds to the wave function of the singlet state and Equation (59, 60 and 64) correspond to wave functions of the triplet states. Since the Hamiltonian (Eq. (58)) does not have a explicit spin-dependent potential, the state of total spin is conserved before and after the collision. In such a way, the total wave function of the system (incident electron + confined electron) can be written as:

$$\Psi(\mathbf{r}_1, \mathbf{r}_2) = \Phi^1(\mathbf{r}_1)\psi_{11}(\mathbf{r}_2) \pm \Phi^1(\mathbf{r}_2)\psi_{11}(\mathbf{r}_1), \tag{65}$$

where the positive (negative) sign refers to the spin singlet (triplet) state. In order to determine the potential for the scattered electron, we have to calculate the following equation:

$$< \Phi^1(\mathbf{r}_1)|H(\mathbf{r}_1, \mathbf{r}_2)|\Psi(\mathbf{r}_1, \mathbf{r}_2) >= E < \Phi^1(\mathbf{r}_1)|\Psi(\mathbf{r}_1, \mathbf{r}_2) >, \tag{66}$$

where

$$E = \varepsilon_1 + \frac{\hbar^2 k_1^2}{2m^*}. \tag{67}$$

The left hand side of Eq. (66) can be rewritten as:

$$< \Phi^1|H|\Psi >=< \Phi^1|H^1_{QD}|\Psi > + < \Phi^1|H^2_{QD}|\Psi > + < \Phi^1|V^{1,2}|\Psi >, \tag{68}$$

where the superscript is related to each electron the operator is operating on. The first term of Equation (68) is equal to:

$$< \Phi^1|H^1_{QD}|\Psi >= \varepsilon_1 \left[< \Phi^1|\Phi^1 > \psi_{11} \pm < \Phi^1|\psi_{11} > \Phi^1 \right]. \tag{69}$$

While the second term of Eq. (68) is given by

$$< \Phi^1|H^2_{QD}|\Psi >=< \Phi^1|\Phi^1 > H^2_{QD}\psi_{11} \pm \varepsilon_1 < \Phi^1|\Phi^1 > \psi_{11}. \tag{70}$$

The third term of Eq. (68) can be written as:

$$< \Phi^1|V^{1,2}|\Psi >=< \Phi^1|V^{1,2}|\Phi^1 > \psi_{11} \pm < \Phi^1|V^{1,2}|\psi_{11} > \Phi^1. \tag{71}$$

By substituting Eqs.(69, 70, and 71) into Eq. (68), we obtain the following result:

$$\begin{aligned} &(H^2_{QD} - \frac{\hbar^2 k_1^2}{2m^*})\psi_{11} + < \Phi^1|V^{1,2}|\Phi^1 > \psi_{11} \pm \\ &\pm \left(< \Phi^1|V^{1,2}|\psi_{11} > \Phi^1 + (\varepsilon_1 - \frac{\hbar^2 k_1^2}{2m^*}) < \Phi^1|\psi_{11} > \Phi^1 \right) = 0. \end{aligned} \tag{72}$$

The previous equation can be further simplified as

$$-\frac{\hbar^2}{2m^*}(\nabla^2 + k_1^2)\psi_{11}(\mathbf{r}) + \left[V^{\mathrm{st}}(\mathbf{r}) \pm V^{\mathrm{ex}}(\mathbf{r})\right]\psi_{11}(\mathbf{r}) = 0, \tag{73}$$

where

$$V^{\mathrm{st}}(\mathbf{r}) =< \Phi^1|V^{1,2}|\Phi^1 > + V_{QD}(\mathbf{r}), \tag{74}$$

and

$$V^{\mathrm{ex}}_{11}(\mathbf{r})\psi_{11}(\mathbf{r}) = \Phi^1(\mathbf{r}) \left[< \Phi^1|V^{1,2}|\psi_{11} > + (\varepsilon_1 - \frac{\hbar^2 k_1^2}{2m^*}) < \Phi^1|\psi_{11} > \right]. \tag{75}$$

Finally, the Lippmann-Schwinger equation corresponding to Eq. (73) is given by

$$\psi_{11}(\mathbf{r}) = \varphi_1(\mathbf{r}) + \int d\mathbf{r}' G^{(0)}\mathbf{k}_1, \mathbf{r}, \mathbf{r}') \left[V^{st}(\mathbf{r}') \pm V^{ex}(\mathbf{r}')\right]\psi_{11}(\mathbf{r}'), \tag{76}$$

which can be numerically solved by the method of continuous fractions.

From Eq. (76), we observe that the exchange potential is different when the two electrons form a singlet spin state (plus sign) or triplet spin state (minus sign). In order to calculate the spin-dependent scattering, we have to calculate separately the scattering considering the singlet state and the triplet state. Moreover, the cross sections are given by

$$\sigma^s_{11}(\theta) = \frac{1}{k_1}|f^s_{11}(\theta)|^2, \tag{77}$$

for the singlet state, and

$$\sigma^t_{11}(\theta) = \frac{1}{k_1}|f^t_{11}(\theta)|^2, \tag{78}$$

for the triplet state, where

$$f^s_{11}(\theta) = -\frac{1}{4}\sqrt{\frac{2}{\pi}}\int d^2\mathbf{r}' e^{-i\mathbf{k}'_1.\mathbf{r}'}[V^{st}(\mathbf{r}') + V^{ex}(\mathbf{r}')]\psi_{11}(\mathbf{r}'), \tag{79}$$

$$f^t_{11}(\theta) = -\frac{1}{4}\sqrt{\frac{2}{\pi}}\int d^2\mathbf{r}' e^{-i\mathbf{k}'_1.\mathbf{r}'}[V^{st}(\mathbf{r}') - V^{ex}(\mathbf{r}')]\psi_{11}(\mathbf{r}'). \tag{80}$$

A quantity that we can obtain is the spin-unpolarized cross section (su-ICS), which is given by the statistical average of possible configurations, *i.e.*,

$$\sigma^{su}_{11}(\theta) = \frac{1}{4k_1}(|f^s_{11}(\theta)|^2 + 3|f^t_{11}(\theta)|^2) \tag{81}$$

where the factor three that multiplies the squared modulus of the scattering amplitude of the triplet state is due to the existence of three different triplet states, which are scattered with equal probability. Another quantity that we can extract from the calculation is the spin-flip (sf) cross-section (da Paixão et al., 1996; Hegemann et al., 1991), which measures the probability of the incident electron changes its spin after being scattered,

$$\sigma^{sf}_{11}(\theta) = \frac{1}{4k_1}|f^t_{11}(\theta) - f^s_{11}(\theta)|^2 \tag{82}$$

In the last expression the factor three does not appear because only one of the triplet states can change its spin $\frac{1}{\sqrt{2}}(|\uparrow\downarrow> + |\downarrow\uparrow>)$.

5.2 Multi-channel scattering

A very important process that we can study by using the previous formalism is the multi-channel scattering, which reveals the probability of an incident electron to promote an excitation or the decay of electrons within the quantum dot. A priori the number of channels of excitation and decay are infinite, but obviously when doing calculations, this number must be truncated. In the case of the parabolic potential of a 2D quantum dot, the first excited energy level is doubly degenerate with an angular momentum $l = \pm 1$. To consider the possible channels of scattering described by the ground state ϵ_1 and by the first excited state ϵ_2, we must consider three coupled channels because the degeneracy of the first excited energy level must be included in the calculation. As the probability of finding the electron in the ground state initially is higher and as the excitation to the first excited state is more likely, we consider only three coupled channels. The calculation details can be found in Ref.(Castelano, 2006). Here we will present some numerical results in Section 6.2.

5.3 Scattering through the QD confined in the Q1D structure

As already discussed in Section 4, the Lippmann-Schwinger equation for the confined QD includes several sub-bands. However, as a first example, we consider only the lowest transversal sub-band due to the confinement in the y-direction. Thus, the L-S equation for a single sub-band is given by:

$$\psi_1(x) = \frac{e^{ik_1x}}{\sqrt{k_1}} + \frac{2m^*}{\hbar^2}\int dx' G_1(x,x')V_{1,1}(x')\psi_1(x'), \tag{83}$$

where $\frac{\hbar^2k^2}{2m^*} = \frac{\hbar^2k_1^2}{2m^*} + \varepsilon_1$. The potential and Green's function for the one sub-band case are respectively given by

$$V_{1,1}(x) = \int dy\chi_1^*(y)V(\mathbf{r})\chi_1(y), \tag{84}$$

and

$$G_1(x,x') = \frac{-i}{2k_1}e^{ik_1|x-x'|}. \tag{85}$$

We also consider only one confined electron in the QD. The electron wave function of the ground state of the QD can be approximated as

$$\Phi_1(x,y) = \frac{1}{\sqrt{\pi l_x l_y}}\exp\left(-\frac{x^2}{2l_x^2} - \frac{y^2}{2l_y^2}\right), \tag{86}$$

where $l_x = \sqrt{\hbar/m^*\omega_x}$ and $l_y = \sqrt{\hbar/m^*\omega_y}$. Here we consider the QD confining potential in the x-direction as a finite parabolic one with confinement frequency ω_x. The calculation of the exchange potential is more complicated in the Q1D system. However, when we use the wave function of the harmonic oscillator, we can partially obtain analytical expressions for the exchange potential.

Just as in the elastic scattering in the 2D case without extra confinement, the exchange potential is different when the two electrons form a singlet or a triple spin state. So, we have to calculate separately the scattering for the different spin sates. The T matrices can be obtained by the following equations:

$$T_{11}^s = \int_{-\infty}^{\infty} dx' \frac{e^{-ik_1x'}}{\sqrt{k_1}}\left[V_{1,1}^{st}(x') + V_{1,1}^{ex}(x')\right]\psi_{11}(x'), \tag{87}$$

and

$$T_{11}^t = \int_{-\infty}^{\infty} dx' \frac{e^{-ik_1x'}}{\sqrt{k_1}}\left[V_{1,1}^{st}(x') - V_{1,1}^{ex}(x')\right]\psi_{11}(x'), \tag{88}$$

where

$$V_{1,1}^{st}(x) = \int dy\chi_1^*(y)V^{st}(\mathbf{r})\chi_1(y), \tag{89}$$

and

$$V_{1,1}^{ex}(x)\psi_{11}(x) = \int dy\chi_1^*(y)V^{ex}(\mathbf{r})\chi_1(y)\psi_{11}(x). \tag{90}$$

The static potential $V_{1,1}^{st}(x)$ and the exchange potential $V_{1,1}^{ex}(x)$ are analog to the Eqs. (10) and (11). The transmission probability for the electrons behaving as singlet and triplet states can also be obtained, see Section 4.

6. Results and analysis

In this section, we present the numerical results of scattering of the incident electron through a quantum dot containing just one confined electron. The first step we have to do is to calculate the eigenfunctions and the eigenenergies of the Hamiltonian Eq. (4) for N = 1, which has the following form:

$$\left[-\frac{\hbar^2}{2m^*}\nabla^2 + V_{QD}(r)\right]\Phi^n = H_{QD}\Phi^n = \varepsilon_n\Phi^n, \tag{91}$$

where

$$V_{QD}(r) = \begin{cases} \frac{1}{2}m^*\omega_0^2(r^2 - R_0^2), & r < R_0, \\ 0, & r > R_0. \end{cases} \tag{92}$$

Usually, the QD is modeled by an infinite parabolic potential $V_{QD}(r) = \frac{1}{2}m^*\omega_0^2 r^2$. However,

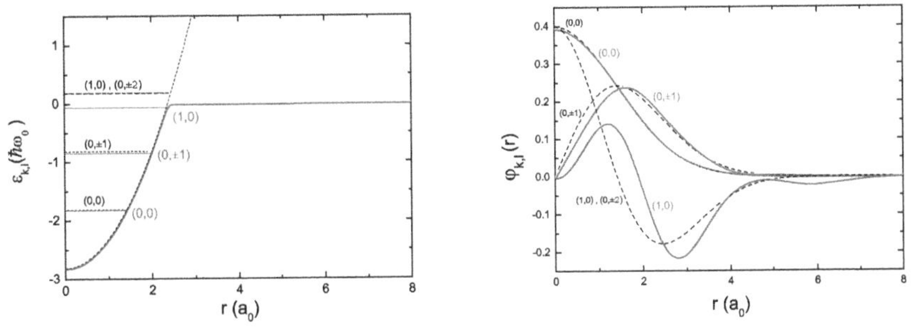

(a) Energy Levels (b) Wave Functions

Fig. 2. (a) Energy levels and (b) wave functions of the infinite (black dashed curves) and finite (blue solid curves) parabolic potential for R_0 =2.37 a_0. The indexes (k, l) denote the radial and angular quantum numbers, respectively.

we are dealing with the scattering processes of an incident electron through the QD, we must employ a potential that goes to zero at the infinity. Therefore, we use the finite parabolic potential (Eq. (92)) as the QD potential. We solve Eq. (91) by expanding the wave function Φ^n in the Fock-Darwin basis (Darwin, 1930; Fock, 1928). The eigenenergies and eigenfunctions are determined by diagonalizing the matrix within the Fock-Darwin basis. Figure 2 (a) compares the energy levels of the infinite and finite parabolic potential. From Figs. 2 (a) and (b) we can see that the ground state $(0,0)$ and the first excited state $(0,\pm1)$ of the finite parabolic potential are not very different from the infinite parabolic potential for R_0 =2.37 a_0, where $a_0 = \sqrt{\hbar/m^*\omega_0}$. However, the state $(1,0)$ is quite different for the two potentials. We also see that the finite parabolic potential with $R_0 = 2.37a_0$ supports three discrete levels only.

6.1 Elastic scattering

The differential cross sections (DCS) for elastic scattering are shown in Figs. 3 (a), (b), and (c) for different incident electron energies $E_0 = \hbar^2 k_0^2/2m^* = 0.6$ meV, 1.7 meV, and 4.2 meV, respectively . In order to understand the role of the electron spin in the scattering, we compare the DCS due to the static potential (blue solid curve) with the spin unpolarized scattering

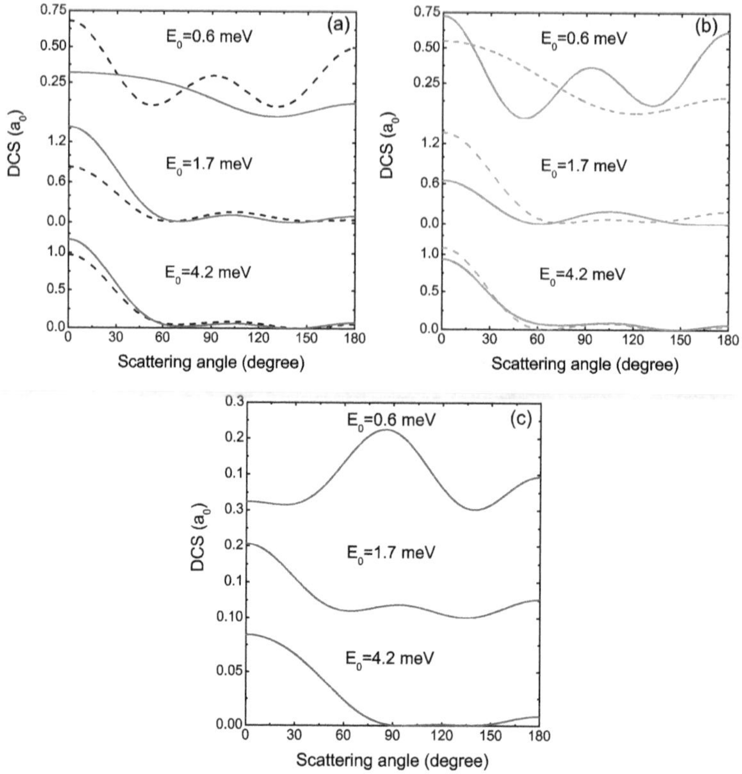

Fig. 3. The elastic DCS's obtained within the one-channel model for electron scattering by the one-electron QD of $\hbar\omega_0 = 5$ meV and $r_0 = 35$ nm. The incident electron energies are indicated in the figures. (a) The spin-unpolarized DCS with (the dashed curves) and without (the solid curves) the exchange potential; (b) The DCS due to the singlet state (the dashed curves) and the triplet state (the solid curves); and (c) The spin-flip DCS.

(black dashed curve) in Figure 3 (a). It is evident that the electron spin is of significant contribution to low energy (E_0=0.6 meV) and/or small scattering angles. The exchange effect on the scattering originates from the two different coupling states between the incident and the QD electron (*i.e.*, the singlet and the triplet states) during the collision. The difference due to the spin states for low-energy and small scattering angles is evident in Figure 3 (b), which compares the DCS of the singlet (orange dashed curve) to that of the triplet (green solid curve) state. For higher energies E_0=1.7 meV and E_0=4.2 meV, the DCS for singlet and for triplet are similar. We observe that the spin-flip DCS in Figure 3 (c) reaches to maximum for angles close to $\pi/2$ for E_0=0.6 meV, while for E_0=1.7 meV and E_0=4.2 meV its maximum appears at angles close to zero.

Figure 4 (a) shows the integral cross section (ICS) for the elastic scattering by static potential (blue solid curve) and spin unpolarized (black dashed curve). Once again, the importance of the dependence on the spin emerges at low energies. The ICS without including the exchange potential is very different from that considering the electron exchange effects for

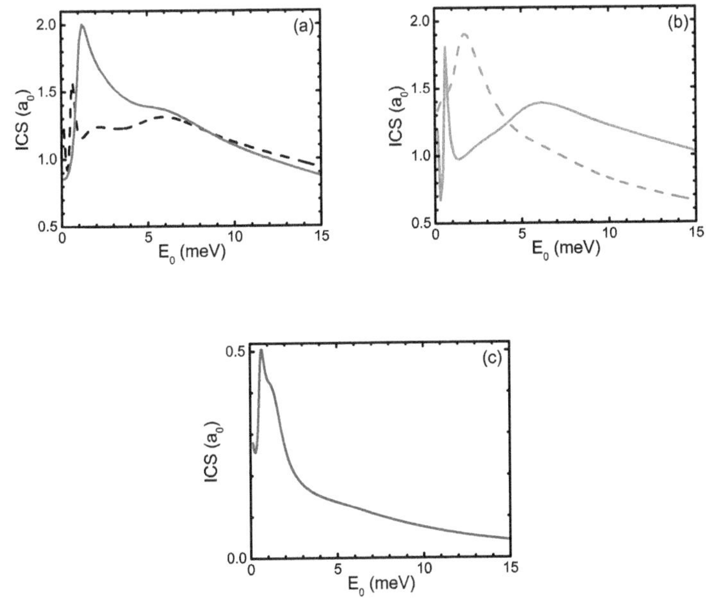

Fig. 4. The elastic ICS as a function of the incident electron energy. (a) The spin-unpolarized ICS with (the dashed curves) and without (the solid curves) the exchange potential; (b) The ICS due to the singlet state (the dashed curves) and the triplet state (the solid curves); and (c) The spin-flip ICS. The parameters for the QD are $\hbar\omega_0 = 5$ meV and $r_0 = 35$ nm

small incident electron energy. However, the ICS is dominated by the static potential at higher energies.

The integral cross section for the singlet (orange dashed curve) and for the triplet (green solid curve) are shown in Figure 4 (b). Note that in the both cases, as well as in the ICS of the static potential (Fig. 4 (a)), a resonant scattering occurs. These resonances can be explained by the analyzing the phase shifts as shown in Figs. 6 (a), (b), and (c). Generally, the phase shifts are functions that smoothly vary as a function of energy. However, under certain circumstances a sudden change of the phase shifts happens in a energy range and a dramatic change in the cross section takes places for these energies, as can be verified by Eq.(43). A physical explanation to this fact can be found when we consider the Schrödinger equation for a central potential, in the basis of angular momentum (equivalent to Eq. (38)),

$$\left[\frac{1}{r}\frac{d}{dr}\left(r\frac{d}{dr}\right) + \frac{l^2}{r^2} + \frac{2m^*}{\hbar^2}V(r)\right]\psi^{l,l}(k,r) + k^2\psi^{l,l}(k,r) = 0. \tag{93}$$

We may identify in Eq. (93) an effective potential $V_{ef} = \frac{l^2}{r^2} + \frac{2m^*}{\hbar^2}V(r)$, for each different angular momentum l. Figure 5 shows this effective potential as a function of r for $V(r) = V_{QD}(r)$. By assuming that there is a metastable state with energy E_r, as sketched in Figure 5, one can prove that when the electron energy E_0 reaches E_r, a rapidly varying phase shift

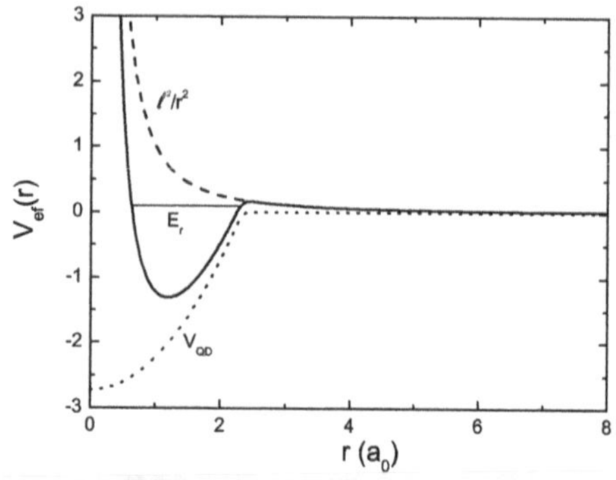

Fig. 5. The effective potential as a function of r is represented by the solid curve. Also, we plot the QD potential (dotted curve) and centrifugal barrier (dashed curve).

occurs and a resonance appears in the ICS (Joachain, 1975). Because this resonance depends on the potential's shape $V(r)$, it is usually called shape resonance.

So, the resonance that appears in the ICS for the static potential (Fig. 4 (a)) with energy $E_0 = 1.22$ meV corresponds to the rapid fluctuation of the phase shift Δ_2 shown in Figure 6 (a). In this case $\Delta_2 \approx \pi/2$ for the value of the energy $E_0 = 1.22$ meV. We also note the appearance of another broader resonance at $E_0 = 6.14$ meV, which corresponds to the rapid increasing of Δ_3 seen in Figure 6 (a). The singlet resonance ($E_0 = 1.72$ meV) and the triplet resonance ($E_0 = 0.57$ meV) shown in Fig. 4 (b) are also resulting from the rapidly varying Δ_2 plotted in Figs.6 (b) and (c).

In Figure 4 (c) we present the integral elastic spin-flip cross section. The spin-flip cross section depends on the scattering amplitudes of the singlet and the triplet as shown by equation (82). As the ICS modulus (the square of the scattering amplitudes) of singlet and triplet states are very distinct at small energies, the spin-flip cross section is of maximum at the same energy range, as shown in Figure 4 (c).

Figure (7) shows the elastic ICS varying the size of the QD, for the scattering by the static potential (Coulomb without exchange). When the radius R_0 is increased, the potential of QD becomes more negative and the resonance energy E_r (Fig. 5) decreases. Thus, we see that the resonance peak in Fig. (7) shifts to lower energy values when R_0 increases. Therefore, the potential becomes deeper and the metastable state E_r decreases, consequently, the shape resonance shifts to lower energies. The shape resonance may disappear when the radius is further increased. In this situation, the state E_r becomes a real bound state instead of a metastable sate.

6.2 Multi-channel scattering

When we consider the three-channel scattering, we have nine possibilities of scattering, *i.e.*, the incident electron has initially energy $E_0 = \hbar^2 k_i^2/2m^*$ and can be scattered with the energy

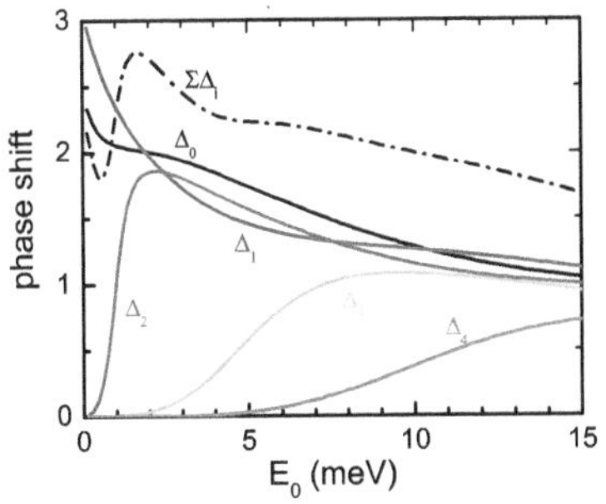

(a) Static potential

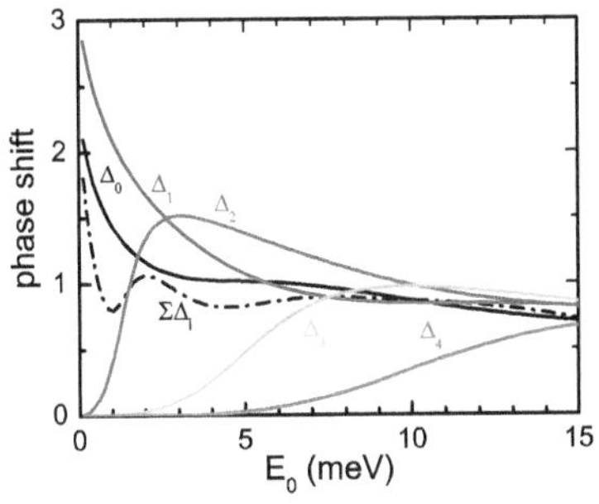

(b) Singlet state

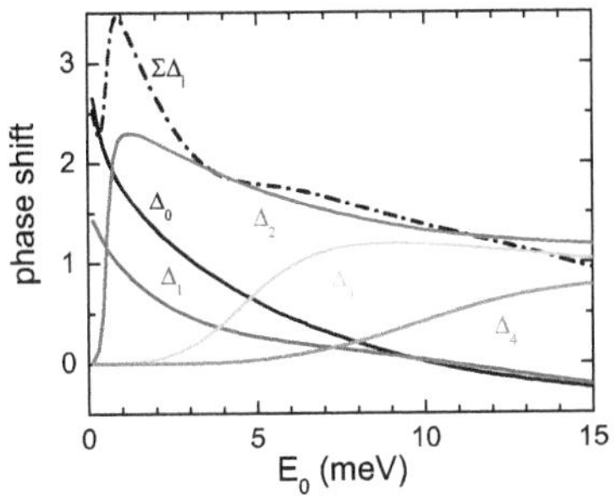

(c) Triplet state.

Fig. 6. The phase shift Δ_l as a function of the incident electron energy for the partial waves with angular momentum l =0, 1, 2, 3, and 4.

$\frac{\hbar^2 k_f^2}{2m} = \frac{\hbar^2 k_i^2}{2m} + \varepsilon_i - \varepsilon_f$, with i and f =1, 2, and 3. If $\varepsilon_f = \varepsilon_i$ the scattering is elastic and if $\varepsilon_f > \varepsilon_i$ there is a excitation. Finally, if $\varepsilon_f < \varepsilon_i$ there is a decay.

In our case, as $\varepsilon_2 = \varepsilon_3$ the probability of exciting or decaying for either of one of these two states is exactly the same. Thus, we calculate the cross section considering elastic and inelastic scattering $\varepsilon_1 \rightarrow \varepsilon_2$. Fig. 8 shows the ICS for the elastic channel (a) and for the excitation channel (b). The black (blue) solid curve represents the spin-unpolarized potential (static), while the dashed curves represent the respective ICS when only one channel is considered. For $E_0 \approx$7 meV, we notice that the ICS for the excitation channel in Fig. 8 (b) has a maximum, while the elastic channel in Fig. 8(a) exhibits a minimum, which is obvious from the probability current conservation. For E_0 >8 meV, the behavior of the ICS shown in Fig. 8 (a) is very similar to the results considering the elastic scattering (dashed curves).

In Figure 9, the ICS is shown for scattering by three-channels, where the green (orange) solid curve represents the scattering by the potential of the triplet (singlet) and the dashed curves represent their ICS when only one channel is considered. In the case of scattering for the triplet state, we verify that when E_0 >9 meV the elastic scattering (Figure 9 (a)) for three-channels is equal to one-channel and therefore, the probability of excitation is practically null (Figure

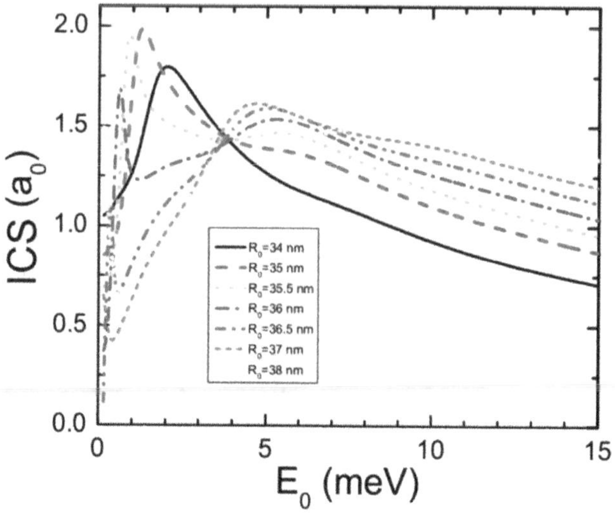

Fig. 7. The elastic ICS as a function of the incident electron energy, considering the static potential for different sizes (R_0) of the QD with $\hbar\omega_0 = 5$ meV.

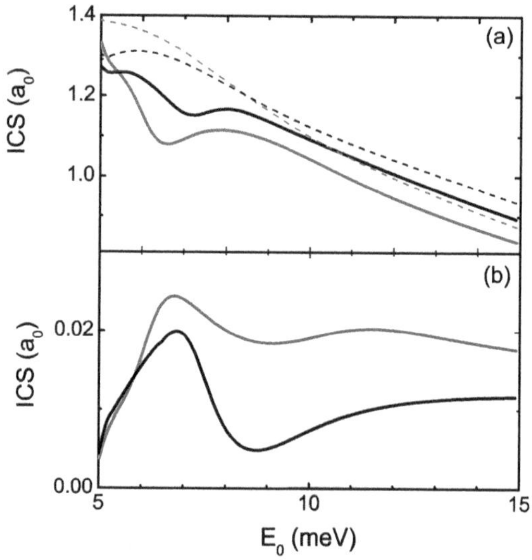

Fig. 8. The ICS for the three-channel scattering as a function of electron energy (R_0 =35 nm and $\hbar\omega_0 = 5$ meV). (a) Elastic channel and (b) excitation channel from the ground state to the first excited state ($l = \pm 1$). Black (blue) curve shows the ICS for the spin-unpolarized (static) scattering. Dashed curves are the respective ICS within the one-channel scattering approximation.

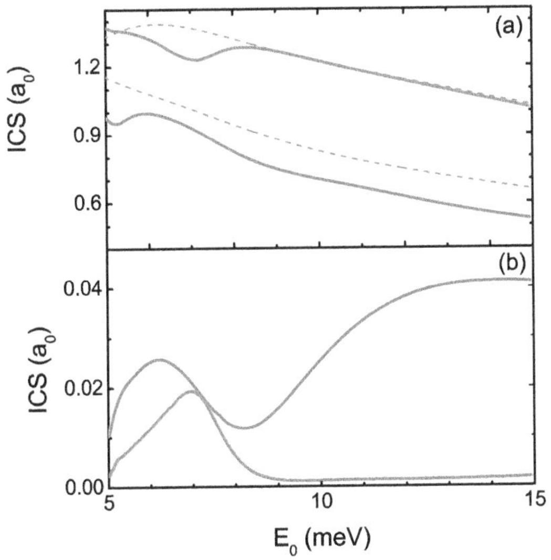

Fig. 9. The same as Fig.8 but now the orange (green) curve shows the ICS for the triplet (singlet) spin state. Dashed curves are the respective ICS within the one-channel scattering approximation.

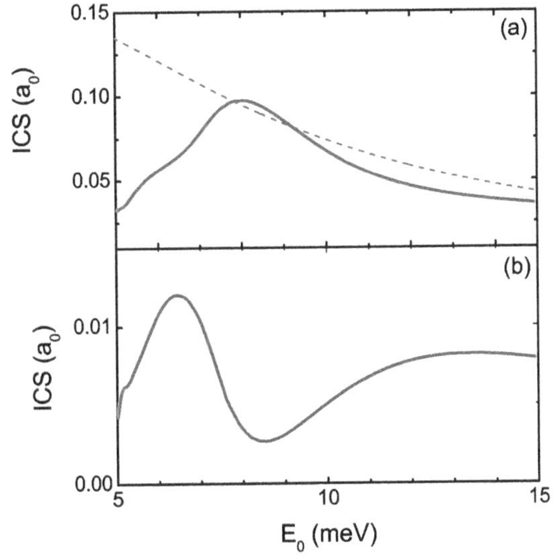

Fig. 10. The same as Fig.8 but now for the spin-flip ICS. The dashed curve is the respective spin-flip ICS within the one-channel approximation.

9 (b)). In the case of the singlet state, the behavior is contrary, *i.e.*, when E_0 >9 meV the excitation probability begins to increase (Fig 9 (b)), thus showing that the scattering may be completely different depending on the spin state of electrons.

In Figure 10, the spin-flip ICS is shown considering three-channels of scattering, where the solid curve represents the scattering from (a) elastic and (b) excitation channel. The dashed curve represents the spin-flip ICS considering only one channel of scattering. The spin-flip ICS for one-channel presents a maximum for $E_0 \approx 8$ meV and a minimum in the same energy range for the excitation channel. We also found that for $E_0 > 9$ meV, the spin-flip cross section of three-channels is similar to the elastic scattering (dashed curve).

6.3 Scattering in the quasi-one-dimensional system

In this section, we apply the MCF to solve the Lippmann-Schwinger equation for the electron scattering through the QD confined in the Q1D structure. The convergency of MCF is very accurate in this case and achieves a precision of 10^{-4} for the transmission probability in approximately 20 interactions. To probe our numerical method, we consider one electron confined in the QD with radius R_0=45 nm. Moreover, two different cases for confined potential in x-direction with $\hbar\omega_x$ =5 meV and $\hbar\omega_x$ =3 meV are tested. The obtained results are shown in Fig. 11. In both cases, the screening length is fixed as $\lambda^{-1} = l_x$ and the confinement frequency the y-direction is set different from that in the x-direction with $\omega_y = 1.7\omega_x$. In Fig. 11 (a) and (b), we plot the transmission probability as a function of the incident electron energy assuming different scattering situations: (i) the QD potential only (black dotted curve), (ii) the static potential (red dash-dotted curve), (ii) the singlet state (orange dashed curve), and (iv) the triplet state (green solid curve). From the results, we see that the confinement potential (or frequency ω_x) of the QD strongly affects the transmission probability through the QD. Furthermore, the Coulomb potential alters considerably the scattering. The electron-electron exchange potential splits the resonant peak into two due to different spin states of the system. It shows that, when the incident electron has anti-parallel (parallel) spin with the confined electron in the QD, the transmission probability is enhanced (suppressed) significantly at low energy. This is a kind of spin filter effect if we could control the spin state of the confined electron.

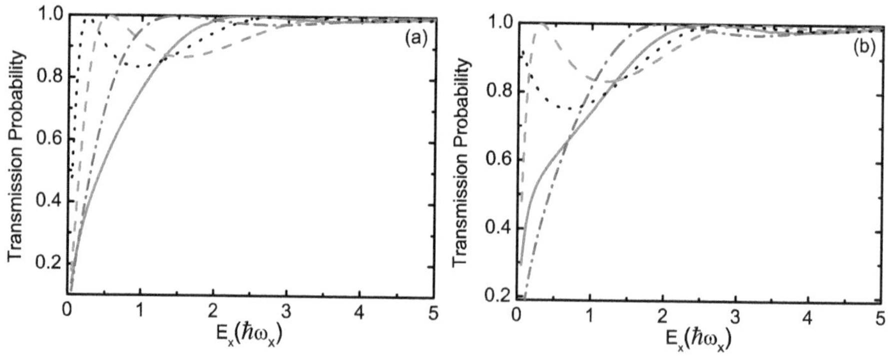

(a) $\hbar\omega_x = 3$ meV

(b) $\hbar\omega_x = 5$meV.

Fig. 11. Transmission probability as a function of the incident electron's energy assuming different scattering situations: only the QD potential (black dotted curve), the static potential (red dash-dotted curve), the singlet state (orange dashed curve), and the triplet state (green solid curve).

7. Conclusion remarks

We presented a theoretical approach to calculate the electron scattering and transport through an N-electron QD embedded in a 2D and a Q1D semiconductor structure. The multichannel L-S equations are solved numerically using the iterative method of continued fractions considering the electron-electron interactions. From this method, we can study the multichannel scattering including the excited states of the QD. The electron transport property due to elastic and inelastic scattering, as well as its dependence on the spin states of the system can be obtained in great precision.

We applied this method to the case where only one electron is confined in the QD. The results indicate a rapid convergency of the numerical method for the electron scattering in both 2D and 1D systems. We found that the electron-electron exchange effects are relevant when the kinetic energy of incident electron is small. For a QD of more electrons, we need firstly to find the eigenstates of the QD with electron-electron interactions. In principal, the scattering processes can be calculated according to the total wave-function defined by Eq. (6).

8. Acknowledgments

This work was supported by FAPESP and CNPq. (Brazil).

9. References

Adhikari, S. K. (1986). *Am. J. Phys.* 54: 362.
Bransden, B. H. & McDowell, M. R. C. (1977). *Phys. Rep.* 30: 207.
Burkard, G., Engel, H. A. & Loss, D. (2000). *Fortschr. Phys.* 48: 965.
Buttiker, M., Imry, Y., Landauer, R. & Pinhas, S. (1985). *Phys. Rev. B* 31: 6207.
Castelano, L. K. (2006). PhD thesis, University of Sao Paulo.
Castelano, L. K., Hai, G.-Q. & Lee, M.-T. (2007a). *Phys. Rev. B* 76: 165306.
Castelano, L. K., Hai, G.-Q. & Lee, M.-T. (2007b). *Phys. Stat. Sol. (c)* 4: 466.
da Paixão, F. J., Lima, M. & McKoy, V. (1996). *Phys. Rev. A* 53: 1400.
Darwin, C. (1930). *Proc. Cambridge Philos. Soc.* 27: 86.
Engel, H.-A. & Loss, D. (2002). *Phys. Rev. B* 65: 2002.
Fisher, D. S. & Lee, P. A. (1981). *Phys. Rev. B* 23: 6851.
Fock, V. (1928). *Z. Phys.* 47: 446.
Fransson, J., Holmstrom, E., Eriksson, O. & Sandalov, I. (2003). *Phys. Rev. B* 67: 205310.
Gundogdu, K., Hall, K. C., Boggess, T. F., Deppe, D. G. & Shchekin, O. B. (2004). *Appl. Phys. Lett.* 84: 2793.
Hegemann, T., Oberste-Vorth, M., Vogts, R. & Hanne, G. F. (1991). *Phys. Rev. Lett.* 66: 2968.
Horacek, J. & Sasakawa, T. (1984). *Phys. Rev. A* 28: 2151.
Imry, Y. & Landauer, R. (1999). *Rev. Mod. Phys.* 71: s306.
Joachain, C. J. (1975). *Quantum collision theory*, North-Holland, Amsterdam.
Konig, J. & Martinek, J. (2003). *Phys. Rev. Lett.* 90: 166602.
Koppens, F. H. L., Buizert, C., Tielrooij, K. J., Vink, I. T., Nowack, K. C., Meunier, T., Kowenhoven, L. P. & Vandersypen, L. M. K. (2006). *Nature* 442: 766.
Landauer, R. (1957). *IBM J. Res.* 1: 223.
Landauer, R. (1970). *Phil. Mag.* 21: 863.
Landauer, R. (1975). *Z. Phys.* B24: 247.
Merzbacher, E. (1970). *Quantum Mechanics*, John Wiley and Sons, Inc., New York.

Morse, P. M. & Feshbach, H. (1953). *Methods of Theoretical Physics*, McGraw-Hill, New York.
Qu, F. Y. & Vasilopoulos, P. (2006). *Phys. Rev. B* 74: 245308.
Sarma, S. D., Fabian, J., Hu, X., & Zutic, I. (2001). *Solid State Commun.* 119: 207.
Seneor, P., Bernand-Mantel, A. & Petroff, F. (2007). *J. Phys.: Condens. Matter* 19: 165222.
Szabo, A. & Ostlund, N. (1982). *Modern Quantum Chemistry*, Macmillan Publishing, New York.
Vargiamidis, V., Valassiades, O. & Kyriakos, D. S. (2003). *Phys. Stat. Sol. (b)* 236: 597.
Wolf, S. A., Awschalom, D. D., Buhrman, R. A., Daughton, J. M., von Molnár, S., Roukes, M. L., Chtchelkanova, A. Y. & Treger, D. M. (2001). *Science* 294: 1488.
Zhang, P., Xue1, Q.-K., Wang, Y. & Xie, X. C. (2002). *Phys. Rev. Lett.* 89: 286803.

6

Coherent Spin Dependent Transport in QD-DTJ Systems

Minjie Ma[1], Mansoor Bin Abdul Jalil[1,2] and Seng Ghee Tan[1,3]
[1]*Computational Nanoelectronics and Nano-Device Laboratory, Electrical and Computer Engineering Department, National University of Singapore*
[2]*Information Storage Materials Laboratory, Department of Electrical and Computer Engineering, National University of Singapore*
[3]*Data Storage Institute, A *STAR (Agency of Science, Technology and Research) Singapore*

1. Introduction

As the dimension of devices reduces to nano-scale regime, the spin-dependent transport (SDT) and spin effects in quantum dot (QD) based systems become significant. These QD based systems have attracted much interest, and can potentially be utilized for spintronic device applications. In this chapter, we consider nano-scale spintronic devices consisting of a QD with a double barrier tunnel junction (QD-DTJ)(schematically shown in Fig. 1). The DTJ couples the QD to two adjacent leads which can be ferromagnetic (FM) or non-magnetic (NM).

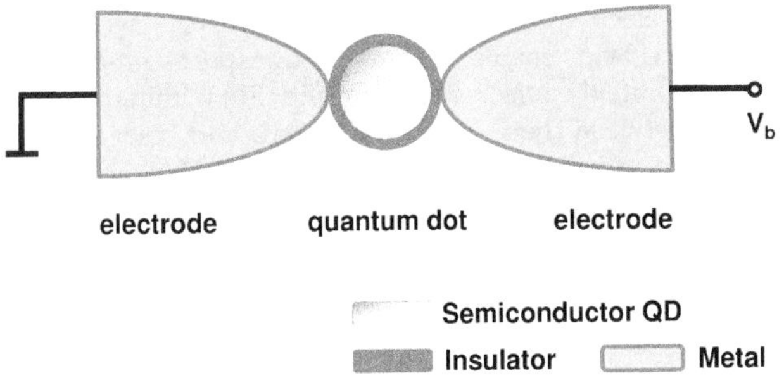

Fig. 1. QD-DTJ system consists of a QD coupling to two electrodes via double tunnel junctions. V_b is the bias voltage, under which the electrons tunnel through the QD one by one.

In a QD-DTJ system, the electron tunneling is affected by the quantized energy levels of the QD, and can thus be referred to as single electron tunneling. The single electron tunneling process becomes spin-dependent when the leads or the QD is a spin polarizer, where the density of states (DOS) for spin-up and spin-down electrons are different. The interplay of SDT with quantum and/or single electron charging effects makes the QD-DTJ systems interesting. In such QD-DTJ systems, it is possible to observe several quantum spin phenomena, such as spin blockade (Shaji et al. (2008)), Coulomb blockade (CB) (Bruus & Flensberg (2004)), cotunneling (Weymann & Barnaś (2007)), tunnel magnetoresistance (TMR) (Rudziński & Barnaś (2001)), spin transfer torque (Mu et al. (2006)) and Kondo effect (Katsura (2007); Lobo et al. (2006); Potok et al. (2007)). The complex spin and charge transport properties of QD-DTJ systems have attracted extensive theoretical (Bao et al. (2008); Braig & Brouwer (2005); Jauho et al. (1994); Kuo & Chang (2007); Ma et al. (2008); Meir & Wingreen (1992); Meir et al. (1991; 1993); Mu et al. (2006); Qi et al. (2008); Qu & Vasilopoulos (2006); Souza et al. (2004); Zhang et al. (2002); Zhu & Balatsky (2002)) and experimental ((Deshmukh & Ralph, 2002; Hamaya et al., 2007; Pasupathy et al., 2004; Potok et al., 2007)) investigations recently. These studies may ultimately lead to the utilization of such devices in diverse applications such as single spin detector (Wabnig & Lovett (2009)) and STM microscopy (Manassen et al. (2001)).

The theoretical study of the SDT through these DTJ systems are mainly based on two approaches, namely the master equation (ME) approach and the Keldysh nonequilibrium Green's function (NEGF) approach. For coherent transport across QD-DTJ devices, quantum transport methods are applied, such as the linear response (Kubo) method applicable for small bias voltage, and its generalization, the NEGF method for arbitrary bias voltage. Since the objective of the study in this Chapter is for device application over a wide voltage range, we focus on the latter. The NEGF method has been employed to analyze various transport properties of QD-DTJ systems, such as TMR, tunneling current (Weymann & Barnaś (2007)) and conductance. These analyses were conducted based on the Anderson model (Meir et al. (1993); Qi et al. (2008)), for collinear or noncollinear (Mu et al. (2006); Sergueev et al. (2002); Weymann & Barnaś (2007)) configurations of the magnetization of the two FM leads, or in the presence of spin-flip scattering in the QD (Lin & D.-S.Chuu (2005); Souza et al. (2004); Zhang et al. (2002)).

In this Chapter, based on the NEGF approach, we study the SDT through two QD-DTJ systems. In Section. 2, the electronic SDT through a single energy-level QD-DTJ is theoretically studied, where the two FM leads enable the electron transport spin-dependent. In the study, we systematically incorporate the effect of the spin-flip (SF) within the QD and the SF during tunneling the junction between the QD and each lead, and consider possible asymmetry between the coupling strengths of the two tunnel junctions. Based on the theoretical model, we first investigate the effects of both types of SF events on the tunneling current and TMR; subsequently, we analyze the effect of coupling asymmetry on the QD's electron occupancies and the charge and spin currents through the system (Ma et al. (2010)).

In Section. 3, we studied the SDT through a QD-DTJ system with finite Zeeman splitting (ZS) in the QD, where the two leads which sandwich the QD are NM. The spin-dependence of the electron transport is induced by the ZS caused by the FM gate attached to the QD. A fully polarized tunneling current is expected through this QD-DTJ system. The charge and spin currents are to be analyzed for the QD-DTJ systems with or without ZS.

2. Single energy level QD

The QD-DTJ device under consideration is shown in Fig. 2. It consists of two FM leads and a central QD in which a single energy level is involved in the electron tunneling process. The SDT through the QD-DTJ is to be theoretically modeled via the Keldysh NEGF approach (Caroli et al. (1971); Meir & Wingreen (1992)). In the transport model, the limit of small correlation energy is assumed, in the case where the energy due to electron-electron interaction in the QD is much smaller than the thermal energy or the separation between the discrete energy levels in the QD (Fransson & Zhu (2008)).

2.1 Theory

For the QD-DTJ device shown in Fig. 2, the full Hamiltonian consists of the lead Hamiltonian H_α, the QD Hamiltonian H_d, and the tunneling Hamiltonian H_t. The explicit form of the Hamiltonian is given by

$$H = \sum_\alpha H_\alpha + H_d + H_t, \tag{1}$$

where

$$H_\alpha = \sum_{k\sigma} \epsilon_{\alpha k\sigma} a^\dagger_{\alpha k\sigma} a_{\alpha k\sigma}, \tag{2}$$

$$H_d = \sum_\sigma \epsilon_{\sigma\sigma} a^\dagger_\sigma a_\sigma + \sum_\sigma \epsilon_{\sigma\bar{\sigma}} a^\dagger_{\bar{\sigma}} a_\sigma, \tag{3}$$

$$H_t = \sum_{\alpha k\sigma} \left(t_{\alpha k\sigma,\sigma} a^\dagger_\sigma a_{\alpha k\sigma} + t^*_{\alpha k\sigma,\sigma} a^\dagger_{\alpha k\sigma} a_\sigma \right) + \sum_{\alpha k\sigma} \left(t_{\alpha k\sigma,\bar{\sigma}} a^\dagger_{\bar{\sigma}} a_{\alpha k\sigma} + t^*_{\alpha k\sigma,\bar{\sigma}} a^\dagger_{\alpha k\sigma} a_{\bar{\sigma}} \right). \tag{4}$$

In the above, $\epsilon_{\sigma\sigma}$ is the single energy level in the QD, $\epsilon_{\sigma\bar{\sigma}}$ denotes the coupling energy of the spin-flip within quantum dot (SF-QD) from spin-σ to spin-$\bar{\sigma}$ state, $t_{\alpha k\sigma,\sigma}$ ($t_{\alpha k\sigma,\bar{\sigma}}$) is the coupling between electrons of the same (opposite) spin states in the lead and the QD. $\alpha = \{L, R\}$ is the lead index for the left and right leads, $\sigma = \{\uparrow, \downarrow\}$ stands for up- and down-spin, and k is the momentum, $\epsilon_{\alpha k\sigma}$ represents the energy in the leads. The operators $a^\dagger_\nu$ (a_ν) and $a^\dagger_\sigma$ (a_σ) are the creation (annihilation) operators for the electrons in the leads and the QD, respectively.

2.1.1 Tunneling current and tunnel magnetoresistance

The tunneling current through the QD-DTJ system can be expressed as the rate of change of the occupation number $N = \sum_\sigma a^\dagger_\sigma a_\sigma$ in the QD,

$$I = e\dot{N} = \frac{ie}{\hbar}\langle [H, N] \rangle. \tag{5}$$

Without loss of generality, we can calculate the tunneling current in Eq. (5) by considering the tunneling current I_L through the left junction between the left lead and the QD. Evaluating the commutator in Eq. (5) in terms of creation and annihilation operators gives

$$I = I_L = \frac{ie}{\hbar} \sum_{Lk\sigma,\sigma'} \left(t_{Lk\sigma,\sigma'} \langle a^\dagger_{Lk\sigma} a_{\sigma'} \rangle - t^*_{Lk\sigma,\sigma'} \langle a^\dagger_{\sigma'} a_{Lk\sigma} \rangle \right). \tag{6}$$

In Eq. (6), one may replace the creation and annihilation operators by the lesser Green's functions, which are defined as $G^<_{\sigma',Lk\sigma}(t) = i\langle a^\dagger_{Lk\sigma} a_{\sigma'}(t) \rangle$ and $G^<_{Lk\sigma,\sigma'}(t) = i\langle a^\dagger_{\sigma'} a_{Lk\sigma}(t) \rangle$ (Meir & Wingreen (1992)). Eq. (6) then takes the form of

$$I_L = \frac{e}{\hbar} \sum_{Lk\sigma,\sigma'} \left(t_{Lk\sigma,\sigma'} G^<_{\sigma',Lk\sigma}(t) - t^*_{Lk\sigma,\sigma'} G^<_{Lk\sigma,\sigma'}(t) \right). \tag{7}$$

After performing a Fourier transform on Eq. (7), $G^<_{Lk\sigma,\sigma'}(\epsilon)$ can be expressed in form of the left lead's and QD's Green's functions, under the assumption of non-interacting

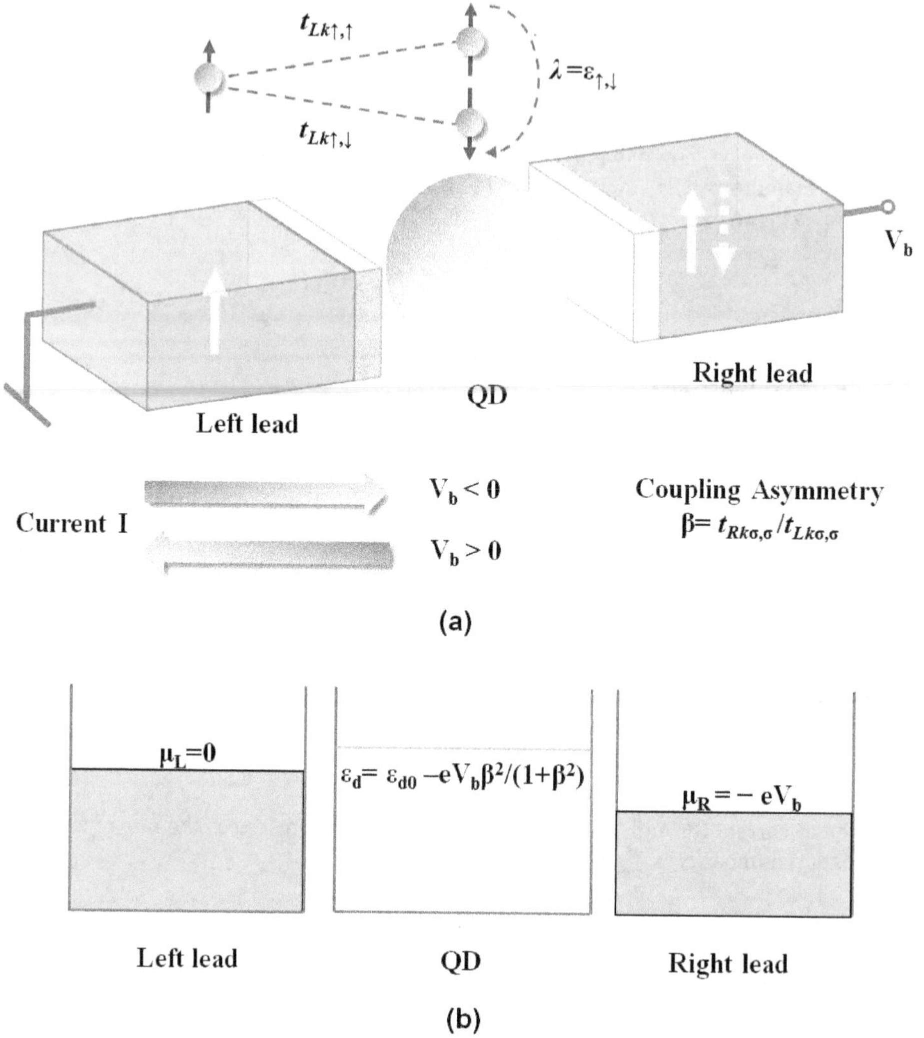

Fig. 2. (a) Schematic diagram of the QD-DTJ structure consisting of a QD sandwiched by two FM leads; (b) the schematic energy diagram for the system in (a). In (a), the arrows in the leads indicate magnetization directions, which can either be in parallel (solid) or antiparallel (dashed) configuration, V_b denotes the bias between the two leads, λ characterizes the strength of the SF-QD, $t_{Lk\uparrow,\downarrow}$ describes the SF-TJ from the up-spin state in left lead and the down-spin state in the QD, $t_{Lk\uparrow,\uparrow}$ shows the coupling between the same electron spin states in left lead and QD, and $\beta = t_{Rk\sigma,\sigma}/t_{Lk\sigma,\sigma}$ represents the coupling asymmetry between the left and right tunneling junctions. In (b), μ_L and μ_R are the chemical potentials of left and right leads respectively, and ϵ_d (ϵ_{d0}) denotes the single energy level of the QD with or without bias voltage.

leads. Taking into account the contour-ordered integration over the time loop, the corresponding Dyson's equations for $G^{<}_{Lk\sigma,\sigma'}(\epsilon)$ can then be obtained (Mahan (1990)), i.e., $G^{<}_{Lk\sigma,\sigma'}(\epsilon) = \sum_{\sigma''} t_{Lk\sigma,\sigma''}[g^{t}_{Lk\sigma,Lk\sigma}(\epsilon)G^{<}_{\sigma'',\sigma'}(\epsilon) - g^{<}_{Lk\sigma,Lk\sigma}(\epsilon)G^{\bar{t}}_{\sigma'',\sigma'}(\epsilon)]$ and $G^{<}_{\sigma',Lk\sigma}(\epsilon) = \sum_{\sigma''} t^{*}_{Lk\sigma,\sigma''}[g^{<}_{Lk\sigma,Lk\sigma'}(\epsilon)G^{t}_{\sigma'',\sigma'}(\epsilon) - g^{\bar{t}}_{Lk\sigma,Lk\sigma}(\epsilon)G^{<}_{\sigma''\sigma'}(\epsilon)]$, where $G^{t} = \theta(t)G^{>} + \theta(-t)G^{<}$ and $G^{\bar{t}} = \theta(-t)G^{>} + \theta(t)G^{<}$ are the time-ordered and anti-time-ordered Green's functions respectively, $G^{<}_{\sigma'',\sigma'}(t) = -i\langle a^{\dagger}_{\sigma'}a_{\sigma''}(t)\rangle$, and the g's are the corresponding unperturbed Green's functions of the leads, whose lesser Green's function and greater Green's function are in form of $g^{<}_{Lk\sigma,Lk\sigma}(\epsilon) = i2\pi f_{L\sigma}(\epsilon)\delta(\epsilon - \epsilon_{L\sigma})$ and $g^{>}_{Lk\sigma,Lk\sigma}(\epsilon) = -i2\pi[1 - f_{L\sigma}(\epsilon)]\delta(\epsilon - \epsilon_{L\sigma})$, where $f_{L\sigma}(\epsilon) = (1 + \exp(\frac{\epsilon - \mu_L}{k_B T}))^{-1}$ is the Fermi-Dirac function, μ_L is the chemical potential, $\epsilon_{L\sigma}$ is the energy for electrons with spin σ in the left lead, k_B is the Boltzmann constant and T is the temperature of the device. With this, the current in Eq. (7) can be expressed in terms of the Green's functions wholly of the leads and the QD, i.e.,

$$I_L = \frac{e}{\hbar}\sum_{Lk\sigma,\sigma'}\int_{-\infty}^{\infty}\frac{d\epsilon}{2\pi}\{t_{Lk\sigma,\sigma'}\sum_{\sigma''} t^{*}_{Lk\sigma,\sigma''}[g^{<}_{Lk\sigma,Lk\sigma'}(\epsilon)G^{t}_{\sigma'',\sigma'}(\epsilon) - g^{\bar{t}}_{Lk\sigma,Lk\sigma}(\epsilon)G^{<}_{\sigma''\sigma'}(\epsilon)]$$
$$-t^{*}_{Lk\sigma,\sigma'}\sum_{\sigma''} t_{Lk\sigma,\sigma''}[g^{t}_{Lk\sigma,Lk\sigma}(\epsilon)G^{<}_{\sigma'',\sigma'}(\epsilon) - g^{<}_{Lk\sigma,Lk\sigma}(\epsilon)G^{\bar{t}}_{\sigma'',\sigma'}(\epsilon)]\}. \tag{8}$$

By applying the identities $G^{t} + G^{\bar{t}} = G^{<} + G^{>}$ and $G^{>} - G^{<} = G^{r} - G^{a}$ to Eq. (8), we obtain after some algebra (Mahan (1990)):

$$I_L = \frac{ie}{\hbar}\sum_{Lk\sigma,\sigma'}\int_{-\infty}^{\infty}\frac{d\epsilon}{2\pi} t_{Lk\sigma,\sigma'}|_{\epsilon=\epsilon_v} t^{*}_{Lk\sigma,\sigma''}|_{\epsilon=\epsilon_v}\{f_v(\epsilon)[G^{r}_{\sigma',\sigma''}(\epsilon) - G^{a}_{\sigma',\sigma''}(\epsilon)] + G^{<}_{\sigma',\sigma''}(\epsilon)\}. \tag{9}$$

We now introduce the density of states for the electrons in the FM leads, denoted by $\rho_{\alpha\sigma}(\epsilon)$. For the electrons in the left FM lead, the density of states is $\rho_{L\sigma}(\epsilon) = [1 + (-1)^{\sigma} p_L]\,\rho_{L0}(\epsilon)$, while for the electrons in the right FM lead, it is $\rho_{R\sigma}(\epsilon) = \left[1 + (-1)^{a+\sigma} p_R\right]\rho_{R0}(\epsilon)$, where $\sigma = \{0, 1\}$ for spin-up/down electrons, $a = \{0, 1\}$ for parallel/antiparallel alignment of the two FM leads' magnetization, $\rho_{\alpha 0} = (\rho_{\alpha\uparrow} + \rho_{\alpha\downarrow})/2$, and p_α is the polarization of the lead α. For the summation over k in Eq.(9), one may apply the continuous limit approximation $\sum_{\{Lk\sigma\}} \rightarrow \sum_{\{L\sigma\}}\int d\epsilon\, \rho_{L\sigma}(\epsilon)$. The current then can be expressed as

$$I_L = \frac{ie}{\hbar}\sum_{v=\{L\sigma\}}\int d\epsilon\, \mathrm{tr}\left\{f_v(\epsilon)\mathbf{\Gamma}_v\left[\mathbf{G}^{r}(\epsilon) - \mathbf{G}^{a}(\epsilon)\right] + \mathbf{\Gamma}_v\mathbf{G}^{<}(\epsilon)\right\}, \tag{10}$$

where $\mathbf{\Gamma}_v$ and $\mathbf{G}^{(r,a,<)}(\epsilon)$ are (2×2) coupling and Green's function matrices, given by

$$\mathbf{\Gamma}_{L\sigma}(\epsilon) = 2\pi\rho_{L\sigma}(\epsilon)\begin{pmatrix} |t_{L\sigma,\sigma}(\epsilon)|^2 & |t^{*}_{L\sigma,\sigma}(\epsilon)t_{L\sigma,\bar{\sigma}}(\epsilon)| \\ |t^{*}_{L\sigma,\bar{\sigma}}(\epsilon)t_{L\sigma,\sigma}(\epsilon)| & |t^{*}_{L\sigma,\bar{\sigma}}(\epsilon)t_{L\sigma,\bar{\sigma}}(\epsilon)| \end{pmatrix}, \tag{11}$$

$$\mathbf{G}^{(r,a,<)}(\epsilon) = \begin{pmatrix} G^{(r,a,<)}_{\sigma,\sigma}(\epsilon) & G^{(r,a,<)}_{\bar{\sigma},\sigma}(\epsilon) \\ G^{(r,a,<)}_{\sigma,\bar{\sigma}}(\epsilon) & G^{(r,a,<)}_{\bar{\sigma},\bar{\sigma}}(\epsilon) \end{pmatrix}. \tag{12}$$

In Eq.(11), $t_{L\sigma,\sigma}(t_{L\sigma,\bar{\sigma}})$ applies for the case of spin-σ electron tunneling to the spin-σ ($\bar{\sigma}$) state with (without) spin-flip. In low-bias approximation, $\Gamma_{L\sigma}(\epsilon)$=$2\pi\rho_{L\sigma}(\epsilon)|t^{*}_{L\sigma,\sigma'}(\epsilon)$ $t_{L\sigma,\sigma''}(\epsilon)|$ is taken to be constant (zero) within (beyond) the energy range close to the lead's electrochemical potential where most of the transport occurs, i.e., $\epsilon \in [\mu_{\alpha} - D, \mu_{\alpha} + D]$, where D is constant (Bruus & Flensberg (2004)). Based on the kinetic equation (Meir & Wingreen (1992)), the lesser Green's function $G^{<}_{\sigma',\sigma''}(\epsilon)$ can be written as $G^{<}_{\sigma',\sigma''}(\epsilon) = iG^{r}_{\sigma',\sigma''}(\epsilon)G^{a}_{\sigma',\sigma''}(\epsilon)[\Gamma_{L\sigma}f_{L\sigma}(\epsilon) + \Gamma_{R\sigma}f_{R\sigma}(\epsilon)]$, where $G^{r}_{\sigma',\sigma''}(t) = -i\theta(t)\langle\{a_{\sigma'},^{+} a_{\sigma''}(t)\}\rangle$ and the advanced Green's function $G^{a}_{\sigma,\sigma}(\epsilon) = [G^{r}_{\sigma,\sigma}(\epsilon)]^{*}$. $\Gamma_{\alpha\sigma}$ is the aforementioned coupling strength, and $f_{\alpha\sigma} = \left(1 + \exp(\frac{\epsilon - \mu_{\alpha\sigma}}{k_B T})\right)^{-1}$ is the Fermi-Dirac function of lead α, with $\mu_{\alpha\sigma}$ being the chemical potential of that lead. When a bias voltage of V_b is between the two leads, the leads' electrochemical potentials are, respectively, given by $\mu_{L\sigma} = 0$ and $\mu_{R\sigma} = -eV_b$.

Considering that the current from the left lead to the QD is equal to the current from the QD to the right lead, one may calculate the current in a symmetric form, i.e., $I = \frac{I_L + I_R}{2}$. The final form for the total current is then given by

$$I = \frac{e}{h}\sum_{\sigma}\int d\epsilon \left[f_{L\sigma}(\epsilon) - f_{R\sigma}(\epsilon)\right] \mathrm{tr}\{\mathbf{G}^{a}\mathbf{\Gamma}_{R\sigma}\mathbf{G}^{r}\mathbf{\Gamma}_{L\sigma}\}. \tag{13}$$

In this QD-DTJ system, there exists the tunnel magnetoresistance (TMR) effect, which is caused by the difference between the resistance in parallel and antiparallel configurations of the two FM leads' magnetization. The TMR is given by

$$\mathrm{TMR} = \frac{R^{AP} - R^{P}}{R^{P}} = \frac{I^{P} - I^{AP}}{I^{AP}}, \tag{14}$$

where I^{P} (I^{AP}) is the tunneling current in parallel (antiparallel) configuration of the two leads' magnetization.

During the course of analyses, we would also consider the state of the QD, which is characterized by its occupancy. The QD's occupancy with electrons of spin-σ can be obtained by considering the lesser Green's function of the QD, i.e.,

$$\langle n_{\sigma}\rangle = \frac{1}{2\pi}\mathrm{Im}\int d\epsilon G^{<}_{\sigma,\sigma}(\epsilon). \tag{15}$$

2.1.2 Retarded Green's function

To calculate the tunneling current in Eq. (13), one has to obtain the explicit expression for the retarded Green's functions $G^{r}_{\sigma\sigma'}(\epsilon)$ of the QD. This can be done by means of the (equation-of-motion) EOM method. By definition, the general form of a retarded Green's function is given by $G^{r}_{\sigma,\sigma'}(t) = -i\theta(t)\langle\{a_{\sigma}(t), a^{\dagger}_{\sigma'}\}\rangle$. In the EOM method, the analytical expression for $G^{r}_{\sigma,\sigma'}(t)$ is obtained by firstly differentiating $G^{r}_{\sigma,\sigma'}(t)$ with respect to time. This yields

$$\begin{aligned} i\partial_t G^{r}_{\sigma,\sigma'}(t) &= \delta(t - t')\delta_{\sigma\sigma'} - i\theta(t - t')\langle\{i\partial_t a_{\sigma}(t), a^{\dagger}_{\sigma'}\}\rangle \\ &= \delta(t - t')\delta_{\sigma\sigma'} - i\theta(t - t')\langle\{-[H, a_{\sigma}], a^{\dagger}_{\sigma'}\}\rangle. \end{aligned} \tag{16}$$

Based on Eq. (16), for the QD-DTJ system with Hamiltonian in Eq. (1), one may obtain a closed set of equations involving $G^r_{\sigma,\sigma'}(\epsilon)$ after Fourier transform,

$$1 = (\epsilon + i\eta - \epsilon_{\sigma\sigma})G^r_{\sigma,\sigma} - \sum_{\nu=\{\alpha,k,\sigma'\}} t_{\nu,\sigma}G^r_{\nu,\sigma} - \epsilon_{\bar{\sigma}\sigma}G^r_{\bar{\sigma},\sigma'} \tag{17}$$

$$0 = (\epsilon + i\eta - \epsilon_{\bar{\sigma}\bar{\sigma}})G^r_{\bar{\sigma},\sigma} - \sum_{\nu=\{\alpha,k,\sigma'\}} t_{\nu,\bar{\sigma}}G^r_{\nu,\sigma} - \epsilon_{\sigma\bar{\sigma}}G^r_{\sigma,\sigma'} \tag{18}$$

$$0 = (\epsilon + i\eta - \epsilon_{\nu})G^r_{\nu,\sigma} - \sum_{\sigma'=\{\sigma,\bar{\sigma}\}} t^*_{\nu,\sigma'}G^r_{\sigma',\sigma'} \text{ where } \nu = \{\alpha, k, \sigma\}, \tag{19}$$

$$0 = (\epsilon + i\eta - \epsilon_{\nu})G^r_{\nu,\sigma} - \sum_{\sigma'=\{\sigma,\bar{\sigma}\}} t^*_{\nu,\sigma'}G^r_{\sigma',\sigma'} \text{ where } \nu = \{\alpha, k, \bar{\sigma}\}. \tag{20}$$

By solving the equation array of Eqs. (17) to (20), one reaches the explicit expressions for the retarded Green's functions (those in Eq. 12) of the QD:

$$G^r_{\sigma,\sigma} = \frac{1}{\epsilon + i\eta - \epsilon_{\sigma\sigma} - \Sigma^{\sigma}_{\sigma\sigma}(\epsilon) - \Sigma^{\bar{\sigma}}_{\sigma\sigma}(\epsilon) - \frac{\prod\limits_{\sigma'=\{\sigma,\bar{\sigma}\}}\left(\epsilon_{\bar{\sigma}'\sigma'} + \Sigma^{\sigma'}_{\sigma'\bar{\sigma}'}(\epsilon) + \Sigma^{\bar{\sigma}'}_{\bar{\sigma}'\sigma'}(\epsilon)\right)}{\epsilon + i\eta - \epsilon_{\bar{\sigma}\bar{\sigma}} - \Sigma^{\sigma}_{\bar{\sigma}\bar{\sigma}}(\epsilon) - \Sigma^{\bar{\sigma}}_{\bar{\sigma}\bar{\sigma}}(\epsilon)}}, \tag{21}$$

$$G^r_{\bar{\sigma},\sigma} = \frac{1}{-\epsilon_{\bar{\sigma}\sigma} - \Sigma^{\sigma}_{\sigma\bar{\sigma}}(\epsilon) - \Sigma^{\bar{\sigma}*}_{\sigma\bar{\sigma}}(\epsilon) + \frac{\prod\limits_{\sigma'=\{\sigma,\bar{\sigma}\}}\left(\epsilon + i\eta - \epsilon_{\bar{\sigma}'\sigma'} - \Sigma^{\sigma'}_{\bar{\sigma}'\bar{\sigma}'}(\epsilon) - \Sigma^{\bar{\sigma}'}_{\bar{\sigma}'\bar{\sigma}'}(\epsilon)\right)}{\epsilon_{\sigma\bar{\sigma}} + \Sigma^{\sigma}_{\bar{\sigma}\sigma}(\epsilon) + \Sigma^{\bar{\sigma}*}_{\sigma\bar{\sigma}}(\omega)}}, \tag{22}$$

where the self energy $\Sigma^{\sigma}_{\sigma'\sigma''}(\epsilon) = \sum_{\{\alpha,k\}} \frac{t_{\alpha k\sigma,\sigma'}t^*_{\alpha k\sigma,\sigma''}}{\epsilon+i\eta-\epsilon_{\alpha k\sigma}}$, $\Sigma^{\sigma*}_{\sigma'\sigma''}(\epsilon) = \sum_{\{\alpha,k\}} \frac{t^*_{\alpha k\sigma,\sigma'}t_{\alpha k\sigma,\sigma''}}{\epsilon+i\eta-\epsilon_{\alpha k\sigma}}$, with $\sigma, \sigma', \sigma'' \in \{\uparrow, \downarrow\}$.

2.2 Results and discussion

Based on the electron transport model developed in Sec. 2.1, one may analyze the SDT properties, such as the spectral functions, the tunneling charge current, spin current, the TMR and the electron occupancies of the QD. The SDT model enables one to investigate the effects of the SF-QD and SF-TJ events and the effect of the coupling asymmetry (CA) on the SDT properties as well.

To focus on the above effects, one may assume that, i) proportional and spin independent lead-QD coupling across the two junctions, i.e., $t_{\alpha k\uparrow,\uparrow}$=$t_{\alpha k\downarrow,\downarrow}$=$t_{\alpha k\sigma,\sigma}$=$t_\alpha$, and t_R=βt_L=t; ii) junction and spin independent strength of SF-TJ, i.e., $t_{\alpha k\uparrow,\downarrow}$=$t_{\alpha k\downarrow,\uparrow}$=$t_{\alpha k\sigma,\bar{\sigma}}$=$v_\alpha$, and v_R=$\beta v_L = v$; and iii) spin independence of SF-QD, i.e., $\epsilon_{\uparrow\downarrow}$=$\epsilon_{\downarrow\uparrow}$=$\lambda$, iv) the chemical potential of the left and right leads are μ_L=0, $\mu_R = -eV_b$; and v) spin independence of the energy level of the QD, i.e., $\epsilon_{\sigma\sigma}$=$\epsilon_{\bar{\sigma}\bar{\sigma}}$=$\epsilon_d$= $\epsilon_{d0} - eV_b\frac{\beta^2}{1+\beta^2}$, where ϵ_{d0} is the QD's energy level without bias voltage. Based on the assumptions i)-v) and Eq. (22), one can readily deduce the spin symmetry of the off-diagonal Green's functions, i.e., $G^r_{\uparrow,\downarrow} = G^r_{\downarrow,\uparrow} = G^r_{\sigma,\bar{\sigma}}$. For simplicity, in the following discussion, the form of $G^r_{\sigma\sigma'}$ is used to replace the form of $G^r_{\sigma,\sigma'}$ for retarded Green's function.

2.2.1 Spin-flip effects

Firstly, one may evaluate the four elements of the retarded Green's function (GF) matrix [given in Eq. (12)], $G^r_{\uparrow\uparrow}$, $G^r_{\uparrow\downarrow}$, $G^r_{\downarrow\uparrow}$ and $G^r_{\downarrow\downarrow}$. Based on Eqs. (21) and (22), one may obtain the respective spectral functions, $-2\text{Im}G^r_{\uparrow\uparrow}$, $-2\text{Im}G^r_{\uparrow\downarrow}$, $-2\text{Im}G^r_{\downarrow\uparrow}$, and $-2\text{Im}G^r_{\downarrow\downarrow}$. Spectral functions provide information about the nature of the QD's electronic states which are involved in the tunneling process, regardless whether the states are occupied or not. The spectral functions can be considered as a generalized density of states.

If one neglects the SF-QD or SF-TJ events in the QD-DTJ system, there is no mixing of the spin-up and spin-down electron transport channels. In such QD-DTJ system, the two off-diagonal Green's functions, $G^r_{\sigma\bar{\sigma}}(\sigma = \{\uparrow, \downarrow\})$ become zero [this can be confirmed by considering Eq. (22)], and so are their respective spectral functions. Thus, we focus on the spectral functions corresponding to the diagonal components of the retarded GF matrix. Those spectral functions are analyzed as a function of energy under both parallel and antiparallel configuration of the two FM leads' magnetization, in Figs. 3(a) to (d). A broad peak is observed corresponding to the QD's energy level ($\epsilon = \epsilon_d$). The broad peak can be referred to as "QD resonance". The broadening of the QD resonance is caused by the finite coupling between the QD and the leads, since the QD resonance is a δ function for an isolated QD with no coupling to leads. The width of the QD resonance reflects the strength of coupling between QD and leads; the stronger the coupling is, the broader the energy spread is, hence, a wider peak.

Under zero-bias [shown in Figs. 3 (a) and (b)], one may note three distinct features of the spectral functions:

1. A second resonance peak which corresponds to the leads' potentials, $\mu_L = \mu_R = 0$ eV. The peak can be referred to as the "lead resonance".
2. The lead resonance for the spin-up spectral function ($-2\text{Im}G^r_{\uparrow\uparrow}$) has a broader and lower profile compared to that of the spin-down spectral function, when the QD-DTJ system is in the parallel configuration. This indicates that the excitation at the lead energy has a larger energy spread for spin-up carriers due to the polarization of the lead.
3. The spin-up and spin-down spectral functions are identical in the antiparallel alignment, due to the spin symmetry of the system in antiparallel configuration.

The spectral functions under an finite bias voltage ($V_b = 0.2$ eV) are shown in Figs. 3 (c) and (d). It is observed that,

1. the lead resonance splits into two peaks at the respective left lead and right lead potentials, $\epsilon = \mu_L = 0$ and $\epsilon = \mu_R = -eV_b = -0.2$ eV.
2. In the parallel configuration, the lead resonance of the spin-down electrons is higher (lower) than that of the spin-up electrons at μ_L (μ_R). This is due to the spin-dependence of the electron tunneling between leads and QD.
3. The antiparallel alignment of leads' magnetization gives rise to similar magnitude of the two lead resonances for both spin-up and down spectral functions, due to the spin symmetry of the two spin channels.

Next, one may investigate the SF-TJ effects on the electron transport through the QD-DTJ system, where the SF-TJ strength $v \neq 0$. Figure 4 shows the effect of SF-TJ on the spectral function of diagonal GFs. With the SF-TJ effects, both the QD resonance and the lead

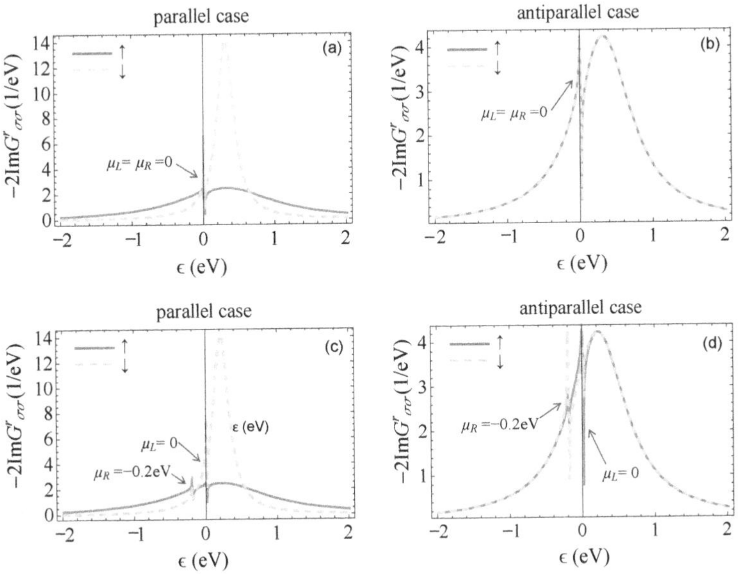

Fig. 3. Spectral functions for spin-up (solid line) and spin-down (dashed line) retarded Green's functions, as a function of electrons' energy, in parallel and antiparallel case. Other parameters are t=0.5 eV, v=0, $\epsilon_{d0} = 0.3$eV, $\rho_0 = 0.7(eV)^{-1}$, $p_L = p_R = 0.7, \lambda = 0, \beta = 1, T =$ 0.3 K.

resonance at $\epsilon = 0$ are enhanced while the lead resonance at $\epsilon = -eV_b$ is suppressed. This indicates that the increasing SF-TJ helps the tunneling to proceed primarily in the vicinity of the QD's energy level, resulting in an effective decrease in the coupling between the same spin states in leads and QD.

Based on the SDT model, one may analyze the effects of the SF-QD events (denoted by λ) on the spectral functions of the diagonal retarded GFs($G^r_{\uparrow\uparrow}$ and $G^r_{\downarrow\downarrow}$) of the QD-DTJ system, for both parallel and antiparallel alignments, as shown in Figure 5. At the QD energy level $\epsilon_d = 0.2$ eV, the presence of the SF-QD causes a symmetric split of the QD resonance, resulting in the suppression of tunneling via the lead resonances. The splitting of the QD resonance indicates that the two effective energy levels within the QD are involved in the tunneling process. This split translates to an additional step in the $I - V$ characteristics, which will be discussed later in Fig. 7.

Considering the off-diagonal GF's ($G^r_{\sigma\bar{\sigma}}$), the spectral functions are ploted in Figure 6, for both parallel and antiparallel alignment, under varying SF-TJ strengths (v) and SF-QD strengths (λ). As shown in Figs. 6(a)-(d), without SF-TJ or SF-QD effects, the off-diagonal spectral functions vanish (the solid lines), i.e., the transport proceeds independently in the spin-up and spin-down channels. The presence of either the SF-TJ ($v > 0$) or the SF-QD ($\lambda \neq 0$) enhances the magnitudes of the off-diagonal spectral functions monotonically, indicating stronger mixing of the tunneling transport through the two spin channels.

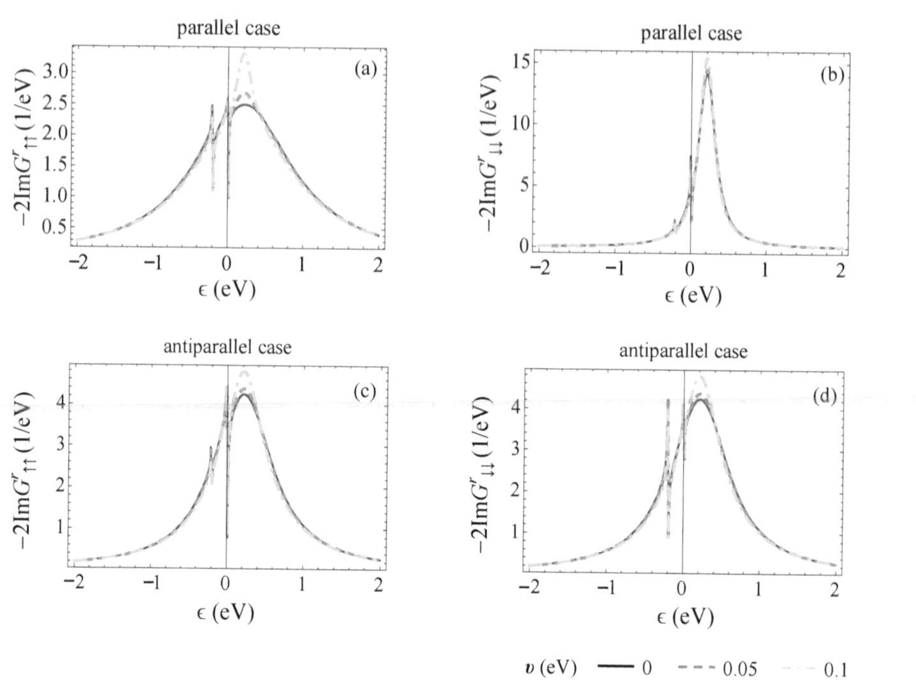

Fig. 4. Spectral functions for the diagonal retarded Green's functions, as a function of electrons' energy, with varied SF-TJ strength (v) between leads and QD, in (a)-(b) parallel and (c)-(d) antiparallel alignment of two leads' magnetization, where $\lambda = 0$ eV, V_b=0.2 V. Other parameters are the same with those in Fig. 3.

The individual effects from SF-TJ or SF-QD on the tunneling current and the TMR are then investigated, as shown in Figs. 7.

The $I - V_b$ characteristics in Figs. 7(a)-(b) and Figs. (d)-(e) show a step at the threshold voltage V_{th}, which is required to overcome the Coulomb blockade (CB). The threshold voltage is given by $V_{th} = 2\epsilon_{d0}$. Considering the bias voltage regions, one may find the following features of the $I - V_b$ characteristics,

1. Within the sub-threshold bias range ($V < V_{th}$), the current is still finite due to thermally assisted tunneling at finite temperature.
2. The sub-threshold current is particularly large in the parallel configuration, due to the stronger lead-QD coupling and hence a greater energy broadening of the QD's level.
3. Overall, the parallel current exceeds the antiparallel current for the entire voltage range considered, due to the nonzero spin polarization of the FM lead.
4. Beyond the threshold voltage (i.e. $V_b \gg V_{th}$), the tunneling current saturates since only a single QD level is assumed to participate in the tunneling transport.

In the presence of SF-TJ, the tunneling currents in the parallel and antiparallel configurations are found to be significantly enhanced for bias voltage exceeding the threshold ($V_b > V_{th}$), as shown in Figs. 7(a) and (b). The enhancement in current stems from the overall stronger coupling between the lead and the QD. Additionally, the degree of enhancement of the

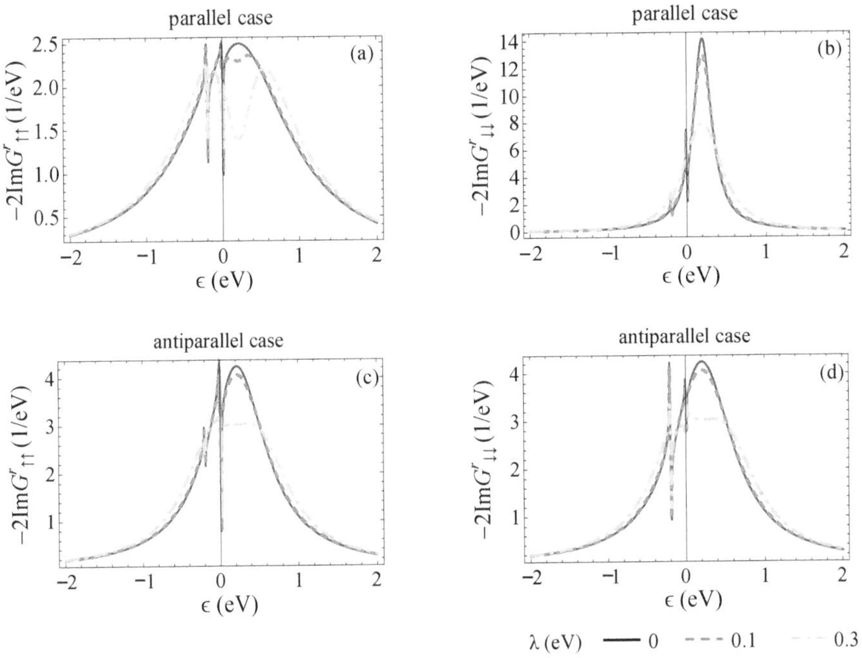

Fig. 5. Spectral function as a function of electrons' energy, with varied SF-QD strength (λ), in (a)-(b) parallel and (c)-(d) antiparallel cases, where $v = 0$ eV. Other parameters are the same with those in Fig. 3.

tunneling current is more pronounced for the parallel alignment of the FM leads. This results in an enhancement of the TMR for the voltage range above the threshold, as shown in Fig. 7(c).

When SF-QD events are present in the system, two new features show up in the $I - V_b$ characteristics, in Figs. 7 (d) and (e). First, the current step at the threshold bias V_{th} splits into two, at $V_b = V_{th} \pm \lambda$, respectively. The presence of the additional step is due to the splitting in the QD resonances observed in the spectral functions of Fig. 5. Secondly, the presence of SF-QD suppresses the current saturation value at large bias voltage (i.e., $V_b \gg V_{th} + \lambda$). The decrease is more pronounced in the antiparallel configuration, resulting in the enhancement of the TMR with the increasing SF-QD probability, as shown in Fig. 7(f).

When both SF processes (Fig. 8) exist in the QD-DTJ system, the two types of SF have competing effects on the tunneling current at large bias voltage exceeding the threshold. The SF-TJ (SF-QD) tends to enhance (suppress) the tunneling current within the bias voltage region exceeding the threshold voltage. This competitive effect is shown for the overall I-V_b characteristics in Figs. 8 (a)-(b). Evidently, the effect caused by one SF mechanism is mitigated by the other for both parallel and antiparallel alignments. However, both SF mechanism contribute to the asymmetry of tunneling current between the parallel and antiparallel cases, leading to an additive effect on the TMR for voltage bias region beyond the threshold voltage, as shown in Fig. 8 (c). The competitive effect on current and collaborative effect on TMR make it possible to attain simultaneously a high TMR and tunneling current density.

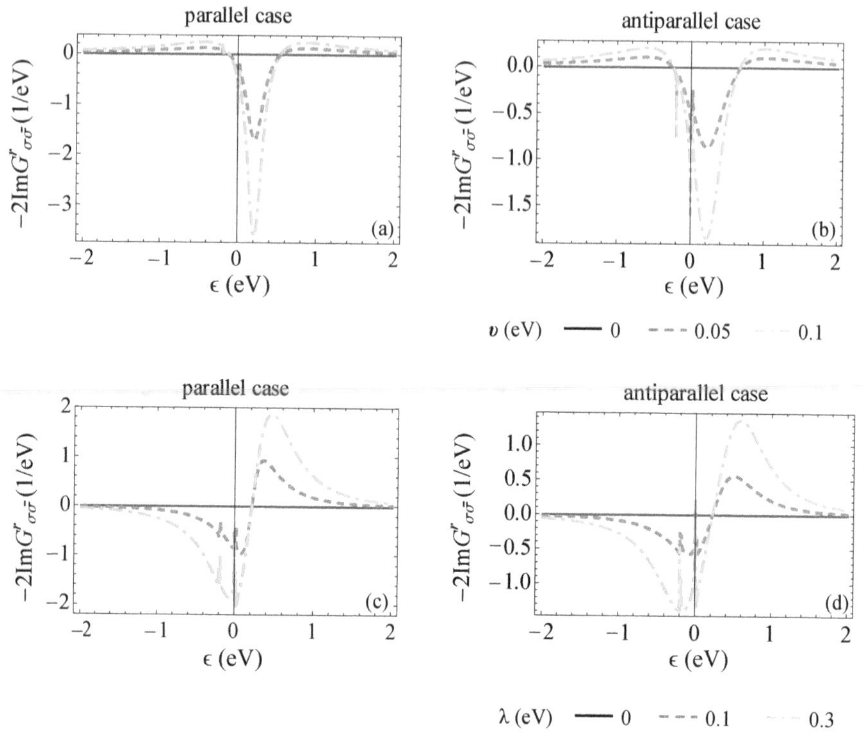

Fig. 6. (a),(b): Off-diagonal spectral functions as a function of energy, for varying SF-TJ strength (v) between leads and QD, in the absence of SF-QD (i.e., $\lambda = 0$ eV).(c),(d):Off-diagonal Spectral functions as a function of electrons' energy, with varied SF-QD strength (λ), in parallel (left) and antiparallel (right) case, where v=0 eV. Other parameters are the same with those in Fig. 3.

2.2.2 Coupling asymmetry effects

Recent experimental studies (Hamaya et al. (2009; 2007)) of QD-DTJ structures revealed that the SDT characteristics are strongly dependent on the coupling asymmetry (CA) between the two junctions. Such asymmetry is inherent in the sandwich structure, given the exponential dependence of the coupling strength on the tunnel barrier width.

One may study the effect of the junction CA on the overall spin and charge current characteristics of the QD-DTJ system. The degree of the CA is characterized by the ratio of the right and left junction coupling parameter. The CA is denoted by β and $\beta = t_{Rk\sigma,\sigma}/t_{Lk\sigma,\sigma}$. The spin-up (spin-down) components of the tunneling current can be represented as $I_\uparrow$ $(I_\downarrow)$, based on which the spin current is defined to be the difference between the two components, $I_s = I_\uparrow - I_\downarrow$. In the following, one may focus on the parallel alignment of the magnetization of the two leads of the QD-DTJ system, since the magnitude of the spin current is the greatest in this case (see Mu et al. (2006)).

For simplicity but without loss of generality, one may assume β to be spin-independent, i.e, $\beta = t_{Rk\uparrow,\uparrow}/t_{Lk\uparrow,\uparrow} = t_{Rk\downarrow,\downarrow}/t_{Lk\downarrow,\downarrow} = t_{Rk\sigma,\sigma}/t_{Lk\sigma,\sigma}$. In Sec. 2.1, the coupling strength is defined as

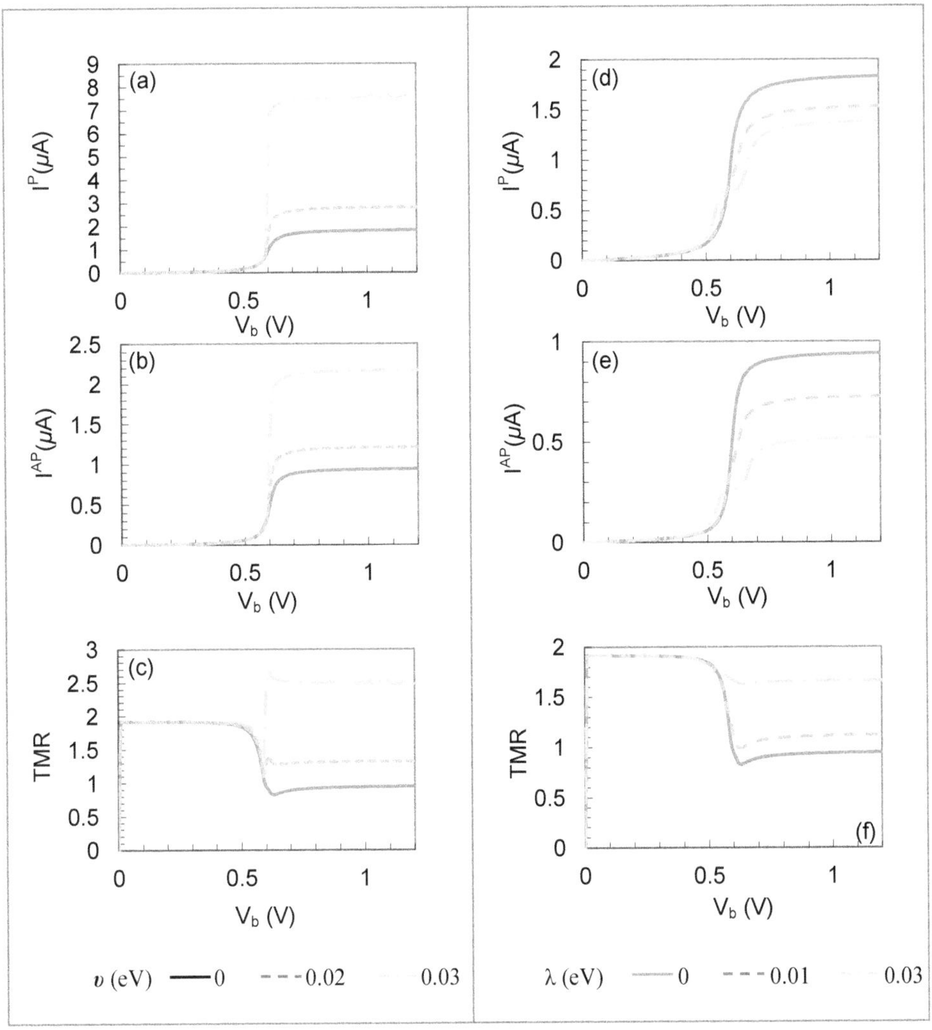

Fig. 7. Current as a function of bias voltage, with varying SF-TJ strength (υ) between the lead and the dot, in (a) parallel and (b) antiparallel cases, or with varying SF-QD strength (λ) in (d) parallel and (e) antiparallel cases. (c)/(f): Tunnel magnetoresistance (TMR) as a function of bias voltage under increasing SF-TJ/SF-QD strength, respectively. In plots (a)-(c), λ=0 eV, and in plots (d)-(f), υ=0, while other parameters are the same with those in Fig. 3.

$\Gamma_{\alpha\sigma} = 2\pi\rho_{\alpha\sigma}|t^*_{\alpha k\sigma,\sigma}\, t_{\alpha k\sigma,\sigma}| = [1+(-1)^\sigma p_\alpha]\, 2\pi\rho_{\alpha 0}|t^*_{\alpha k\sigma,\sigma}\, t_{\alpha k\sigma,\sigma}| = [1+(-1)^\sigma p_\alpha]\,\Gamma_{\alpha 0}$. If assuming identical intrinsic electron density of states and identical polarization of the two leads, i.e., $\rho_{\alpha 0} = \rho_0$, $p_\alpha = p$, one may obtain that $\beta = \sqrt{\Gamma_{R\sigma}/\Gamma_{L\sigma}} = \sqrt{\Gamma_{R0}/\Gamma_{L0}}$.

We consider the *I-V* characteristics for the charge current and spin current, shown in Fig. 9 for two different CA values. These two values were chosen so that $\beta_1 = 1/\beta_2$, meaning

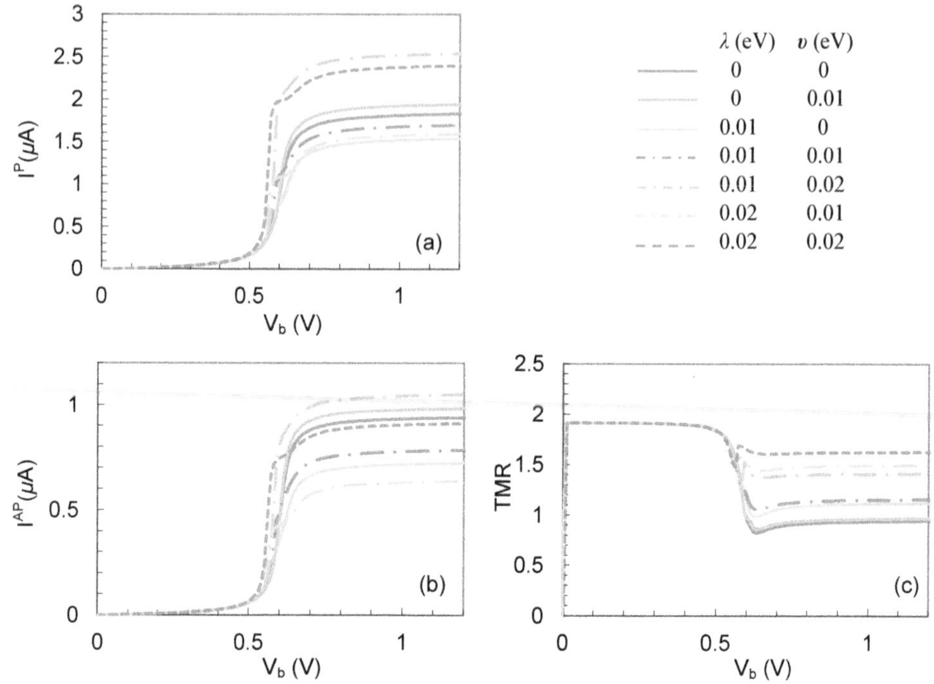

Fig. 8. (a),(b): Current as a function of bias voltage in parallel case and antiparallel case, and (c) TMR as a function of bias voltage, with varying SF-TJ strength (v) and varying SF-QD (λ). Other parameters are the same as those in Fig. 3.

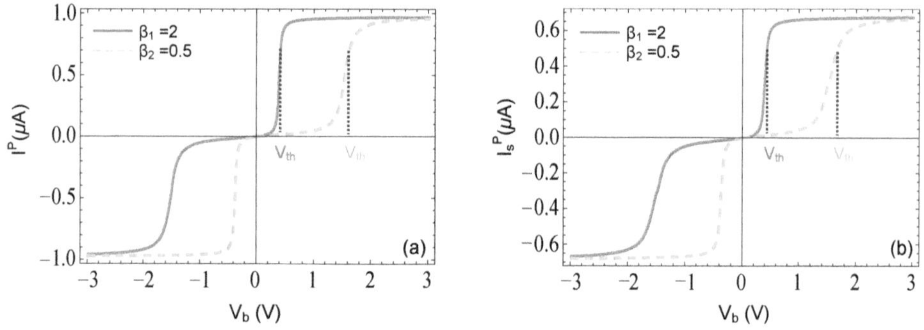

Fig. 9. (a) Charge current I as a function of bias voltage V_b and (b) spin current (I_s) as a function of bias voltage V_b, with two different coupling asymmetry β, in the parallel alignment of the leads' magnetization. The coupling asymmetry is denoted by $\beta = t_{Rk\sigma,\sigma} / t_{Lk\sigma,\sigma} = \sqrt{\Gamma_{R\sigma}/\Gamma_{L\sigma}}$, where $\Gamma_{L\sigma} = (1 \pm p_L)\Gamma_{L0}$ and $\Gamma_{R\sigma} = (1 \pm p_R)\Gamma_{R0}$. Other parameters are $\Gamma_{L0} = 0.012$ eV and $\Gamma_{R0} = \Gamma_{L0} \times {\beta_1}^2$ for β_1 case, $\Gamma_{L0} = 0.006$ eV and $\Gamma_{R0} = \Gamma_{L0} \times {\beta_1}^2$ for β_2 case, $\epsilon_{d0} = 0.3$eV, $p_L = p_R = 0.7$, $T = 100$ K, v=0 eV, λ=0 eV.

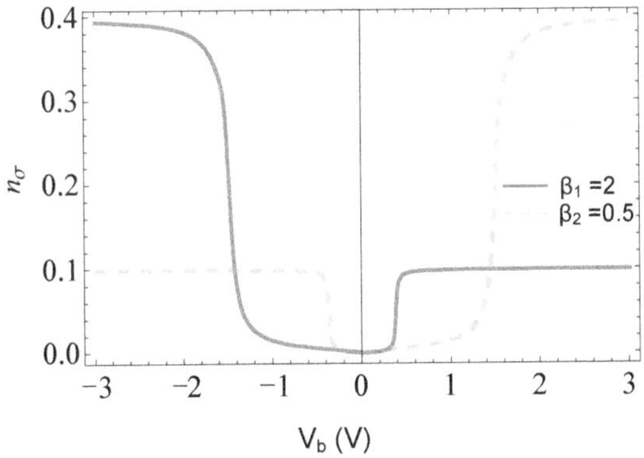

Fig. 10. The occupancy of the quantum dot as a function of bias voltage, with two different coupling asymmetry β, in the parallel alignment of the leadsąŕ magnetization. Other parameters are the same with those Fig. 9.

that the left (right) junction in β_1 system is the right (left) junction for β_2 system. It is found that when the coupling strength of the right junction is four times as strong as that of the left junction, i.e., $\beta = 2$, both the magnitude of the charge and spin currents beyond the threshold voltage are the same as those for the reverse case ($\beta = 0.5$). This is due to the fact that the total resistance of the two QD-DTJ system is maintained regardless of the coupling asymmetry reversal. However, the CA affects the threshold voltage V_{th}. This is due to the different shifts of the QD energy level under positive and negative bias voltage, i.e., $V_{th} = 2\epsilon_d$, where $\epsilon_d = \epsilon_{d0} - \frac{\beta^2}{1+\beta^2} eV_b$. The CA effect on the charge current I-V characteristics is consistent with the experimental results observed by K. Hamaya *et al.* for an asymmetric Co/InAs/Co QD-DTJ system (Fig. 2(a) of Ref. Hamaya et al. (2007)).

Next, one may investigate the CA effect on the QD occupancies, which are obtained by integrating the spectral function in Eq. (15). The QD occupancies for both spin-up and spin-down electrons are shown in Fig. 10. The occupancies for spin-up and spin-down electrons in the QD actually coincide since the QD-DTJ system is operated in the parallel configuration of the leads' magnetization. Moreover, as β is increased from 0.5 to 2, the QD occupancies of both spin orientations decrease. This decrease is reasonable since as Γ_L is decreased with respect to Γ_R, the coupling which allows the electron to tunnel to the QD from the source (left lead) is reduced, while the coupling which allows the electron to tunnel out of the QD to the drain (right lead) is enhanced. In this circumstance, electrons start to have a higher occupancy in the QD for $\beta < 1$ case, where $\Gamma_L > \Gamma_R$.

2.3 Summary

In summary, the SDT through a QD-DTJ system is theoretically studied. In the SDT model described in Sec.2.1, well-separated QD levels are assumed such that only a single energy level are involved in the SDT process, and the correlation between different energy levels is then neglected. The spectral functions, QD electron occupancies, tunneling charge current, spin current as well as TMR are evaluated based on the Keldysh NEGF formulism and EOM method, with consideration of the effects of the SF-TJ events, SF-QD events, and the CA between the two tunnel junctions on the SDT of the system.

3. QD with Zeeman splitting

In the last section, the SDT is studied for the QD-DTJ system where the spin dependence of the electron transport is caused by the spin polarization in the FM leads. In this section, one may analyze the SDT through the QD-DTJ system where the leads sandwiching the QD are non-magnetic (NM), and a FM gate is applied above the QD. The electron transport through this QD-DTJ system is spin-dependent due to the Zeeman splitting (ZS) generated in the QD. In this QD-DTJ system, one may expect a fully polarized current to tunnel through (Recher et al. (2000)). A fully spin-polarized current is important for detecting or generating single spin states (Prinz (1995; 1998)), and thus is of great importance in the realization of quantum computing (Hanson et al. (2007); Kroutvar et al. (2004); Loss & DiVincenzo (1998); Moodera et al. (2007); Petta et al. (2005); Wabnig & Lovett (2009)).

The QD-DTJ system is schematically shown in Fig. 11. The magnetic field generated by the FM gate is assumed to be spatially localized such that it gives rise to a ZS of the discrete energy levels of the QD, but negligible ZS in the energy levels of the NM electrodes. When the bias voltage V_b between the two NM electrodes, and the size of the ZS in the QD are appropriately tuned, a fully polarized spin current is observed in this QD-DTJ system. The polarization of the current depends on the magnetization direction of the FM gate. Here, the down (up)-spin electrons have spins which are aligned parallel (antiparallel) to the magnetization direction of the FM gate.

3.1 Theory

The Hamiltonian of the QD-DTJ system is in form of

$$H = \sum_{\alpha} H_{\alpha} + H_d + H_t, \tag{23}$$

where

$$H_{\alpha} = \sum_{k\sigma} \epsilon_{\alpha k\sigma} a^{\dagger}_{\alpha k\sigma} a_{\alpha k\sigma}, \tag{24}$$

$$H_d = \sum_{\sigma} \epsilon_{\sigma} a^{\dagger}_{\sigma} a_{\sigma} + U n_{\uparrow} n_{\downarrow}, \tag{25}$$

$$H_t = \sum_{\alpha k\sigma} \left(t_{\alpha k\sigma,\sigma} a^{\dagger}_{\sigma} a_{\alpha k\sigma} + t^{*}_{\alpha k\sigma,\sigma} a^{\dagger}_{\alpha k\sigma} a_{\sigma} \right), \tag{26}$$

where $\alpha = \{L, R\}$ is the lead index for the left and right leads, k is the momentum, $\sigma = \{\uparrow, \downarrow\}$ is the spin-up and spin-down index, $a^{\dagger}$ and a are the electron creation and annihilation operators, ϵ_{σ} is the energy level in the QD for electrons with spin-σ, U is the Coulomb blockade energy when the QD is doubly occupied by two electrons with opposite spins, and $t_{\alpha k\sigma,\sigma}$ describes the coupling between the electron states with spin-σ in the lead α and the QD.

In our model, we consider only the lowest unoccupied energy level of the QD ϵ_{σ} since most of the overall transport occurs via that level. With the presence of an applied magnetic field B, the lowest unoccupied energy level is given by $\epsilon_{\sigma} = \epsilon_d + \frac{(-1)^{\sigma}}{2} g\mu_B B$, where $\sigma = 0$ ($\sigma = 1$) for up-spin (down-spin) electrons, g is the electron spin g-factor, μ_B is the Bohr magneton, and $g\mu_B B$ is the ZS between the two spin states $\epsilon_{\downarrow}$ and $\epsilon_{\uparrow}$. Under an applied bias V_b between the

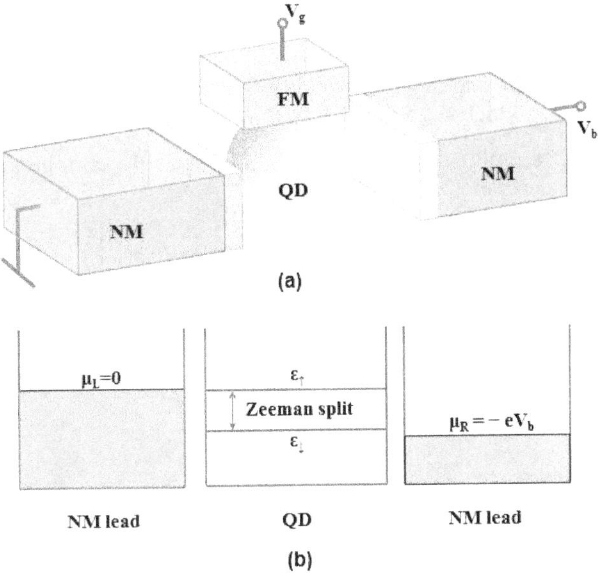

Fig. 11. (a) Schematic diagram of the QD-DTJ set up, which consists of two NM electrodes, one QD and one FM gate. (b) schematic energy diagram of the QD-DTJ system in (a), where V_b is the bias voltage, $\epsilon_\downarrow = \epsilon_d - g\mu_B B/2$ ($\epsilon_\uparrow = \epsilon_d + g\mu_B B/2$) is the energy level for spin-down(up) electrons, respectively, $g\mu_B B$ is the Zeeman splitting between $\epsilon_\downarrow$ and $\epsilon_\uparrow$, g is the electron spin g-factor, μ_B is the Bohr magneton, B is the applied magnetic field generated by the FM gate, and $\epsilon_d = \epsilon_{d0} - eV_b\beta^2/(1+\beta^2)$ is the single energy level of the QD without applied magnetic field, with ϵ_{d0} being the single energy level under zero bias voltage and β being the coupling asymmetry between the two tunnel junctions. We assume a symmetrical QD-DTJ system where $\beta = 1$.

leads and in the absence of B-field, the QD's energy level is modified as $\epsilon_d = \epsilon_{d0} - eV_b\beta^2/(1+\beta^2)$, where ϵ_{d0} is the energy level at zero bias voltage, and $\beta = t_{Rk\sigma,\sigma}/t_{Lk\sigma,\sigma}$ denotes the asymmetry of the coupling in the left and right tunnel junctions. In the following, a symmetric QD-DTJ system is assumed where $\beta = 1$.

Based on the Hamiltonian, the tunneling current is evaluated via the NEGF formalism introduced in Section. 2.1. The charge and spin current are defined as $I_c = I_\downarrow + I_\uparrow$ and $I_s = I_\downarrow - I_\uparrow$, respectively, where the tunneling current of spin-σ electron tunneling through the QD-DTJ system is given by

$$I_\sigma = \frac{e}{h}\int d\epsilon \left[f_{L\sigma}(\epsilon) - f_{R\sigma}(\epsilon)\right] G^a_{\sigma,\sigma}\Gamma_{R\sigma}G^r_{\sigma,\sigma}\Gamma_{L\sigma}. \tag{27}$$

Here, $\Gamma_{\alpha\sigma}(\epsilon) = 2\pi\rho_{\alpha\sigma}(\epsilon)t_{\alpha\sigma,\sigma}(\epsilon)t^*_{\alpha\sigma,\sigma}(\epsilon)$, $G^a_{\sigma,\sigma}(\epsilon) = [G^r_{\sigma,\sigma}(\epsilon)]^*$, $G^r_{\sigma,\sigma}(t) = -i\theta(t)\langle\{a_\sigma(t), a^\dagger_\sigma\}\rangle$.

The explicit form of $G^r_{\sigma,\sigma}(\epsilon)$ is given by

$$G^r_{\sigma,\sigma}(\epsilon) = \frac{1 - n_{\bar{\sigma}}}{\epsilon + i\eta - \epsilon_\sigma - \Sigma(\epsilon)} + \frac{n_{\bar{\sigma}}}{\epsilon + i\eta - \epsilon_{\bar{\sigma}} - \Sigma(\epsilon)}, \tag{28}$$

where the self energy terms are $\Sigma(\epsilon) = \sum_{\{\alpha,k\}} \frac{t_{\alpha k\sigma,\sigma} t^*_{\alpha k\sigma,\sigma}}{\epsilon + i\eta - \epsilon_{\alpha k\sigma}}$. The coupling coefficients $t_{\alpha k\sigma,\sigma}$ and $t^*_{\alpha k\sigma,\sigma}$ are spin-independent since the two leads are NM.

Based on Eqs. (27) and (28), one can then calculate the spin-σ current $I_\uparrow$ and $I_\downarrow$, and hence the charge and spin current, which are defined as $I_c = I_\downarrow + I_\uparrow$ and $I_s = I_\downarrow - I_\uparrow$, respectively.

In the EOM method, the following Hartree-Fock decoupling decoupling approximation of (Lacroix (1981); Sergueev et al. (2002)) is applied,

$$\left\langle\left\{a^\dagger_{k\alpha\sigma} a_\sigma a_{k\alpha\bar{\sigma}}, a^\dagger_{\bar{\sigma}}\right\}\right\rangle = \left\langle a^\dagger_{k\alpha\sigma} a_\sigma\right\rangle \left\langle\left\{a_{k\alpha\bar{\sigma}} a^\dagger_{\bar{\sigma}}\right\}\right\rangle, \tag{29}$$

$$\left\langle\left\{a^\dagger_{\sigma} a_{k\alpha\sigma} a_{k\alpha\bar{\sigma}}, a^\dagger_{\bar{\sigma}}\right\}\right\rangle = \left\langle a^\dagger_{\sigma} a_{k\alpha\sigma}\right\rangle \left\langle\left\{a_{k\alpha\bar{\sigma}} a^\dagger_{\bar{\sigma}}\right\}\right\rangle, \tag{30}$$

$$\left\langle\left\{a_\sigma a^\dagger_{k\alpha\sigma} a_{\bar{\sigma}}, a^\dagger_{\bar{\sigma}}\right\}\right\rangle \simeq 0, \tag{31}$$

$$\left\langle\left\{a_{k\alpha\sigma} a^\dagger_{\sigma} a_{\bar{\sigma}}, a^\dagger_{\bar{\sigma}}\right\}\right\rangle \simeq 0, \tag{32}$$

$$\left\langle a^\dagger_{k\alpha\sigma} a_\sigma\right\rangle = \left\langle a^\dagger_{\sigma} a_{k\alpha\sigma}\right\rangle \simeq 0. \tag{33}$$

3.2 Spin polarized current

Based on the SDT model in Sec. 3.1, one may obtain the $I-V$ characteristics of the system for both spin current I_s and charge current I_c, as shown in Fig. 12.

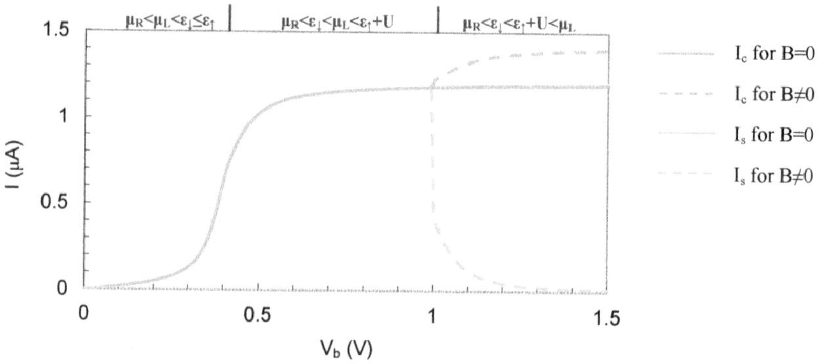

Fig. 12. Charge current (I_c) and spin current (I_s) as a function of the bias voltage, with ($B \neq 0$) or without ($B = 0$) ZS. $V_g = 0$. The following parameters are assumed: lowest unoccupied energy level in the QD under zero bias voltage ϵ_{d0}=0.2eV, the Coulomb blockade energy $U = 0.26$ eV, the Zeeman splitting due to the FM gate is $g\mu_B B = 0(0.36$ meV) for $B = 0$ ($B \neq 0$) case, the gate voltage $V_g = 0$, and temperature $T = 3$ K.

In the absence of a FM gate, i.e., with zero magnetic field ($B = 0$) applied to the QD, the magnitude of the charge current I_c is the same as that of the system with a FM gate, within the bias region $\mu_R < \epsilon_\downarrow < \mu_L < \epsilon_\uparrow + U$. In this region, the spin current I_s is zero for the system without a FM gate, where the device is spin-symmetric and the transport across it is spin-independent.

For the system with a FM gate, both the charge current I_c and spin current I_s of the system show the three distinct regions with respect to the bias voltage,

1. $\mu_R < \mu_L < \epsilon_\downarrow < \epsilon_\uparrow$, where both I_c and I_s are negligible due to the suppression of electron tunneling by Coulomb blockade;
2. $\mu_R < \epsilon_\downarrow < \mu_L < \epsilon_\uparrow + U$, where due to spin blockade, only the spin-down channel contributes to the transport across the system, resulting in a fully spin-down polarized current with $I_s = I_c$;
3. and $\mu_R < \epsilon_\downarrow < \epsilon_\uparrow + U < \mu_L$, where it is energetically favorable for both types of spins to tunnel across the device, leading to zero spin current.

The sign of the spin polarization of the tunneling current can be *electrically* modulated, i.e., by means of a gate voltage V_g. The gate voltage modulation of the QD energy level ϵ_d can result in the switching of the spin polarization of current, without requiring any corresponding change to the magnetization of the FM gate. If the energy diagram of the system satisfies $\epsilon_\downarrow - eV_g < \mu_R < \epsilon_\uparrow - eV_g < \mu_L$, a fully spin-up polarized current will thus flow continuously through the system.

3.3 Summary

In summary, the SDT through a QD-DTJ system is analyzed with NM leads and FM gate. Under the applied magnetic field from the FM gate, the energy level in the QD splits to two due to ZS effect. The two energy levels can be modulated by the gate voltage applied to the FM gate. Based on the SDT model developed by NEGF formulism and EOM method, the $I - V_b$ characteristics is analyzed, and a fully spin-down polarized current is obtained when the system is operated under a proper bias voltage between the two leads. Additionally, by utilizing the gate voltage modulation instead of switching the magnetization of the FM gate, the polarization of the current can be reversed from spin-down to spin-up by electrical means.

4. Conclusion

In conclusion, the SDT is theoretically studied for the QD-DTJ systems where a QD is sandwiched by two adjacent leads. The tunneling current through the systems is shown to be rigorously derived via the Keldysh NEGF approach and EOM method. The SF events, CA, ZS and FM gating are systematically incorporated in the SDT models. Considering these effects, one may analyze the SDT properties of QD-DTJ systems, including the tunneling current (charge current and spin current), the TMR, the spectral functions and the occupancies of the QD. The SF-TJ and the SF-QD events are found to have competitive effects on the tunneling current. The presence of CA effectively modifies the threshold voltage, and gives rise to additional bias voltage dependence of the QD's electron occupancy and the charge and spin currents. The FM gate attached to the QD can be utilized to generate a bipolar spin polarization of the current through QD-DTJ systems. The above investigations done have yielded a better understanding of the SDT in QD-DTJ systems.

5. Acknowledgement

We gratefully acknowledge the SERC Grant No. 092 101 0060 (R-398-000-061-305) for financially supporting this work.

6. List of Abbreviations

CA	coupling asymmetry
CB	Coulomb blockade
DOS	density of states
DTJ	double tunnel junction
EOM	equation of motion
FM	ferromagnetic
GF	Green's function
ME	master equation
NEGF	nonequilibrium Green's function
NM	non-magnetic
QD	quantum dot
SDT	spin dependent transport
SF	spin-flip
TMR	tunnel magnetoresistance
ZS	Zeeman splitting

7. References

Bao, Y. J., Tong, N. H., Sun, Q. F. & Shen, S. Q. (2008). Conductance plateau in quantum spin transport through an interacting quantum dot, *Europhys. Lett.* 83: 37007:1–5.

Braig, S. & Brouwer, P. W. (2005). Rate equations for coulomb blockade with ferromagnetic leads, *Phys. Rev. B* 71: 195324:1–9.

Bruus, H. & Flensberg, K. (2004). *Many-body Quantum Theory in Condensed Matter Physics*, Oxford University Press, New York.

Caroli, C., Combescot, R., Nozieres, P. & Saint-James, D. (1971). Direct calculation of the tunneling current, *J. Phys. C* 4: 916.

Deshmukh, M. M. & Ralph, D. C. (2002). Using single quantum states as spin filters to study spin polarization in ferromagnets, *Phys. Rev. Lett.* 89: 266803:1–4.

Fransson, J. & Zhu, J.-X. (2008). Spin dynamics in a tunnel junction between ferromagnet, *New J. Phys.* 10: 013017:1–9.

Hamaya, K., Kitabatake, M., Shibata, K., Jung, M., Ishida, S., Taniyama, T., Hirakawa, K., Arakawa, Y. & Machida, T. (2009). Spin-related current suppression in a semiconductor quantum dot spin-diode structure, *Phys. Rev. Lett.* 102: 236806:1–4.

Hamaya, K., Masubuchi, S., Kawamura, M., Machida, T., Jung, M., Shibata, K., Hirakawa, K., Taniyama, T., Ishida, S. & Arakawa, Y. (2007). Spin transport through a single self-assembled inas quantum dot with ferromagnetic leads, *Appl. Phys. Lett.* 90: 053108.

Hanson, R., Kouwenhoven, L. P., Petta, J. R., Tarucha, S. & Vandersypen, L. M. K. (2007). *Rev. Mod. Phys.* 79: 1217.

Jauho, A. P., Wingreen, N. S. & Meir, Y. (1994). Time-dependent transport in interacting and noninteracting resonant-tunneling systems, *Phys. Rev. B* 50: 5528–5544.

Katsura, H. (2007). Nonequilibrium kondo problem with spin-dependent chemical potentials: Exact results, *J. Phys. Soc. Jpn* 76: 054710.

Kroutvar, M., Ducommun, Y., Heiss, D., Bichler, M., Schuh, D., Abstreiter, G. & Finley, J. J. (2004). Optically programmable electron spin memory using semiconductor quantum dots, *Nature* 432: 81–84.

Kuo, D. M. T. & Chang, Y. (2007). Tunneling current spectroscopy of a nanostructure junction involving multiple energy levels, *Phys. Rev. Lett.* 99: 086803:1–4.

Lacroix, C. (1981). Density of states for the anderson model, *J. Phys. F: Met. Phys.* 11: 2389.

Lin, K.-C. & D.-S.Chuu (2005). Anderson model with spin-flip-associated tunneling, *Phys. Rev. B* 72: 125314:1–11.

Lobo, T., Figueira, M. S. & Foglio, M. E. (2006). Kondo effect in a quantum dot-the atomic approach, *Nanotechnology* 17: 6016.

Loss, D. & DiVincenzo, D. P. (1998). Quantum computation with quantum dots, *Phys. Rev. A* 57: 120–126.

Ma, M. J., Jalil, M. B. A. & Tan, S. G. (2008). Sequential tunneling through a two-level semiconductor quantum dot system coupled to magnetic leads, *J. Appl. Phys.* 104: 053902:1–6.

Ma, M. J., Jalil, M. B. A. & Tan, S. G. (2010). Coupling asymmetry effect on the coherent spin current through a ferromagnetic lead/quantum dot/ferromagnetic lead system, *IEEE Trans. Magnetics* 46: 1495.

Mahan, G. D. (1990). *Many-Particle Physics*, 2nd edn, Plenum, New York.

Manassen, Y., Mukhopadhay, I. & Rao, N. R. (2001). Electron-spin-resonance stm on iron atoms in silicon, *Phys. Rev. B* 61: 16223–16228.

Meir, Y. & Wingreen, N. S. (1992). Landauer formula for the current through an interacting electron region, *Phys. Rev. Lett.* 68: 2512–2515.

Meir, Y., Wingreen, N. S. & Lee, P. A. (1991). Transport through a strongly interacting electron system: Theory of periodic conductance oscillations, *Phys. Rev. Lett.* 66: 3048–3051.

Meir, Y., Wingreen, N. S. & Lee, P. A. (1993). Low-temperature transport through a quantum dot: The anderson model out of equilibrium, *Phys. Rev. Lett.* 70: 2601–2604.

Moodera, J. S., Santos, T. S. & Nagahama, T. (2007). The phenomena of spin-filter tunnelling, *J. Phys.: Condens. Matter* 19: 165202.

Mu, H. F., Su, G. & Zheng, Q. R. (2006). Spin current and current-induced spin transfer torque in ferromagnet-quantum dot-ferromagnet coupled systems, *Phys. Rev. B* 73: 054414:1–9.

Pasupathy, A., Bialczak, R., Martinek, J., Grose, J., Donev, L., McEuen, P. & Ralph, D. C. (2004). The kondo effect in the presence of ferromagnetism, *Science* 306: 86–89.

Petta, J. R., Johnson, A. C., Taylor, J. M., Laird, E. A., Yacoby, A., Lukin, M. D., Marcus, C. M., Hanson, M. P. & Gossard, A. C. (2005). Coherent manipulation of coupled electron spins in semiconductor quantum dots, *Science* 309: 2180–2184.

Potok, R. M., Rau, I. G., Shtrikman, H., Oreg, Y. & Goldhaber-Gordon, D. (2007). Observation of the two-channel kondo effect, *Nature* 446: 167–171.

Prinz, G. (1995). Spin-polarized transport, *Phys. Today* 48: 58.

Prinz, G. (1998). Magnetoelectronics, *Science* 282: 1660–1663.

Qi, Y., Zhu, J.-X., Zhang, S. & S.Ting, C. (2008). Kondo resonance in the presence of spin-polarized currents, *Phys. Rev. B* 78: 045305:1–5.

Qu, F. & Vasilopoulos, P. (2006). Spin transport across a quantum dot doped with a magnetic ion, *Appl. Phys. Lett.* 89: 122512:1–3.

Recher, P., Sukhorukov, E. V. & Loss, D. (2000). Quantum dot as spin filter and spin memory, *Phys. Rev. Lett.* 85: 1962–1965.

Rudziński, W. & Barnaś, J. (2001). Tunnel magnetoresistance in ferromagnetic junctions: Tunneling through a single discrete level, *Phys. Rev. B* 64: 085318:1–10.

Sergueev, N., Sun, Q. F., Guo, H., Wang, B. G. & Wang, J. (2002). Spin-polarized transport through a quantum dot: Anderson model with on-site coulomb repulsion, *Phys. Rev. B* 65: 165303.

Shaji, N., Simmons, C. B., Thalakulam, M., Klein, L. J., Qin, H., Luo, H., Savage, D. E., Lagally, M. G., Rimberg, A. J., Joynt, R., Friesen, M., Blick, R. H., Coppersmith, S. N. & Eriksson, M. A. (2008). Spin blockade and lifetime-enhanced transport in a few-electron si/sige double quantum dot, *Nature Physics* 4: 540–544.

Souza, F. M., Egues, J. C. & Jauho, A. P. (2004). Tmr effect in a fm-qd-fm system, *Brazilian Journal of Physics* 34: 565–567.

Wabnig, J. & Lovett, B. W. (2009). A quantum dot single spin meter, *New J. Phys.* 11: 043031.

Weymann, I. & Barnaś, J. (2007). Cotunneling through quantum dots coupled to magnetic leads: Zero-bias anomaly for noncollinear magnetic configurations, *Phys. Rev. B* 75: 155308:1–12.

Zhang, P., Xue, Q. K., Wang, Y. P. & Xie, X. C. (2002). Spin-dependent transport through an interacting quantum dot, *Phys. Rev. Lett.* 89: 286803:1–4.

Zhu, J.-X. & Balatsky, A. V. (2002). Quantum electronic transport through a precessing spin, *Phys. Rev. Lett.* 89: 286802:1–4.

Non-Equilibrium Green Functions of Electrons in Single-Level Quantum Dots at Finite Temperature

Nguyen Bich Ha
Institute of Materials Science, Vietnam Academy of Science and Technology, Cau Giay, Hanoi Vietnam

1. Introduction

In the quantum field theory with the vacuum being the ground state the Green functions are the vacuum expectation values of the chronological, retarded or advanced products of the field operators (Bjoken & Drell, 1964; Itzykson & Zuber, 1985; Peskin & Schroeder, 1995). They are the generalized functions of the real time variables t_i (and also other spatial coordinates). For the application of the Green function technique to the study of the time-independent phenomena in equilibrium many-body systems at a finite temperature, the Matsubara imaginary time Green functions were introduced and widely used (Abrikosov et al., 1975; Bruuns & Flensberg, 2004; Haken, 1976). They are the mean values over a statistical ensemble at a finite temperature of the chronological products of the imaginary time-dependent operators. Both these types of Green functions are inadequate for the application to the study of the time-dependent phenomena in the many-body systems with a finite density and at a finite temperature, in particular the non-equilibrium systems. For the application to the study of the time-dependent dynamical processes in non-equilibrium many-body systems Keldysh (Keldysh, 1965) has introduced a more general class of time-dependent Green functions at finite temperature and density. They are the mean values of the time-ordered products of quantum operators in the Heisenberg picture over statistical ensembles of many-body systems with finite densities and at finite temperatures (which may be non-vanishing). The simplest example is the two-point Green function

$$G_{ab}(t) = -i\langle T[a(t)b(0)]\rangle = -i\frac{Tr\{e^{-\beta H}T[a(t)b(0)]\}}{Tr\{e^{-\beta H}\}} , \tag{1.1}$$

where $a(t)$ and $b(t)$ are two quantum operators in the Heisenberg picture, H is the total Hamiltonian of the system, β and T are the Boltzmann constant and the temperature.

Having shown that the Green functions at finite density and temperature of the form (1.1) can be analytically continued with respect to the time variable t to become the functions of a complex variable z analytical in the stripe $-\beta < \mathrm{Im}\ z < 0$ parallel to the real axis, Keldysh (Keldysh, 1965) has proposed to consider these functions as the quantum statistical averages of the linear combinations of the products of ordered operators depending on complex variables as complex times. For the definition of the ordering of the complex variables z, z' it was proposed to use some contour C in above-mentioned stripe with some initial point t_0 on the real axis and the final point $t_0 - i\beta$ such that all the complex numbers z, z'... belong to this contour. Then the "chronological" ordering T_C of the complex times z, z' ... is defined as the ordering along the contour C. The complex time-dependent operators $a(z)$, $b(z)$ and $\bar{a}(z), \bar{b}(z)$, for example, are defined in the analogy with the operators in the Heisenberg picture

$$\begin{aligned} a(z) &= e^{iHz}a(0)e^{-iHz}, \\ b(z) &= e^{iHz}b(0)e^{-iHz}, \\ \bar{a}(z) &= e^{iHz}a^{+}(0)e^{-iHz}, \\ \bar{b}(z) &= e^{iHz}b^{+}(0)e^{-iHz}. \end{aligned} \tag{1.2}$$

As the generalization of formula (1.1) one defines the two-point Green function of two operators $a(z)$ and $b(z')$, for example, depending on two complex times $z, z' \in C$, as follows:

$$G_{ab}(z - z')_C = -i\langle T_C[a(z)b(z')]\rangle = -i\frac{Tr\{e^{-\beta H}T_C[a(z)b(z')]\}}{Tr\{e^{-\beta H}\}}, \tag{1.3}$$

T_C denoting the "chronological" ordering along the contour C. The Green functions of the form (1.3), usually called the Keldysh complex time-dependent Green functions at finite density and temperature, some time also simply called non-equilibrium Green functions, are widely used in quantum statistical physics and many-body theories (Chou et al., 1985; Kapusta, 1989; Le Bellac, 1996).

In practice we need to know the Green functions at the real values of the time variables. For the convenience we chose the contour C to consists of four parts $C = C_1 \cup C_2 \cup C_3 \cup C_\infty$, C_1 being the part of the straight line over and infinitely close to the real axis from some point $t_0 + io$ to infinity $+\infty + io$, C_2 being the part of the straight line under and infinitely close to the real axis from infinity $+\infty - io$ to the point $t_0 - io$, C_3 and C_∞ being the segments $[t_0, t_0 - i\beta]$ and $[+\infty + io, +\infty - io]$ parallel to the axis Oy (figure 1).

The contributions of the segment $[+\infty + io, +\infty - io]$ to all physical observables are negligibly small, because of its vanishing length. Therefore this segment plays no role, and the contour C can be considered to consist of only three parts C_1, C_2 and C_3. Then the function $G_{ab}(z - z')_C$ with the complex time variables z and z' on the contour C effectively becomes a set of nine functions of two variables, each of which has the values on one among three segments C_1, C_2 and C_3. When both variables z and z' belong to the line C_1, the function (1.3) is the quantum statistical average of the usual chronological product of two quantum operators $a(t)$ and $b(t')$ in the Heisenberg picture over a statistical ensemble of a many-body system at finite density and temperature, and can be denoted by $G_{ab}(t - t')_{11}$. When both variables z and z' belong to the line C_3, the function (1.3) is reduced to the Matsubara imaginary time Green function and can be denoted $G_{ab}(-i\tau + i\tau')_{33}$.

In the study of stationary physical processes one often uses the complex time Green functions of the form (1.3) in the limit $t_0 \to -\infty$. Because the interaction must satisfy the "adiabatic hypothesis" and therefore vanishes at this limit, the segment C_3 also gives no contribution to the stationary physical processes. In this case the contour C can be considered to consist of only two segments C_1 and C_2, and the complex time Green function (1.3) effectively becomes a set of four functions of real variables $G_{ab}(t-t')_{11}$, $G_{ab}(t-t')_{12}$, $G_{ab}(t-t')_{21}$, $G_{ab}(t-t')_{22}$.

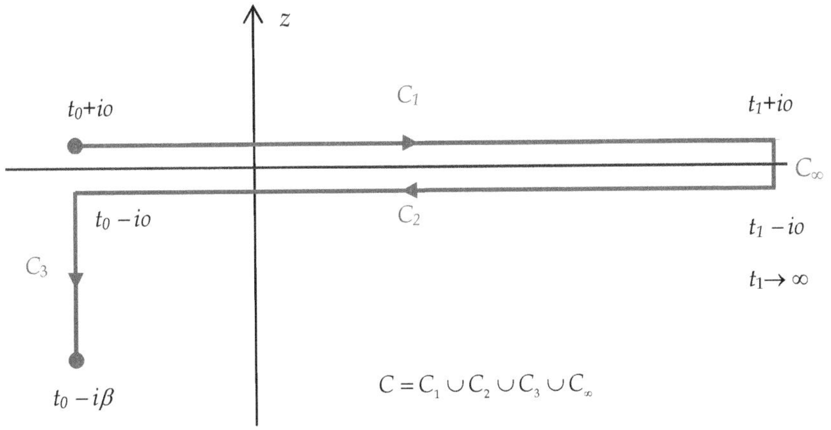

Fig. 1. Contour C consists of four parts $C=C_1\cup C_2\cup C_3\cup C_\infty$.

The electrons transport through a single-level quantum dot (QD) connected with two conducting leads has been the subject for theoretical and experimental studies in many works since the early days of nanophysics (Choi et al., 2004; Costi et al., 1994; Craco & Kang, 1999; Fujii & Ueda, 2003; Hershfield et al., 1991; Inoshita et al., 1993; Izumida et al., 1997, 1998, 2001; Konig & Gefen, 2005; Meir et al., 1991, 1993; Ng, 1993; Nguyen Van Hieu & Nguyen Bich Ha, 2005, 2006; Nguyen Van Hieu et al., 2006a, 2006b; Pustilnik & Glasman, 2004; Sakai et al., 1999; Swirkowicz et al., 2003, 2006; Takagi & Saso, 1999a, 1999b; Torio et al., 2002; Wingreen & Meir, 1994; Yeyati et al., 1993). Two observable physical quantities, which can be measured in experiments on electron transport, are the electron current through the QD and the time-averaged value of the electron number in the QD. Both can be expressed in terms of the single-electron Green functions. In the pioneering theoretical works (Meir et al., 1991, 1993) on the electron transport through a single-level QD, the differential equations for the non-equilibrium Green functions were derived with the use of the Heisenberg equations of motion for the electron destruction and creation operators. Due to the presence of the strong Coulomb interaction between electrons in the QD, the differential equations for the single-electron Green functions contain multi-electron Green functions, and all the coupled equations for these Green functions form an infinite system of differential equations. In order to have a finite closed system of equations, one can assume some approximation to decouple the infinite system of equations. Moreover, since the electron transport is a non-equilibrium process, one should work with the Keldysh formalism of non-equilibrium complex time Green functions.

As the simplest explanation of the calculation methods for establishing the differential equations of non-equilibrium Green functions and deriving their exact solutions, in Section 2 we present the theory of non-equilibrium Green functions of free electron in a single-level quantum system. In Section 3 we study non-equilibrium Green functions of interacting electron in an isolated single-level QD. The elaborated calculation methods are then applied in Section 4 to the study of non-equilibrium Green functions of electrons in a single-level QD connected with two conducting leads. Due to the electron tunneling between QD and conducting leads there does not exist a closed finite system of differential equations for some finite number of Green functions. In order to truncate the infinite system of differential equations for the infinite number of Green functions we can apply some suitable approximation. In Section 5 the mean-field approximation was used to truncate the infinite system of differential equations for the Green functions. As the result we establish a closed system of Dyson equations for a finite number of Green functions. This system of differential equations can be exactly solved. The asymptotic analytical expressions of these Green functions at the resonances, Kondo and Fano resonances, are derived in Section 6. Section 7 is the Conclusion.

2. Non-equilibrium Green function of free electrons in a single-level quantum system

For the demonstration of the calculation methods to derive the differential equations and the expressions of the non-equilibrium Green functions let us consider a simplest quantum system - that of free electrons at a single energy level E. Denote by c_σ and c_σ^+ the destruction and creation operators of the electron with the spin projection $\sigma = \uparrow, \downarrow$ in the Schrödinger picture and by H_0 the Hamiltonian of this system. We have

$$H_0 = E \sum_\sigma c_\sigma^+ c_\sigma \,. \tag{2.1}$$

The non-equilibrium Green function of electron system with Hamiltonian (2.1) is defined as follows:

$$S_{\sigma\sigma'}^{E}(z-z')_C = -i\langle T_C[c_\sigma(z)\overline{c}_{\sigma'}(z')]\rangle = -i\frac{Tr\{e^{-\beta H_0}T_C[c_\sigma(z)\overline{c}_{\sigma'}(z')]\}}{Tr\{e^{-\beta H_0}\}}\,, \tag{2.2}$$

where

$$\begin{aligned} c_\sigma(z) &= e^{iH_0 z} c_\sigma e^{-iH_0 z}\,, \\ \overline{c}_\sigma(z') &= e^{iH_0 z'} c_\sigma^+ e^{-iH_0 z'}\,. \end{aligned} \tag{2.3}$$

Note that at the real values t of the time variable we have $\overline{c}_\sigma(t) = c_\sigma^+(t) = c_\sigma(t)^+$. Complex time-dependent operators $c_\sigma(z)$ and $\overline{c}_\sigma(z)$ satisfy Heisenberg quantum equation of motion

$$\begin{aligned} i\frac{dc_\sigma(z)}{dz} &= -[H_0, c_\sigma(z)]\,, \\ i\frac{d\overline{c}_\sigma(z)}{dz} &= -[H_0, \overline{c}_\sigma(z)]\,. \end{aligned} \tag{2.4}$$

From the canonical anti-commutation relations

$$\begin{aligned} \{c_\sigma, c_{\sigma'}^+\} &= \delta_{\sigma\sigma'}\,, \\ \{c_\sigma, c_{\sigma'}\} &= \{c_\sigma^+, c_{\sigma'}^+\} = 0 \end{aligned} \tag{2.5}$$

it follows that

$$i\frac{dc_\sigma(z)}{dz} = Ec_\sigma(z)\,,$$
$$i\frac{d\overline{c}_\sigma(z)}{dz} = -E\overline{c}_\sigma(z)\,. \tag{2.6}$$

Green function $S^E_{\sigma\sigma'}(z-z')_C$ with both variables z and z' ranging over contour C is a set of nine functions $S^E_{\sigma\sigma'}(z-z')_{ij}$ with the variable $z(z')$ ranging over the segment $C_i(C_j)$. All they have the form

$$S^E_{\sigma\sigma'}(z-z')_C = \delta_{\sigma\sigma'}S^E(z-z')_C\,,$$
$$S^E_{\sigma\sigma'}(z-z')_{ij} = \delta_{\sigma\sigma'}S^E(z-z')_{ij}\,. \tag{2.7}$$

First consider three cases when both variables z and z' belong to one and the same segment C_i, i = 1, 2, 3. For $z,z'\in C_1, ,z=t+io, ,z'=t'+io,$ T_C is the usual chronological ordering T of the real times t and t':

$$T[c_\sigma(t)c^+_{\sigma'}(t')] = \theta(t-t')c_\sigma(t)c^+_{\sigma'}(t') - \theta(t'-t)c^+_{\sigma'}(t')c_\sigma(t).$$

Using one of equations (2.6) and one of anti-commutation relations (2.5), we derive the differential equation for $S^E_{\sigma\sigma'}(z-z')_{11} = S^E_{\sigma\sigma'}(t-t')_{11}$ and obtain

$$\left[i\frac{d}{dt} - E\right]S^E(t-t')_{11} = \delta(t-t'). \tag{2.8.1}$$

For $z,z'\in C_2, ,z=t-io, ,z'=t'-io,$ T_C is the anti-chronological ordering T^{-1} reverse to the usual chronological ordering T of the real times t and t':

$$T^{-1}[c_\sigma(t)c^+_{\sigma'}(t')] = \theta(t'-t)c_\sigma(t)c^+_{\sigma'}(t') - \theta(t-t')c^+_{\sigma'}(t')c_\sigma(t).$$

In this case we have the differential equation

$$\left[i\frac{d}{dt} - E\right]S^E(t-t')_{22} = -\delta(t-t'). \tag{2.8.2}$$

For $z,z'\in C_3, ,z=t_0-i\tau, ,z'=t_0-i\tau',$ T_C becomes the usual chronological ordering T_τ of the real values τ and τ' in the imaginary times $i\tau$ and $i\tau'$, $0\le\tau,\tau'\le\beta$, and we have

$$S^E_{\sigma\sigma'}(z-z')_{33} = S^E_{\sigma\sigma'}(-i\tau+i\tau')_{33} = -i\mathfrak{S}^E_{\sigma\sigma'}(\tau-\tau'), \tag{2.9}$$

where $\mathfrak{S}^E_{\sigma\sigma'}(\tau-\tau')$ is the Matsubara imaginary time-dependent two-point Green function in statistical physics

$$\mathfrak{S}^E_{\sigma\sigma'}(\tau-\tau') = \delta_{\sigma\sigma'}\mathfrak{S}^E(\tau-\tau') = \left\langle T_\tau\left[\gamma_\sigma(\tau)\overline{\gamma}_{\sigma'}(\tau')\right]\right\rangle, \tag{2.10}$$

where

$$\gamma_\sigma(\tau) = e^{H_0\tau}c_\sigma e^{-H_0\tau}\,,$$
$$\overline{\gamma}_\sigma(\tau) = e^{H_0\tau}c^+_\sigma e^{-H_0\tau} \tag{2.11}$$

and

$$T_\tau\left[\gamma_\sigma(\tau)\overline{\gamma}_{\sigma'}(\tau')\right]=\theta(\tau-\tau')\gamma_\sigma(\tau)\overline{\gamma}_{\sigma'}(\tau')-\theta(\tau'-\tau)\overline{\gamma}_{\sigma'}(\tau')\gamma_\sigma(\tau). \tag{2.12}$$

The Heisenberg quantum equation of motion for imaginary time-dependent operators (2.11) has the form

$$\begin{aligned}\frac{d\gamma_\sigma(\tau)}{d\tau}&=[H_0,\gamma_\sigma(\tau)],\\ \frac{d\overline{\gamma}_\sigma(\tau)}{d\tau}&=[H_0,\overline{\gamma}_\sigma(\tau)].\end{aligned} \tag{2.13}$$

From the anti-commutation relations (2.5) it follows that

$$\begin{aligned}\frac{d\gamma_\sigma(\tau)}{d\tau}&=-E\gamma_\sigma(\tau),\\ \frac{d\overline{\gamma}_\sigma(\tau)}{d\tau}&=E\overline{\gamma}_\sigma(\tau),\end{aligned} \tag{2.14}$$

and therefore

$$\left[\frac{d}{d\tau}+E\right]\mathfrak{S}^E(\tau-\tau')=\delta(\tau-\tau'). \tag{2.15}$$

In the analogy with relations (2.9) we set

$$S^E(-i\tau+i\tau')_{33}=-i\mathfrak{S}^E(\tau-\tau') \tag{2.16}$$

and rewrite equation (2.15) in the form similar to equations (2.8.1) and (2.8.2):

$$\left[i\frac{d}{d(-i\tau)}-E\right]S^E(-i\tau+i\tau')_{33}=i\delta(\tau-\tau'). \tag{2.17}$$

Now consider six other cases when two variables z and z' belong to different segments C_i and C_j with $i\neq j$. For $z=t+io\in C_1$ and $z'=t'-io\in C_2$ the values of z always precede those of z' with respect to the ordering along the contour C and therefore

$$T_C[c_\sigma(t+io)\overline{c}_{\sigma'}(t'-io)]=-\overline{c}_{\sigma'}(t'-io)c_\sigma(t+io).$$

Similarly, for $z=t+io\in C_1$ or $z=t-io\in C_2$ and $z'=t_0-i\tau\in C_3$ we have

$$T_C[c_\sigma(t\pm io)\overline{c}_{\sigma'}(t_0-i\tau)]=-\overline{c}_{\sigma'}(t_0-i\tau)c_\sigma(t\pm io).$$

On the contrary, for $z=t-io\in C_2$ and $z'=t'+io\in C_1$ the values of z' always precede those of z with respect to the ordering along the contour C and therefore

$$T_C[c_\sigma(t-io)\overline{c}_{\sigma'}(t'+io)]=c_\sigma(t-io)\overline{c}_{\sigma'}(t'+io).$$

Similarly, for $z=t_0-i\tau\in C_3$ and $z'=t'+io\in C_1$ or $z'=t'-io\in C_2$ we have

$$T_C[c_\sigma(t_0 - i\tau)\overline{c}_{\sigma'}(t' \pm io)] = c_\sigma(t_0 - i\tau)\overline{c}_{\sigma'}(t' \pm io).$$

In all six later cases the differential equations for corresponding functions $S^E(z-z')_{ij}$, $i \neq j$, are six homogeneous ones:

$$\begin{aligned}&\left[i\frac{d}{dt} - E\right]S^E(t-t')_{12} = \left[i\frac{d}{dt} - E\right]S^E(t-t')_{21} \\ &= \left[i\frac{d}{dt} - E\right]S^E(t-t_0+i\tau)_{13} = \left[i\frac{d}{dt} - E\right]S^E(t-t_0+i\tau)_{23} \\ &= \left[i\frac{d}{d(-i\tau)} - E\right]S^E(t_0-i\tau-t)_{31} = \left[i\frac{d}{d(-i\tau)} - E\right]S^E(t_0-i\tau-t)_{32} = 0.\end{aligned} \tag{2.18}$$

By introducing a new notation

$$\delta(z-z')_C = \begin{cases} \delta(t-t') & \text{for } z,z' \in C_1, \\ -\delta(t-t') & \text{for } z,z' \in C_2, \\ i\delta(\tau-\tau') & \text{for } z,z' \in C_3, \\ 0 & \text{for } z \in C_i, z' \in C_j, i \neq j \end{cases} \tag{2.19}$$

we rewrite equations (2.8.1), (2.8.2), (2.17) and (2.18) in the unified form

$$\left[i\frac{d}{dz} - E\right]S^E(z-z')_C = \delta(z-z')_C\,. \tag{2.20}$$

From above presented reasonnings and relations determining nine functions $S^E(z-z')_{ij}$, and formula (2.1) for total Hamiltonian, it is straightforward to derive explicit expressions of these functions. They depend on the average electron number with a definite spin projection

$$n = \langle n_\sigma \rangle = \langle c_\sigma^+ c_\sigma \rangle = \frac{e^{-\beta H_0}}{1 + e^{-\beta H_0}}. \tag{2.21}$$

We obtain following results:

$$S^E(z-z')_{11} = -i[\theta(t-t') - n]e^{-iE(t-t')}, \tag{2.22.1}$$

$$S^E(z-z')_{22} = -i[\theta(t'-t) - n]e^{-iE(t-t')}, \tag{2.22.2}$$

$$S^E(z-z')_{33} = -i[\theta(\tau-\tau') - n]e^{-E(\tau-\tau')}, \tag{2.22.3}$$

$$S^E(z-z')_{12} = ine^{-iE(t-t')}, \tag{2.22.4}$$

$$S^E(z-z')_{21} = -i(1-n)e^{-iE(t-t')}, \tag{2.22.5}$$

$$S^E(z-z')_{13} = S^E(z-z')_{23} = ine^{E\tau}e^{-iE(t-t_0)}, \tag{2.22.6}$$

$$S^E(z-z')_{31} = S^E(z-z')_{32} = -i(1-n)e^{-E\tau}e^{-iE(t_0-t)}. \tag{2.22.7}$$

They satisfy above presented differential equations (2.8.1), (2.8.2), (2.17) and (2.18), respectively.

For concluding this Section we consider the Fourier transformation of the functions $S^E(z-z')_{ij}$:

$$S^E(z-z')_{ij} = \frac{1}{2\pi}\int d\omega e^{-i\omega(t-t')}\tilde{S}^E(\omega)_{ij}, \qquad i,j=1,2, \tag{2.23.1}$$

$$S^E(z-z')_{i3} = \frac{1}{2\pi}\int d\omega e^{-i\omega(t-t_0+i\tau)}\tilde{S}^E(\omega)_{i3}, \qquad i=1,2, \tag{2.23.2}$$

$$S^E(z-z')_{3i} = \frac{1}{2\pi}\int d\omega e^{-i\omega(t_0-t-i\tau)}\tilde{S}^E(\omega)_{3i}, \qquad i=1,2, \tag{2.23.3}$$

$$S^E(z-z')_{33} = -i\frac{1}{\beta}\sum_{\nu} e^{i\varepsilon_\nu(\tau-\tau')}\tilde{S}^E_\nu, \tag{2.23.4}$$

$$\varepsilon_\nu = (2\nu+1)\frac{\pi}{\beta}, \qquad \nu = 0,\pm 1,\pm 2,...$$

From the expressions (2.22.1)-(2.22.7) of the functions $S^E(z-z')_{ij}$ it follows that

$$\begin{aligned}\tilde{S}^E(\omega)_{11} &= \frac{1}{\omega - E + io} + i2\pi n\delta(\omega - E)\\ &= P\frac{1}{\omega - E} - i\pi\frac{1-e^{-\beta E}}{1+e^{-\beta E}}\delta(\omega - E),\end{aligned} \tag{2.24.1}$$

$$\begin{aligned}\tilde{S}^E(\omega)_{22} &= -\frac{1}{\omega - E - io} + i2\pi n\delta(\omega - E)\\ &= -P\frac{1}{\omega - E} - i\pi\frac{1-e^{-\beta E}}{1+e^{-\beta E}}\delta(\omega - E),\end{aligned} \tag{2.24.2}$$

where P means the principal value,

$$\tilde{S}^E(\omega)_{12} = i2\pi n\delta(\omega - E), \tag{2.24.3}$$

$$\tilde{S}^E(\omega)_{21} = -i2\pi(1-n)\delta(\omega - E), \tag{2.24.4}$$

$$\tilde{S}^E(\omega)_{i3} = i2\pi n\delta(\omega - E), \qquad i=1,2, \tag{2.24.5}$$

$$\tilde{S}^E(\omega)_{3i} = -i2\pi(1-n)\delta(\omega - E), \qquad i=1,2, \tag{2.24.6}$$

$$\tilde{S}^E_\nu = \frac{1}{i\varepsilon_\nu + E}. \tag{2.24.7}$$

The explicit expressions of Green functions of free electrons presented in this Section are often used in the theoretical studies of non-equilibrium processes by means of the perturbation theory.

3. Non-equilibrium Green functions of electrons in isolated single-level quantum dot

The calculation methods and reasonnings presented in the preceding Section are now applied to the study of the Keldysh non-equilibrium Green functions of interacting electrons in the simplest nanosystem - the isolated single-level quantum dot (QD) with total Hamiltonian

$$H = E\sum_{\sigma} c_{\sigma}^{+}c_{\sigma} + UN_{\uparrow}N_{\downarrow} , \tag{3.1}$$

where U is the value of a potential energy, $\sigma=\uparrow,\downarrow$ denotes the spin projection (if $\sigma=\uparrow$ then $-\sigma=\downarrow$ and vice versa) and

$$N_{\sigma} = c_{\sigma}^{+}c_{\sigma} \tag{3.2}$$

is the number of electrons with the spin projection σ. The second term in Hamiltonian (3.1) is the potential energy of the Coulomb electron-electron interaction (two electrons with different spin projections in one and the same energy level). The interacting nanosystem with total Hamiltonian (3.1) is an exactly solvable model. There are four exactly determined eigenstates and eigenvalues of H: the vacuum with vanishing energy, two degenerate single-electron states with two different spin projections and the same energy E, and a two-electron state with total energy $2E+U$. The Keldysh complex time-dependent two-point Green function of two operators $c(z)$ and $\overline{c}(z')$ is defined as follows

$$G_{\sigma\sigma'}(z-z')_C = -i\langle T_C[c_{\sigma}(z)\overline{c}_{\sigma'}(z')]\rangle = -i\frac{Tr\{e^{-\beta H}T_C[c_{\sigma}(z)\overline{c}_{\sigma'}(z')]\}}{Tr\{e^{-\beta H}\}} \tag{3.3}$$

with total Hamiltonian (3.1). They have the form

$$G_{\sigma\sigma'}(z-z')_C = \delta_{\sigma\sigma'}G(z-z')_C . \tag{3.4}$$

As in the preceding Section, we choose the contour C to consist of three segments C_1, C_2 and C_3. Then $G(z-z')_C$ becomes the set of nine functions $G(z-z')_{ij}$, $i,j=1,2,3$. The calculations of these functions are straightforward, as they have been done in the preceding Section for free electrons at a single energy level. We obtain following results:

$$G(z-z')_{11} = \frac{i}{Z}\{-\theta(t-t')[e^{-iE(t-t')} + e^{-\beta E}e^{-i(E+U)(t-t')}] \\ +\theta(t'-t)[e^{-\beta E}e^{-iE(t-t')} + e^{-\beta(2E+U)}e^{-i(E+U)(t-t')}]\} , \tag{3.5.1}$$

$$G(z-z')_{22} = \frac{i}{Z}\{-\theta(t'-t)[e^{-iE(t-t')} + e^{-\beta E}e^{-i(E+U)(t-t')}] \\ +\theta(t-t')[e^{-\beta E}e^{-iE(t-t')} + e^{-\beta(2E+U)}e^{-i(E+U)(t-t')}]\} , \tag{3.5.2}$$

$$G(z-z')_{33}=i\mathcal{G}(\tau-\tau'),$$
$$\mathcal{G}(\tau)=\frac{1}{Z}\{[\theta(\tau)e^{-\tau E}-\theta(-\tau)e^{-(\tau+\beta)E}] \quad (3.5.3)$$
$$+e^{-\beta E}[\theta(\tau)e^{-\tau(E+U)}-\theta(-\tau)e^{-(\tau+\beta)(E+U)}]\},$$

$$G(z-z')_{12}=\frac{i}{Z}\{e^{-\beta E}e^{-iE(t-t')}+e^{-\beta(2E+U)}e^{-i(E+U)(t-t')}\}, \quad (3.5.4)$$

$$G(z-z')_{21}=\frac{i}{Z}\{-e^{-iE(t-t')}-e^{-\beta E}e^{-i(E+U)(t-t')}\}, \quad (3.5.5)$$

$$G(z-z')_{13}=G(z-z')_{23}$$
$$=\frac{i}{Z}\{e^{-\beta E}e^{\tau E}e^{-iE(t-t_0)}+e^{-\beta(2E+U)}e^{\tau(E+U)}e^{-i(E+U)(t-t_0)}\}, \quad (3.5.6)$$

and

$$G(z-z')_{31}=G(z-z')_{32}$$
$$=\frac{i}{Z}\{-e^{-\tau E}e^{-iE(t_0-t')}+e^{-\beta E}e^{-\tau(E+U)}e^{-i(E+U)(t_0-t')}\}. \quad (3.5.7)$$

In the study of non-equlibrium dynamical processes by means of the perturbation theory one often needs to use the Fourier transformation of four functions $G(z-z')_{ij}$ with $i,j=1,2$:

$$G(z-z')_{ij}=G(t-t')_{ij}=\frac{1}{2\pi}\int d\omega e^{-i\omega(t-t')}\tilde{G}(\omega)_{ij}. \quad (3.6)$$

We have following exact expressions of their Fourier transforms:

$$\tilde{G}(\omega)_{11}=\frac{1}{Z}\left\{\frac{1}{\omega-E+io}+\frac{e^{-\beta E}}{\omega-E-io}+\frac{e^{-\beta E}}{\omega-E-U+io}+\frac{e^{-\beta(2E+U)}}{\omega-E-U-io}\right\}$$
$$=\frac{1}{Z}\left\{\left[1+e^{-\beta E}\right]P\frac{1}{\omega-E}-i\pi\left[1-e^{-\beta E}\right]\delta(\omega-E)\right. \quad (3.7.1)$$
$$\left.+e^{-\beta E}\left[1+e^{-\beta(E+U)}\right]P\frac{1}{\omega-E-U}-i\pi e^{-\beta E}\left[1-e^{-\beta(E+U)}\right]\delta(\omega-E-U)\right\},$$

$$\tilde{G}(\omega)_{22}=\frac{1}{Z}\left\{-\frac{1}{\omega-E+io}-\frac{e^{-\beta E}}{\omega-E-io}-\frac{e^{-\beta E}}{\omega-E-U-io}-\frac{e^{-\beta(2E+U)}}{\omega-E-U+io}\right\}$$
$$=-\frac{1}{Z}\left\{\left[1+e^{-\beta E}\right]P\frac{1}{\omega-E}+i\pi\left[1-e^{-\beta E}\right]\delta(\omega-E)\right. \quad (3.7.2)$$
$$\left.+e^{-\beta E}\left[1+e^{-\beta(E+U)}\right]P\frac{1}{\omega-E-U}+i\pi e^{-\beta E}\left[1-e^{-\beta(E+U)}\right]\delta(\omega-E-U)\right\},$$

$$\tilde{G}(\omega)_{12}=\frac{2\pi i}{Z}\{\delta(\omega-E)+e^{-\beta E}\delta(\omega-E-U)\}, \quad (3.7.3)$$

$$\tilde{G}(\omega)_{21} = -\frac{2\pi i}{Z}\{ e^{-\beta E}\delta(\omega - E) + e^{-\beta(2E+U)}\delta(\omega - E - U)\} \tag{3.7.4}$$

with

$$Z = 1 + 2e^{-\beta E} + e^{-\beta(2E+U)}.$$

Now we derive the system of differential equations for two-point Green functions $G_{\sigma\sigma'}(t-t')_{ij}$. Consider first the function with $i = j = 1$:

$$G_{\sigma\sigma'}(t-t')_{11} = -i\langle T[c_{\sigma}(t)\overline{c}_{\sigma'}(t')]\rangle . \tag{3.8}$$

We have

$$i\frac{dG_{\sigma\sigma'}(t-t')_{11}}{dt} = \delta_{\sigma\sigma'}\delta(t-t') - i\left\langle T\left[i\frac{dc_{\sigma}(t)}{dt}\overline{c}_{\sigma'}(t')\right]\right\rangle . \tag{3.9}$$

From the Heisenberg quantum equation of motion

$$i\frac{dc_{\sigma}(t)}{dt} = -[H, c_{\sigma}(t)] \tag{3.10}$$

with total Hamiltonian (3.1) it follows that

$$i\frac{dc_{\sigma}(t)}{dt} = Ec_{\sigma}(t) + UN_{-\sigma}c_{\sigma}(t). \tag{3.11}$$

Substituting this expression of $i\frac{dc_{\sigma}(t)}{dt}$ into the r.h.s. of equation (3.9), we obtain

$$\left[i\frac{d}{dt} - E\right] G_{\sigma\sigma'}(t-t') = \delta_{\sigma\sigma'}\delta(t-t') + UH_{\sigma\sigma'}(t-t')_{11}, \tag{3.12}$$

where

$$\begin{aligned} H_{\sigma\sigma'}(t-t')_{11} &= -i\langle T[N_{-\sigma}c_{\sigma}(t)\overline{c}_{\sigma'}(t')]\rangle = \\ &-i\theta(t-t')\langle [N_{-\sigma}c_{\sigma}(t)\overline{c}_{\sigma'}(t')]\rangle + i\theta(t'-t)\langle [\overline{c}_{\sigma'}(t')N_{-\sigma}c_{\sigma}(t)]\rangle . \end{aligned} \tag{3.13}$$

Thus the differential equation for $G_{\sigma\sigma'}(t-t')_{11}$ contains a new Green function $H_{\sigma\sigma'}(t-t')_{11}$. In order to derive the differential equation for this new Green function it is necessary to calculate the time derivatives of both sides of equation (3.13). Note that $N_{-\sigma}$ commutes with H and therefore does not depend on t. Moreover, it has following property

$$N_{-\sigma}^2 = N_{-\sigma}.$$

Multiplying both sides of relation (3.11) with $N_{-\sigma}$ and using these two above-mentioned properties of $N_{-\sigma}$, we obtain

$$i\frac{d}{dt}[N_{-\sigma}c_{\sigma}(t)] = (E+U)[N_{-\sigma}c_{\sigma}(t)] . \tag{3.14}$$

Differentiating both sides of equation (3.13) and using relation (3.14), we derive following differential equation for the new Green function $H_{\sigma\sigma'}(t-t')_{11}$:

$$\left[i\frac{d}{dt}-(E+U)\right]H_{\sigma\sigma'}(t-t')=n\delta_{\sigma\sigma'}\delta(t-t'). \tag{3.15}$$

Thus both $G_{\sigma\sigma'}(t-t')_{11}$ and $H_{\sigma\sigma'}(t-t')_{11}$ have the common form

$$\begin{aligned}G_{\sigma\sigma'}(t-t')_{11}&=\delta_{\sigma\sigma'}G(t-t')_{11},\\ H_{\sigma\sigma'}(t-t')_{11}&=\delta_{\sigma\sigma'}H(t-t')_{11},\end{aligned} \tag{3.16}$$

where $G(t-t')_{11}$ and $H(t-t')_{11}$ must satisfy differential equations

$$\left[i\frac{d}{dt}-E\right]G(t-t')_{11}=\delta(t-t')+UH(t-t')_{11}, \tag{3.17}$$

$$\left[i\frac{d}{dt}-(E+U)\right]H(t-t')_{11}=n\delta(t-t'). \tag{3.18}$$

In preceding Section we have shown that

$$\left[i\frac{d}{dt}-E\right]S^{E}(t-t')_{11}=\delta(t-t')$$

(equation (2.8.1)). Therefore

$$\left[i\frac{d}{dt}-(E+U)\right]S^{E+U}(t-t')_{11}=\delta(t-t'). \tag{3.19}$$

Equations (3.18) and (3.19) show that $\frac{1}{n}H(t-t')_{11}$ satisfies the same inhomogeneous differential equation as $S^{E+U}(t-t')$ does. It follows that

$$H(t-t')_{11}=nS^{E+U}(t-t')_{11}, \tag{3.20}$$

and the differential equation for $G(t-t')_{11}$ becomes

$$\left[i\frac{d}{dt}-E\right]G(t-t')_{11}=\delta(t-t')+nS^{E+U}(t-t')_{11}. \tag{3.21}$$

Similarly, it can be shown that the Green function

$$G_{\sigma\sigma'}(t-t')_{22}=-i\{\theta(t'-t)\langle c_{\sigma}(t)\overline{c}_{\sigma'}(t')\rangle-\theta(t-t')\langle\overline{c}_{\sigma'}(t')c_{\sigma}(t)\rangle\} \tag{3.22}$$

has the form

$$G_{\sigma\sigma'}(t-t')_{22}=\delta_{\sigma\sigma'}G(t-t')_{22}, \tag{3.23}$$

and $G(t-t')_{22}$ satisfies differential equation

$$\left[i\frac{d}{dt}-E\right]G(t-t')_{22}=-\delta(t-t')+nUS^{E+U}(t-t')_{22} \tag{3.24}$$

etc. In general, Keldysh complex time Green function

$$G_{\sigma\sigma'}(z-z')_C=-i\langle T_C[c_\sigma(z)\overline{c}_\sigma(z')\rangle \tag{3.25}$$

has the form

$$G_{\sigma\sigma'}(z-z')_C=\delta_{\sigma\sigma'}G(z-z')_C, \tag{3.26}$$

and $G(z-z')_C$ satisfies differential equation

$$\left[i\frac{d}{dt}-E\right]G(z-z')_C=\delta(z-z')_C+nUS^{E+U}(z-z')_C. \tag{3.27}$$

4. Non-equilibrium Green functions of electrons in single-level quantum dot connected with two conducting leads

Consider the single-electron transistor (SET) consisting of a single-level quantum dot (QD) connected with two conducting leads through two potential barriers. The electron transport through this SET was investigated experimentally and studied theoretically in many works (Choi et al., 2004; Costi et al., 1994; Craco & Kang, 1999; Fujii & Ueda, 2003; Hershfield et al., 1991; Inoshita et al., 1993; Izumida et al., 1997, 1998, 2001; Meir et al., 1991, 1993; Ng, 1993; Pustilnik & Glasman, 2004; Sakai et al., 1999; Swirkowicz et al., 2003, 2006; Takagi & Saso, 1999a, 1999b; Torio et al., 2002; Wingreen & Meir, 1994; Yeyati et al., 1993). It was assumed that the electron system in this SET has following total Hamiltonian

$$\begin{aligned}H=E\sum_\sigma c_\sigma^+c_\sigma+UN_\uparrow N_\downarrow+\sum_k\sum_\sigma\{E_a(\mathrm{k})a_\sigma^+(\mathrm{k})a_\sigma(\mathrm{k})+E_b(\mathrm{k})b_\sigma^+(\mathrm{k})b_\sigma(\mathrm{k})\}\\+\sum_k\sum_\sigma\{V_a(\mathrm{k})a_\sigma^+(\mathrm{k})c_\sigma+V_a(\mathrm{k})^*c_\sigma^+a_\sigma(\mathrm{k})+V_b(\mathrm{k})b_\sigma^+(\mathrm{k})c_\sigma+V_b(\mathrm{k})^*c_\sigma^+b_\sigma(\mathrm{k})\}.\end{aligned} \tag{4.1}$$

In order to define the complex time-dependent Green functions we introduce the complex time-dependent quantum operators

$$\begin{aligned}c_\sigma(z)=e^{iHz}c_\sigma e^{-iHz},\quad \overline{c}_\sigma(z)=e^{iHz}c_\sigma^+e^{-iHz},\\ a_\sigma(\mathrm{k},z)=e^{iHz}a_\sigma(\mathrm{k})e^{-iHz},\quad \overline{a}_\sigma(\mathrm{k},z)=e^{iHz}a_\sigma^+(\mathrm{k})e^{-iHz},\\ b_\sigma(\mathrm{k},z)=e^{iHz}b_\sigma(\mathrm{k})e^{-iHz},\quad \overline{b}_\sigma(\mathrm{k},z)=e^{iHz}b_\sigma^+(\mathrm{k})e^{-iHz}.\end{aligned} \tag{4.2}$$

The Keldysh non-equilibrium Green functions of electrons are defined as follows:

$$G_{\sigma\sigma'}^{c\overline{c}}(z-z')_C=\delta_{\sigma\sigma'}G^{c\overline{c}}(z-z')_C=-i\langle T_C[c_\sigma(z)\overline{c}_{\sigma'}(z')\rangle, \tag{4.3}$$

$$H_{\sigma\sigma'}^{c\overline{c}}(z-z')_C=\delta_{\sigma\sigma'}H^{c\overline{c}}(z-z')_C=-i\langle T_C[N_{-\sigma}(z)c_\sigma(z)\overline{c}_{\sigma'}(z')\rangle, \tag{4.4}$$

$$G^{a\bar{c}}_{\sigma\sigma'}(\mathbf{k};z-z')_C = \delta_{\sigma\sigma'}G^{a\bar{c}}(\mathbf{k};z-z')_C = -i\langle T_C[a_\sigma(\mathbf{k};z)\bar{c}_{\sigma'}(z')]\rangle, \tag{4.5}$$

$$H^{a\bar{c}}_{\sigma\sigma'}(\mathbf{k};z-z')_C = \delta_{\sigma\sigma'}H^{a\bar{c}}(\mathbf{k};z-z')_C = -i\langle T_C[N_{-\sigma}(z)a_\sigma(\mathbf{k};z)\bar{c}_{\sigma'}(z')]\rangle, \tag{4.6}$$

$$\begin{aligned} G^{ac\bar{c}\bar{c}}_{\sigma\sigma'}(\mathbf{k};z-z')_C &= \delta_{\sigma\sigma'}G^{ac\bar{c}\bar{c}}(\mathbf{k};z-z')_C \\ &= -i\langle T_C[a_{-\sigma}(\mathbf{k};z)c_\sigma(z)\bar{c}_{-\sigma}(z)\bar{c}_{\sigma'}(z')]\rangle, \end{aligned} \tag{4.7}$$

$$\begin{aligned} G^{cc\bar{a}\bar{c}}_{\sigma\sigma'}(\mathbf{k};z-z')_C &= \delta_{\sigma\sigma'}G^{cc\bar{a}\bar{c}}(\mathbf{k};z-z')_C \\ &= -i\langle T_C[c_{-\sigma}(z)c_\sigma(z)\bar{a}_{-\sigma}(\mathbf{k};z)\bar{c}_{\sigma'}(z')]\rangle, \end{aligned} \tag{4.8}$$

$$\begin{aligned} G^{aa\bar{c}\bar{c}}_{\sigma\sigma'}(\mathbf{k},\mathbf{l};z-z')_C &= \delta_{\sigma\sigma'}G^{aa\bar{c}\bar{c}}(\mathbf{k},\mathbf{l};z-z')_C \\ &= -i\langle T_C[a_{-\sigma}(\mathbf{k};z)a_\sigma(\mathbf{l};z)\bar{c}_{-\sigma}(z)\bar{c}_{\sigma'}(z')]\rangle, \end{aligned} \tag{4.9}$$

$$\begin{aligned} G^{ac\bar{a}\bar{c}}_{\sigma\sigma'}(\mathbf{k},\mathbf{l};z-z')_C &= \delta_{\sigma\sigma'}G^{ac\bar{a}\bar{c}}(\mathbf{k},\mathbf{l};z-z')_C \\ &= -i\langle T_C[a_{-\sigma}(\mathbf{k};z)c_\sigma(z)\bar{a}_{-\sigma}(\mathbf{l};z)\bar{c}_{\sigma'}(z')]\rangle, \end{aligned} \tag{4.10}$$

and similarly for the others $G^{b\bar{c}}_{\sigma\sigma'}(\mathbf{k};z-z')_C$, $H^{b\bar{c}}_{\sigma\sigma'}(\mathbf{k};z-z')_C$, $G^{bc\bar{c}\bar{c}}_{\sigma\sigma'}(\mathbf{k};z-z')_C$; $G^{cc\bar{b}\bar{c}}_{\sigma\sigma'}(\mathbf{k};z-z')_C$, $G^{ab\bar{c}\bar{c}}_{\sigma\sigma'}(\mathbf{k},\mathbf{l};z-z')_C$, $G^{ac\bar{b}\bar{c}}_{\sigma\sigma'}(\mathbf{k},\mathbf{l};z-z')_C$ etc.

Because there is no magnetic interaction, all Green functions (4.3)-(4.10) and other ones are proportional to $\delta_{\sigma\sigma'}$. From Heisenberg quantum equations of motion and equal-time canonical anti-commutation relations for the electron destruction and creation operators it follows the differential equations for these operators:

$$i\frac{dc_\sigma(z)}{dz} = Ec_\sigma(z) + UN_{-\sigma}(z)c_\sigma(z) + \sum_{\mathbf{k}}\left[V_a^*(\mathbf{k})a_\sigma(\mathbf{k};z) + V_b^*(\mathbf{k})b_\sigma(\mathbf{k};z)\right], \tag{4.11}$$

$$i\frac{d\bar{c}_\sigma(z)}{dz} = -Ec_\sigma(z) - UN_{-\sigma}(z)\bar{c}_\sigma(z) - \sum_{\mathbf{k}}\left[V_a(\mathbf{k})\bar{a}_\sigma(\mathbf{k};z) + V_b(\mathbf{k})\bar{b}_\sigma(\mathbf{k};z)\right], \tag{4.12}$$

$$i\frac{da_\sigma(\mathbf{k};z)}{dz} = E_a(\mathbf{k})a_\sigma(\mathbf{k};z) + V_a(\mathbf{k})c_\sigma(z), \tag{4.13}$$

$$i\frac{d\bar{a}_\sigma(\mathbf{k};z)}{dz} = -E_a(\mathbf{k})\bar{a}_\sigma(\mathbf{k};z) - V_a^*(\mathbf{k})\bar{c}_\sigma(z) \tag{4.14}$$

and similarly for $b_\sigma(\mathbf{k};z)$ and $\bar{b}_\sigma(\mathbf{k};z)$.

By using differential equation (4.11) and the equal-time canonical anti-commutation relation between $c_\sigma(z)$ and $\bar{c}_{\sigma'}(z)$, it is easy to derive the differential equation for the Green function $G^{c\bar{c}}_{\sigma\sigma'}(z-z')_C$

$$\begin{aligned} \left[i\frac{d}{dz} - E\right]G^{c\bar{c}}_{\sigma\sigma'}(z-z')_C &= \delta_{\sigma\sigma'}\delta(z-z')_C + UH^{c\bar{c}}_{\sigma\sigma'}(z-z') \\ &+ \sum_{\mathbf{k}}\left[V_a^*(\mathbf{k})G^{a\bar{c}}_{\sigma\sigma'}(\mathbf{k};z-z')_C + V_b^*(\mathbf{k})G^{b\bar{c}}_{\sigma\sigma'}(\mathbf{k};z-z')_C\right], \end{aligned} \tag{4.15}$$

which contains Green functions $H^{c\bar{c}}_{\sigma\sigma'}(z-z')_C$, $G^{a\bar{c}}_{\sigma\sigma'}(\mathbf{k};z-z')_C$ and $G^{b\bar{c}}_{\sigma\sigma'}(\mathbf{k};z-z')_C$. These new functions must satisfy following differential equations which can be also derived by using differential equations (4.11)-(4.14):

$$\left[i\frac{d}{dz}-(E+U)\right]H^{c\bar{c}}_{\sigma\sigma'}(z-z')_C = n\delta_{\sigma\sigma'}\delta(z-z')_C$$
$$+\sum_{\mathbf{k}}\left[V_a^*(\mathbf{k})H^{a\bar{c}}_{\sigma\sigma'}(\mathbf{k};z-z')_C + V_b^*(\mathbf{k})H^{b\bar{c}}_{\sigma\sigma'}(\mathbf{k};z-z')_C\right.$$
$$\left.-V_a(\mathbf{k})G^{cc\bar{a}\bar{c}}_{\sigma\sigma'}(\mathbf{k};z-z')_C - V_b(\mathbf{k})G^{cc\bar{b}\bar{c}}_{\sigma\sigma'}(\mathbf{k};z-z')_C\right], \tag{4.16}$$

where

$$n=\langle c_\uparrow^+ c_\uparrow\rangle = \langle c_\downarrow^+ c_\downarrow\rangle, \tag{4.17}$$

$$\left[i\frac{d}{dz}-E_a(\mathbf{k})\right]G^{a\bar{c}}_{\sigma\sigma'}(\mathbf{k};z-z')_C = V_a(\mathbf{k})G^{c\bar{c}}_{\sigma\sigma'}(z-z')_C \tag{4.18}$$

and similarly for $G^{b\bar{c}}_{\sigma\sigma'}(\mathbf{k};z-z')_C$.

In Section 2 we have established the differential equation (2.20) for the Keldysh non-equilibrium Green function of a free electron. If the free electron has energy $E_a(\mathbf{k})$, then it is denoted by $S^{Ea(\mathbf{k})}(z-z')_C$ and must satisfy differential equation

$$\left[i\frac{d}{dz}-E_a(\mathbf{k})\right]S^{Ea(\mathbf{k})}(z-z')_C = \delta(z-z')_C. \tag{4.19}$$

Using this function, we obtain following expression of the solution of equation (4.18)

$$G^{a\bar{c}}_{\sigma\sigma'}(\mathbf{k};z-z')_C = V_a(\mathbf{k})\int dz'' S^{Ea(\mathbf{k})}(z-z'')_C G^{c\bar{c}}_{\sigma\sigma'}(z''-z')_C \tag{4.20}$$

and similarly for $G^{b\bar{c}}_{\sigma\sigma'}(\mathbf{k};z-z')_C$. Substituting the expression of the form (4.20) for $G^{a\bar{c}}_{\sigma\sigma'}(\mathbf{k};z-z')_C$ and $G^{b\bar{c}}_{\sigma\sigma'}(\mathbf{k};z-z')_C$ into the r.h.s. of differential equation (4.15) for $G^{c\bar{c}}_{\sigma\sigma'}(z-z')_C$, we rewrite this equation in a new form

$$\left[i\frac{d}{dz}-E\right]G^{c\bar{c}}_{\sigma\sigma'}(z-z')_C = \delta_{\sigma\sigma'}\delta(z-z')_C + UH^{c\bar{c}}_{\sigma\sigma'}(z-z')_C$$
$$+\int_C dz''\Sigma^{(1)}(z-z'')_C G^{c\bar{c}}_{\sigma\sigma'}(z''-z')_C, \tag{4.21}$$

where $\Sigma^{(1)}(z-z')_C$ is following self-energy part

$$\Sigma^{(1)}(z-z')_C = \sum_{\mathbf{k}}\left[|V_a(\mathbf{k})|^2 S^{Ea(\mathbf{k})}(z-z')_C + |V_b(\mathbf{k})|^2 S^{Eb(\mathbf{k})}(z-z')_C\right]. \tag{4.22}$$

The differential equation for $H^{c\bar{c}}_{\sigma\sigma'}(z-z')_C$ contains new functions $H^{a\bar{c}}_{\sigma\sigma'}(\mathbf{k};z-z')_C$, $H^{b\bar{c}}_{\sigma\sigma'}(\mathbf{k};z-z')_C$, $G^{ac\bar{c}\bar{c}}_{\sigma\sigma'}(\mathbf{k};z-z')_C$, $G^{bc\bar{c}\bar{c}}_{\sigma\sigma'}(\mathbf{k};z-z')_C$, $G^{cc\bar{a}\bar{c}}_{\sigma\sigma'}(\mathbf{k};z-z')_C$ and $G^{cc\bar{b}\bar{c}}_{\sigma\sigma'}(\mathbf{k};z-z')_C$, which must satisfy following differential equations

$$\left[i\frac{d}{dz}-E_a(\mathbf{k})\right]H^{a\bar{c}}_{\sigma\sigma'}(\mathbf{k};z-z')_C=V_a(\mathbf{k})H^{c\bar{c}}_{\sigma\sigma'}(z-z')_C \tag{4.23}$$

and similarly for $H^{b\bar{c}}_{\sigma\sigma'}(\mathbf{k};z-z')_C$,

$$\begin{aligned}\left[i\frac{d}{dt}-E_a(\mathbf{k})\right]G^{ac\bar{c}\bar{c}}_{\sigma\sigma'}(\mathbf{k};z-z')_c&=\left\langle\left\{a_{-\sigma}(\mathbf{k})c_\sigma c^+_{-\sigma},c^+_{\sigma'}\right\}\right\rangle\delta(z-z')_C\\&+V_a(\mathbf{k})\left[H^{c\bar{c}}_{\sigma\sigma'}(z-z')_C-G^{c\bar{c}}_{\sigma\sigma'}(z-z')_C\right]\\&+\sum_{\mathbf{l}}\left[V^*_a(\mathbf{l})G^{aac\bar{c}}_{\sigma\sigma'}(\mathbf{k},\mathbf{l};z-z')_C+V^*_b(\mathbf{l})G^{ab\bar{c}\bar{c}}_{\sigma\sigma'}(\mathbf{k},\mathbf{l};z-z')_C\right]\\&-\sum_{\mathbf{l}}\left[V_a(\mathbf{l})G^{ac\bar{a}\bar{c}}_{\sigma\sigma'}(\mathbf{k},\mathbf{l};z-z')_C+V_b(\mathbf{l})G^{ac\bar{b}\bar{c}}_{\sigma\sigma'}(\mathbf{k},\mathbf{l};z-z')_C\right],\end{aligned} \tag{4.24}$$

and similarly for $G^{bc\bar{c}\bar{c}}_{\sigma\sigma'}(\mathbf{k};z-z')$,

$$\begin{aligned}\left\{i\frac{d}{dt}-\left[2E-E_a(\mathbf{k})+U\right]\right\}G^{cc\bar{a}\bar{c}}_{\sigma\sigma'}(\mathbf{k};z-z')_c&=\left\langle\left\{c_{-\sigma}c_\sigma a^+_{-\sigma}(\mathbf{k}),c^+_{\sigma'}\right\}\right\rangle\delta(z-z')_C\\&-V^*_a(\mathbf{k})\left[H^{c\bar{c}}_{\sigma\sigma'}(z-z')_C-G^{c\bar{c}}_{\sigma\sigma'}(z-z')_C\right]+\\&+\sum_{\mathbf{l}}\left\{V^*_a(\mathbf{l})\left[G^{ac\bar{a}\bar{c}}_{\sigma\sigma'}(\mathbf{l},\mathbf{k};z-z')_C+G^{ca\bar{a}\bar{c}}_{\sigma\sigma'}(\mathbf{l},\mathbf{k};z-z')_C\right]\right.\\&\left.+V^*_b(\mathbf{l})\left[G^{bc\bar{a}\bar{c}}_{\sigma\sigma'}(\mathbf{l},\mathbf{k};z-z')_C+G^{cb\bar{a}\bar{c}}_{\sigma\sigma'}(\mathbf{l},\mathbf{k};z-z')_C\right]\right\}\end{aligned} \tag{4.25}$$

and similarly for $G^{cc\bar{b}\bar{c}}_{\sigma\sigma'}(\mathbf{k};z-z')_C$.

The presented calculations for deriving differential equations of Green function showed that there does not exist a closed system of a finite number of differential equations for a finite number of Green functions. Some approximation should be used for truncating the infinite system of all differential equations at some step. The mean-field approximation is the most appropriate one. In order to apply this approximation we rewrite equations (4.23)-(4.25) in the form of integral equations:

$$H^{a\bar{c}}_{\sigma\sigma'}(\mathbf{k};z-z')_C=V_a(\mathbf{k})\int_C dz''S^{Ea(\mathbf{k})}(z-z'')_C H^{c\bar{c}}_{\sigma\sigma'}(z-z'')_C \tag{4.26}$$

and similarly for $H^{b\bar{c}}_{\sigma\sigma'}(\mathbf{k};z-z')_C$,

$$\begin{aligned}G^{ac\bar{c}\bar{c}}_{\sigma\sigma'}(\mathbf{k};z-z')_C&=\left\langle\left\{a_{-\sigma}(\mathbf{k})c_\sigma c^+_{-\sigma},c^+_{\sigma'}\right\}\right\rangle S^{Ea(\mathbf{k})}(z-z')_C\\&+V_a(\mathbf{k})\int_C dz''S^{Ea(\mathbf{k})}(z-z'')_C\left[H^{c\bar{c}}_{\sigma\sigma'}(z''-z')_C-G^{c\bar{c}}_{\sigma\sigma'}(z''-z')_C\right]\\&+\int_C dz''S^{Ea(\mathbf{k})}(z-z'')_C\sum_{\mathbf{l}}\left[V^*_a(\mathbf{l})G^{aa\bar{c}\bar{c}}_{\sigma\sigma'}(\mathbf{k},\mathbf{l};z''-z')_C+V^*_b(\mathbf{l})G^{ab\bar{c}\bar{c}}_{\sigma\sigma'}(\mathbf{k},\mathbf{l};z''-z')_C\right]\\&-\int_C dz''S^{Ea(\mathbf{k})}(z-z'')_C\sum_{\mathbf{l}}\left[V_a(\mathbf{l})G^{ac\bar{a}\bar{c}}_{\sigma\sigma'}(\mathbf{k},\mathbf{l};z''-z')_C+V_b(\mathbf{l})G^{ac\bar{b}\bar{c}}_{\sigma\sigma'}(\mathbf{k},\mathbf{l};z''-z')_C\right],\end{aligned} \tag{4.27}$$

and similarly for $G^{bc\bar{c}\bar{c}}_{\sigma\sigma'}(\mathbf{k};z-z')_C$,

$$\begin{aligned} G^{cc\bar{a}\bar{c}}_{\sigma\sigma'}(\mathbf{k};z-z')_C &= \left\langle\left\{c_{-\sigma}c_{\sigma}a^{+}_{-\sigma}(\mathbf{k}),c^{+}_{\sigma'}\right\}\right\rangle S^{2E+U-Ea(\mathbf{k})}(z-z')_C \\ &-V^{*}_{a}(\mathbf{k})\int_C dz'' S^{2E+U-Ea(\mathbf{k})}(z-z')_C\left[H^{c\bar{c}}_{\sigma\sigma'}(z''-z')_C-G^{c\bar{c}}_{\sigma\sigma'}(z''-z')_C\right] \\ &+\int_C dz'' S^{2E+U-Ea(\mathbf{k})}(z''-z')_C\sum_{\mathbf{l}}\left\{V^{*}_{a}(\mathbf{l})\left[G^{ac\bar{a}\bar{c}}_{\sigma\sigma'}(\mathbf{l},\mathbf{k};z''-z')_C+G^{ca\bar{a}\bar{c}}_{\sigma\sigma'}(\mathbf{l},\mathbf{k};z''-z')_C\right]\right. \\ &\left.+V^{*}_{b}(\mathbf{l})\left[G^{bc\bar{a}\bar{c}}_{\sigma\sigma'}(\mathbf{l},\mathbf{k};z''-z')_C+G^{cb\bar{a}\bar{c}}_{\sigma\sigma'}(\mathbf{l},\mathbf{k};z''-z')_C\right]\right\}, \end{aligned} \tag{4.28}$$

and similarly for $G^{cc\bar{b}\bar{c}}_{\sigma\sigma'}(\mathbf{k};z-z')_C$. Substituting these solutions into the r.h.s. of the differential equation (4.16) for $H^{c\bar{c}}_{\sigma\sigma'}(z-z')_C$, we rewrite this equation in the new form

$$\begin{aligned} &\left[i\frac{d}{dz}-(E+U)\right]H^{c\bar{c}}_{\sigma\sigma'}(z-z')_C = n\delta_{\sigma\sigma'}\delta(z-z')_C+\int_C dz''\Sigma^{(1)}(z-z'')_C H^{c\bar{c}}_{\sigma\sigma'}(z''-z')_C \\ &+\sum_{\mathbf{k}}\left\{V^{*}_{a}(\mathbf{k})\left\langle\left\{a_{-\sigma}(\mathbf{k})c_{\sigma}c^{+}_{-\sigma},c^{+}_{\sigma'}\right\}\right\rangle S^{Ea(\mathbf{k})}(z-z')_C\right. \\ &\qquad\left.+\left|V_a(\mathbf{k})\right|^2\int_C dz'' S^{Ea(\mathbf{k})}(z-z'')_C\left[H^{c\bar{c}}_{\sigma\sigma'}(z''-z')_C-G^{c\bar{c}}_{\sigma\sigma'}(z''-z')_C\right]\right\} \\ &+\int_C dz''\sum_{\mathbf{k}}V^{*}_{a}(\mathbf{k})S^{Ea(\mathbf{k})}(z-z'')_C\sum_{\mathbf{l}}\left[V^{*}_{a}(\mathbf{l})G^{aa\bar{c}\bar{c}}_{\sigma\sigma'}(\mathbf{k},\mathbf{l};z''-z')_C+V^{*}_{b}(\mathbf{l})G^{ab\bar{c}\bar{c}}_{\sigma\sigma'}(\mathbf{k},\mathbf{l};z''-z')_C\right] \\ &-\int_C dz''\sum_{\mathbf{k}}V^{*}_{a}(\mathbf{k})S^{Ea(\mathbf{k})}(z-z'')_C\sum_{\mathbf{l}}\left[V_a(\mathbf{l})G^{ac\bar{a}\bar{c}}_{\sigma\sigma'}(\mathbf{k},\mathbf{l};z''-z')_C+V_b(\mathbf{l})G^{ac\bar{b}\bar{c}}_{\sigma\sigma'}(\mathbf{k},\mathbf{l};z''-z')_C\right] \\ &-\sum_{\mathbf{k}}\left\{V_a(\mathbf{k})\left\langle\left\{c_{-\sigma}c_{\sigma}a^{+}_{-\sigma}(\mathbf{k}),c^{+}_{\sigma'}\right\}\right\rangle S^{2E+U-Ea(\mathbf{k})}(z-z'')_C\right. \\ &\qquad\left.+\left|V_a(\mathbf{k})\right|^2\int_C dz'' S^{2E+U-Ea(\mathbf{k})}(z-z'')_C\left[H^{c\bar{c}}_{\sigma\sigma'}(z''-z')_C-G^{c\bar{c}}_{\sigma\sigma'}(t''-t)\right]\right\} \\ &-\int_C dz''\sum_{\mathbf{k}}V_a(\mathbf{k})S^{2E+U-Ea(\mathbf{k})}(z-z'')_C\sum_{\mathbf{l}}\left\{V^{*}_{a}(\mathbf{l})\left[G^{ac\bar{a}\bar{c}}_{\sigma\sigma'}(\mathbf{l},\mathbf{k};z''-z')_C+G^{ca\bar{a}\bar{c}}_{\sigma\sigma'}(\mathbf{l},\mathbf{k};z''-z')_C\right]\right\} \\ &-\int_C dz''\sum_{\mathbf{k}}V_a(\mathbf{k})S^{2E+U-Ea(\mathbf{k})}(z-z'')_C\sum_{\mathbf{l}}V^{*}_{b}(\mathbf{l})\left[G^{bc\bar{a}\bar{c}}_{\sigma\sigma'}(\mathbf{l},\mathbf{k};z''-z')_C+G^{cb\bar{a}\bar{c}}_{\sigma\sigma'}(\mathbf{l},\mathbf{k};z''-z')_C\right] \end{aligned} \tag{4.29}$$

+ similar terms with suitable interchange $(a\leftrightarrow b)$.

5. Dyson equations for non-equilibrium Green functions of electrons in single-level quantum dot connected with two conducting leads and their solutions

The r.h.s. of equation (4.29) for Green function $H^{c\bar{c}}_{\sigma\sigma'}(z-z')_C$ contains multi-electron Green functions $G^{aa\bar{c}\bar{c}}_{\sigma\sigma'}(\mathbf{k},\mathbf{l};z-z')_C$, $G^{ab\bar{c}\bar{c}}_{\sigma\sigma'}(\mathbf{k},\mathbf{l};z-z')_C$, $G^{ac\bar{a}\bar{c}}_{\sigma\sigma'}(\mathbf{k},\mathbf{l};z-z')_C$, $G^{ac\bar{b}\bar{c}}_{\sigma\sigma'}(\mathbf{k},\mathbf{l};z-z')_C$, $G^{ca\bar{a}\bar{c}}_{\sigma\sigma'}(\mathbf{k},\mathbf{l};z-z')_C$, $G^{bc\bar{a}\bar{c}}_{\sigma\sigma'}(\mathbf{k},\mathbf{l};z-z')_C$, $G^{bc\bar{a}\bar{c}}_{\sigma\sigma'}(\mathbf{k},\mathbf{l};z-z')_C$ and similar ones with suitable interchange (a ↔ b). In order to decouple this equation from those for other multi-electron Green functions we apply the mean-field approximation to the products of four operators. For example

$$\left\langle T_C\left[\bar{a}_{-\sigma}(\mathbf{l};z)a_{-\sigma}(\mathbf{k};z)c_{\sigma}(z)\bar{c}_{\sigma'}(z')\right]\right\rangle\approx\left\langle\bar{a}_{-\sigma}(\mathbf{l};z)a_{-\sigma}(\mathbf{k};z)\right\rangle\left\langle T_C\left[c_{\sigma}(z)\bar{c}_{\sigma'}(z')\right]\right\rangle \tag{5.1}$$

with

$$\left\langle \overline{a}_{-\sigma}(\mathbf{l};z)a_{-\sigma}(\mathbf{k};z)\right\rangle = \delta_{\mathbf{kl}}\left\langle \overline{a}_{-\sigma}(\mathbf{k})a_{-\sigma}(\mathbf{k})\right\rangle = \delta_{\mathbf{kl}}n_a(\mathbf{k}), \tag{5.2}$$

where $n_a(\mathbf{k})$ is the density of electrons with momentum $\mathbf{k}$ and spin projection σ or $-\sigma$ in the lead "a" at the given temperature

$$n_a(\mathbf{k}) = \frac{e^{-\beta Ea(\mathbf{k})}}{1+e^{-\beta Ea(\mathbf{k})}}. \tag{5.3}$$

Note that

$$\left\langle T_C\left[c_\sigma(z)\overline{c}_{\sigma'}(z')\right]\right\rangle = iG^{c\overline{c}}_{\sigma\sigma'}(z-z')_C.$$

As the result we have

$$G^{ac\overline{a}\overline{c}}_{\sigma\sigma'}(\mathbf{k},\mathbf{l};z-z')_C \approx -\delta_{\mathbf{kl}}\left[1-n_a(\mathbf{k})\right]G^{c\overline{c}}_{\sigma\sigma'}(z-z')_C \tag{5.4}$$

and similarly for $G^{bc\overline{b}\overline{c}}_{\sigma\sigma'}(\mathbf{k},\mathbf{l};z-z')_C$. Applying the mean-field approximation to each of others above-mentioned multi-electron Green functions in any manner, we always obtain the vanishing mean value in the lowest order the perturbation theory with respect to the effective tunnelling coupling constants $V_{a,b}(\mathbf{k})$. Note that these functions enter the r. h. s. of the equation (29) with the coefficients of the second order with respect to the effective tunnelling coupling constants. This means that in this second order they do not give contributions. Thus in the second order approximation the equation (4.29) is simplified and becomes

$$\begin{aligned}&\left[i\frac{d}{dz}-(E+U)\right]H^{c\overline{c}}_{\sigma\sigma'}(z-z')_C = n\delta_{\sigma\sigma'}\,\delta(z-z')_C+\Delta(z-z')_C\\&+\int_C dz''\Sigma^{(2)}(z-z')_C\,H^{c\overline{c}}_{\sigma\sigma'}(z''-z')_C-\int_C dz''\Sigma^{(3)}(z-z'')_C\,G^{c\overline{c}}_{\sigma\sigma'}(z''-z')_C,\end{aligned} \tag{5.5}$$

where

$$\begin{aligned}\Delta(z-z')_C = \sum_{\mathbf{k}}\Big[&V_a^*(\mathbf{k})\left\langle\left\{a_{-\sigma}(\mathbf{k})c_\sigma c^+_{-\sigma},c^+_{\sigma'}\right\}\right\rangle S^{Ea(\mathbf{k})}(z-z')_C\\&-V_a(\mathbf{k})\left\langle\left\{c_{-\sigma}c_\sigma a^+_{-\sigma}(\mathbf{k}),c^+_{\sigma'}\right\}\right\rangle S^{2E+U-Ea(\mathbf{k})}(z-z')_C+(a\to b)\Big],\end{aligned} \tag{5.6}$$

$$\Sigma^{(2)}(z-z')_C = \sum_{\mathbf{k}}\left\{\left|V_a(\mathbf{k})\right|^2\left[2S^{Ea(\mathbf{k})}(z-z')_C+S^{2E+U-Ea(\mathbf{k})}(z-z')_C\right]+(a\to b)\right\}, \tag{5.7}$$

$$\Sigma^{(3)}(z-z')_C = \sum_{\mathbf{k}}\left\{n_a(\mathbf{k})\left|V_a(\mathbf{k})\right|^2\left[S^{Ea(\mathbf{k})}(z-z')_C+S^{2E+U-Ea(\mathbf{k})}(z-z')_C\right]+(a\to b)\right\}. \tag{5.8}$$

Note that in the r.h.s. of equations (5.6)-(5.8), there appear the crossing terms containing $S^{2E+U-Ea(\mathbf{k})}(z-z')_C$. They must disappear in the non-crossing approximation. Two equations (4.21) and (5.5) form the closed system of Dyson equations for two Green functions $G^{c\overline{c}}_{\sigma\sigma'}(z-z')_C$ and $H^{c\overline{c}}_{\sigma\sigma'}(z-z')_C$.

To proceed further we note that

$$\left\langle\left\{a_{-\sigma}(\mathbf{k})c_{\sigma}c_{-\sigma}^{+},c_{\sigma'}^{+}\right\}\right\rangle=-\delta_{\sigma\sigma'}\left\langle a_{-\sigma}(\mathbf{k})c_{-\sigma}^{+}\right\rangle,$$
$$\left\langle\left\{c_{-\sigma}c_{\sigma}a_{-\sigma}^{+}(\mathbf{k}),c_{\sigma'}^{+}\right\}\right\rangle=-\delta_{\sigma\sigma'}\left\langle a_{-\sigma}(\mathbf{k})c_{-\sigma}^{+}\right\rangle^{*}, \tag{5.9}$$

where $\left\langle a_{-\sigma}(\mathbf{k})c_{-\sigma}^{+}\right\rangle$ is a limiting value of the Green function $G_{-\sigma-\sigma}^{ac}(\mathbf{k};t)_{11}$:

$$\left\langle a_{-\sigma}(\mathbf{k})c_{-\sigma}^{+}\right\rangle=iG_{-\sigma-\sigma}^{ac}(\mathbf{k};+0)_{11}. \tag{5.10}$$

For evaluating the vertex (5.6) in the second order with respect to the tunnelling coupling constants $V_{a,b}(\mathbf{k})$ we calculate the limiting value (5.10) in the first order. Introduce the Fourier transformations of the Green functions, for example

$$G^{a\bar{c}}(\mathbf{k};t)_{ij}=\frac{1}{2\pi}\int_{-\infty}^{+\infty}d\omega\, e^{-i\omega t}\tilde{G}^{a\bar{c}}(\mathbf{k};\omega)_{ij},$$

$$S^{Ea(\mathbf{k})}(t)_{ij}=\frac{1}{2\pi}\int_{-\infty}^{+\infty}d\omega\, e^{-i\omega t}\tilde{S}^{Ea(\mathbf{k})}(\omega)_{ij}, \tag{5.11}$$

$$G^{c\bar{c}}(t)_{ij}=\frac{1}{2\pi}\int_{-\infty}^{+\infty}d\omega\, e^{-i\omega t}\tilde{G}^{c\bar{c}}(\omega)_{ij}\quad \text{for } i,j=1,2.$$

From the equation (4.20) it follows that

$$\tilde{G}^{a\bar{c}}(\mathbf{k};\omega)_{11}=V_{a}(\mathbf{k})\tilde{S}^{Ea(\mathbf{k})}(\omega)_{11}\tilde{G}^{c\bar{c}}(\omega)_{11}. \tag{5.12}$$

For deriving $\tilde{G}^{a\bar{c}}(\mathbf{k};\omega)_{11}$ in the first order with respect to the constant $V_{a}(\mathbf{k})$ it is enough to use the expression of $\tilde{G}^{c\bar{c}}(\omega)_{11}$ in the case of the vanishing tunnelling coupling constant and have

$$\tilde{G}^{a\bar{c}}(\mathbf{k};\omega)_{11}=\frac{1}{Z}V_{a}(\mathbf{k})\left[\frac{n_{a}(\mathbf{k})}{\omega-io-E_{a}(\mathbf{k})}+\frac{1-n_{a}(\mathbf{k})}{\omega+io-E_{a}(\mathbf{k})}\right]$$
$$\times\left[\frac{e^{-\beta E}}{\omega-io-E}+\frac{e^{-\beta(2E+U)}}{\omega-io-E-U}+\frac{1}{\omega+io-E}+\frac{e^{-\beta E}}{\omega+io-E-U}\right], \tag{5.13}$$

$$Z=1+2e^{-\beta E}+e^{-\beta(2E+U)}.$$

It is easy to calculate the limit

$$G^{a\bar{c}}(\mathbf{k};+0)_{11}=\lim_{\varepsilon\to+0}\frac{1}{2\pi}\int_{-\infty}^{+\infty}d\omega\, e^{-i\omega\varepsilon}\,\tilde{G}^{a\bar{c}}(\mathbf{k};\omega)_{11} \tag{5.14}$$

by using the residue theorem and obtain

$$\left\langle a_{-\sigma}(\mathbf{k})c_{-\sigma}^{+}\right\rangle=-\nu_{a}(\mathbf{k})V_{a}(\mathbf{k}), \tag{5.15}$$

where

$$\nu_a(\mathbf{k})=\frac{1}{Z}\left\{\frac{e^{-\beta E}-\left[1+e^{-\beta E}\right]n_a(\mathbf{k})}{E-E_a(\mathbf{k})}+e^{-\beta E}\frac{e^{-\beta(E+U)}-\left[1+e^{-\beta(E+U)}\right]n_a(\mathbf{k})}{E+U-E_a(\mathbf{k})}\right\}. \tag{5.16}$$

The formula (5.6) becomes

$$\Delta(z-z')_C=\sum_{\mathbf{k}}\left\{\left|V_a(\mathbf{k})\right|^2\nu_a(\mathbf{k})\left[S^{Ea(\mathbf{k})}(z-z')_C-S^{2E+U-Ea(\mathbf{k})}(z-z')_C\right]+(a\to b)\right\}. \tag{5.17}$$

The system of Dyson equations (4.21) and (5.5) is the mathematical tool for the study of the electron transport through a single-level QD. Since this is a stationary process one can apply the Keldysh non-equilibrium Green function formalism in the limit $t_0\to-\infty$. Because the interaction vanishes at this limit, the contour C can be considered to consist of only two segment $C_1=[-\infty+io,+\infty+io]$ and $C_2=[\infty-io,-\infty+io]$. In this case each complex time-dependent Green function $S^E(z-z')_C$, $G^{c\bar{c}}(z-z')_C$, $H^{c\bar{c}}(z-z')_C$, $\Sigma^{(\alpha)}(z-z')_C$, $\alpha=1,2,3$ or $\Delta(z-z')_C$ becomes a set of four real time-dependent functions $S^E(t-t')_{ij}$, $G^{c\bar{c}}(t-t')_{ij}$, $H^{c\bar{c}}(t-t')_{ij}$, $\Sigma^{(\alpha)}(t-t')_{ij}$, $\alpha=1,2,3$ or $\Delta(t-t')_{ij}$ with their Fourier transforms $\tilde{S}^E(\omega)_{ij}$, $\tilde{G}^{c\bar{c}}(\omega)_{ij}$, $\tilde{H}^{c\bar{c}}(\omega)_{ij}$, $\tilde{\Sigma}^{(\alpha)}(\omega)_{ij}$, $\alpha=1,2,3$ or $\tilde{\Delta}(\omega)_{ij}$ with $i,j=1,2$. Considering them as the elements of corresponding 2×2 matrices $\hat{S}^E(\omega)$, $\hat{G}^{c\bar{c}}(\omega)$, $\hat{H}^{c\bar{c}}(\omega)$, $\hat{\Sigma}^{(\alpha)}(\omega)$, $\alpha=1,2,3$ or $\hat{\Delta}(\omega)$, and setting

$$\hat{\eta}=\begin{pmatrix}1 & 0\\ 0 & -1\end{pmatrix}, \tag{5.18}$$

we rewrite the system of Dyson equations (4.21) and (5.5) in the matrix form

$$\hat{G}(\omega)=\hat{S}^E(\omega)+U\hat{S}^E(\omega)\hat{\eta}\hat{H}(\omega)+\hat{S}^E(\omega)\hat{\eta}\hat{\Sigma}^{(1)}(\omega)\hat{\eta}\hat{G}(\omega), \tag{5.19}$$

$$\begin{aligned}\hat{H}(\omega)=n\hat{S}^{E+U}(\omega)+\hat{S}^{E+U}(\omega)\hat{\eta}\hat{\Delta}(\omega)+\hat{S}^{E+U}(\omega)\hat{\eta}\hat{\Sigma}^{(2)}(\omega)\hat{\eta}\hat{H}(\omega)\\ -\hat{S}^{E+U}(\omega)\hat{\eta}\hat{\Sigma}^{(3)}(\omega)\hat{\eta}\hat{G}(\omega).\end{aligned} \tag{5.20}$$

From these matrix equations we derive two systems of algebraic equations, each of which consists of four equations for four functions $\tilde{G}(\omega)_{i1}$ and $\tilde{H}(\omega)_{i1}$ or $\tilde{G}(\omega)_{i2}$ and $\tilde{H}(\omega)_{i2}$, $i=1,2$. The observable physical quantities are expressed in terms of these functions.

For the application let us calculate the Green function $\tilde{G}(\omega)_{11}$. By solving the system of equation (5.19) and (5.20) we obtain following result:

$$\tilde{G}(\omega)_{11}=\frac{Z(\omega)}{Y(\omega)}, \tag{5.21}$$

$$\begin{aligned}Z(\omega)=&\left\{B(\omega)[1+\Omega_{22}^{(1)}(\omega)]-UD_{22}(\omega)\right\}\left\{B(\omega)\tilde{S}^E(\omega)_{11}+UC_1(\omega)\right\}\\ &-\left\{B(\omega)\Omega_{12}^{(1)}(\omega)-UD_{12}(\omega)\right\}\left\{B(\omega)\tilde{S}^E(\omega)_{21}+UC_2(\omega)\right\},\end{aligned} \tag{5.22}$$

$$\begin{aligned}Y(\omega)=&\left\{B(\omega)[1-\Omega_{11}^{(1)}(\omega)]+UD_{11}(\omega)\right\}\left\{B(\omega)[1+\Omega_{22}^{(1)}(\omega)]-UD_{22}(\omega)\right\}\\ &+\left\{B(\omega)\Omega_{12}^{(1)}(\omega)-UD_{12}(\omega)\right\}\left\{B(\omega)\Omega_{21}^{(1)}(\omega)-UD_{21}(\omega)\right\},\end{aligned} \tag{5.23}$$

$$B(\omega)=[1+\Omega_{22}^{(2)}(\omega)][1-\Omega_{11}^{(2)}(\omega)]+\Omega_{12}^{(2)}(\omega)\Omega_{21}^{(2)}(\omega), \tag{5.24}$$

$$\begin{aligned}C_i(\omega)=&\left\{\tilde{S}^E(\omega)_{i1}[1+\Omega_{22}^{(2)}(\omega)]-\tilde{S}^E(\omega)_{i2}\Omega_{21}^{(2)}(\omega)\right\}\lambda_{11}(\omega)\\&-\left\{\tilde{S}^E(\omega)_{i1}\Omega_{12}^{(2)}(\omega)+\tilde{S}^E(\omega)_{i2}[1-\Omega_{11}^{(2)}(\omega)]\right\}\lambda_{21}(\omega),\end{aligned} \tag{5.25}$$

$$i=1,2$$

$$\begin{aligned}D_{ij}(\omega)=&\left\{\tilde{S}^E(\omega)_{i1}[1+\Omega_{22}^{(2)}(\omega)]-\tilde{S}^E(\omega)_{i2}\Omega_{21}^{(2)}(\omega)\right\}\Omega_{1j}^{(3)}(\omega)-\\&-\left\{\tilde{S}^E(\omega)_{i1}\Omega_{12}^{(2)}(\omega)+\tilde{S}^E(\omega)_{i2}[1-\Omega_{11}^{(2)}(\omega)]\right\}\Omega_{2j}^{(3)}(\omega),\end{aligned} \tag{5.26}$$

$$\begin{aligned}\Omega_{ij}^{(1)}(\omega)&=\tilde{S}^E(\omega)_{i1}\tilde{\Sigma}^{(1)}(\omega)_{1j}-\tilde{S}^E(\omega)_{i2}\tilde{\Sigma}^{(1)}(\omega)_{2j},\\\Omega_{ij}^{(2)}(\omega)&=\tilde{S}^{E+U}(\omega)_{i1}\tilde{\Sigma}^{(2)}(\omega)_{1j}-\tilde{S}^{E+U}(\omega)_{i2}\tilde{\Sigma}^{(2)}(\omega)_{2j},\\\Omega_{ij}^{(3)}(\omega)&=\tilde{S}^{E+U}(\omega)_{i1}\tilde{\Sigma}^{(3)}(\omega)_{1j}-\tilde{S}^{E+U}(\omega)_{i2}\tilde{\Sigma}^{(3)}(\omega)_{2j},\end{aligned} \tag{5.27}$$

$$\lambda_{ij}(\omega)=n\tilde{S}^{E+U}(\omega)_{ij}+\tilde{S}^{E+U}(\omega)_{i1}\tilde{\Delta}(\omega)_{1j}-\tilde{S}^{E+U}(\omega)_{i2}\tilde{\Delta}(\omega)_{2j}, \tag{5.28}$$

$$i=1,2,\quad j=1,2.$$

The expressions (5.21)-(5.28) of $\tilde{G}(\omega)_{11}$ and similar ones for the Fourier transforms of other Green functions contain the self-energies $\tilde{\Sigma}^{(\alpha)}(\omega)_{ij},\alpha=1,2,3$. Because the tunnelling coupling constants $V_a(\mathbf{k})$ and $V_b(\mathbf{k})$ have small values, the contributions of these self-energies, in general, give small corrections to the Green functions. However, the self-energies may be divergent at some special values of ω. At some points near these special values the denominator $Y(\omega)$ may vanish and the Green functions have the resonances. The formulae (5.23)-(5.28) would be used for the rigorous study of the behaviour of $\tilde{G}(\omega)_{11}$ at the resonances. This will be done in the subsequent Section.

6. Kondo and Fano resonances in electron transport through single-level quantum dot

In this Section we study the appearance of the resonances in the expressions of the Fourier transforms of the Green functions when the denominator $Y(\omega)$ is vanishing. The expression of $Y(\omega)$ consists of the terms of two types: the finite terms which do not depend on the Fourier transforms $\tilde{\Sigma}^{(\alpha)}(\omega)_{ij}$ of the self-energies and those proportional to $\tilde{\Sigma}^{(\alpha)}(\omega)_{ij},\alpha=1,2,3$. The functions $\tilde{\Sigma}^{(\alpha)}(\omega)_{ij}$ contain the small tunnelling coupling constants $V_a(\mathbf{k})$ and $V_b(\mathbf{k})$. They are determined by following formulae:

$$\tilde{\Sigma}^{(1)}(\omega)_{ij}=\sum_{\mathbf{k}}\left\{|V_a(\mathbf{k})|^2\tilde{S}^{E_a(\mathbf{k})}(\omega)_{ij}+(a\to b)\right\}, \tag{6.1}$$

$$\tilde{\Sigma}^{(2)}(\omega)_{ij}=\sum_{\mathbf{k}}\left\{|V_a(\mathbf{k})|^2[2\tilde{S}^{E_a(\mathbf{k})}(\omega)_{ij}+\tilde{S}^{2E+U-E_a(\mathbf{k})}(\omega)_{ij}]+(a\to b)\right\}, \tag{6.2}$$

$$\tilde{\Sigma}^{(3)}(\omega)=\sum_{\mathbf{k}}\left\{n_a(\mathbf{k})|V_a(\mathbf{k})|^2[\tilde{S}^{E_a(\mathbf{k})}(\omega)_{ij}+\tilde{S}^{2E+U-E_a(\mathbf{k})}(\omega)_{ij}]+(a\to b)\right\}, \tag{6.3}$$

Introducing the spectral functions

$$\Gamma^{(\alpha)}_{a,b}(\omega) = \pi \sum_{\mathbf{k}} \left[\frac{e^{-\beta E_{a,b}(\mathbf{k})}}{1+e^{-\beta E_{a,b}(\mathbf{k})}} \right]^{\alpha} |V_{a,b}(\mathbf{k})|^2 \, \delta[\omega - E_{a,b}(\mathbf{k})], \tag{6.4}$$

$$a = 0, 1, 2,$$

we rewrite them in the new form convenient for the study of their divergence:

$$\begin{aligned} \tilde{\Sigma}^{(1)}(\omega)_{11} = \frac{1}{\pi} P \int d\omega' \frac{1}{\omega-\omega'} \left[\Gamma^{(0)}_a(\omega') + \Gamma^{(0)}_b(\omega') \right] \\ -i\left[\Gamma^{(0)}_a(\omega) + \Gamma^{(0)}_b(\omega) - 2\Gamma^{(1)}_a(\omega) - 2\Gamma^{(1)}_b(\omega) \right], \end{aligned} \tag{6.5}$$

$$\tilde{\Sigma}^{(1)}(\omega)_{12} = 2i\left[\Gamma^{(1)}_a(\omega) + \Gamma^{(1)}_b(\omega) \right],$$

$$\tilde{\Sigma}^{(1)}(\omega)_{21} = -2i\left[\Gamma^{(0)}_a(\omega) + \Gamma^{(0)}_b(\omega) - \Gamma^{(1)}_a(\omega) - \Gamma^{(1)}_b(\omega) \right],$$

$$\begin{aligned} \tilde{\Sigma}^{(1)}(\omega)_{22} = -\frac{1}{\pi} P \int d\omega' \frac{1}{\omega-\omega'} \left[\Gamma^{(0)}_a(\omega') + \Gamma^{(0)}_b(\omega') \right] \\ -i\left[\Gamma^{(0)}_a(\omega) + \Gamma^{(0)}_b(\omega) - 2\Gamma^{(1)}_a(\omega) - 2\Gamma^{(1)}_b(\omega) \right]. \end{aligned}$$

$$\begin{aligned} \tilde{\Sigma}^{(2)}(\omega)_{11} = \frac{1}{\pi} P \int d\omega' \left[\frac{2}{\omega-\omega'} + \frac{1}{\omega-2E-U+\omega'} \right] \left[\Gamma^{(0)}_a(\omega') + \Gamma^{(0)}_b(\omega') \right] - \\ - 2i\left[\Gamma^{(0)}_a(\omega) + \Gamma^{(0)}_b(\omega) - 2\Gamma^{(1)}_a(\omega) - 2\Gamma^{(1)}_b(\omega) \right] - \\ - i\left[\Gamma^{(0)}_a(2E+U-\omega) + \Gamma^{(0)}_b(2E+U-\omega) \right. \\ \left. -2\Gamma^{(1)}_a(2E+U-\omega) - 2\Gamma^{(1)}_b(2E+U-\omega) \right], \end{aligned} \tag{6.6}$$

$$\tilde{\Sigma}^{(2)}(\omega)_{12} = 4i\left[\Gamma^{(1)}_a(\omega) + \Gamma^{(1)}_b(\omega) \right] + 2i\left[\Gamma^{(1)}_a(2E+U-\omega) + \Gamma^{(1)}_b(2E+U-\omega) \right],$$

$$\begin{aligned} \tilde{\Sigma}^{(2)}(\omega)_{21} = -4i\left[\Gamma^{(0)}_a(\omega) + \Gamma^{(0)}_b(\omega) - \Gamma^{(1)}_a(\omega) - \Gamma^{(1)}_b(\omega) \right] - \\ -2i\left[\Gamma^{(0)}_a(2E+U-\omega) + \Gamma^{(0)}_b(2E+U-\omega) \right. \\ \left. -\Gamma^{(1)}_a(2E+U-\omega) - \Gamma^{(1)}_b(2E+U-\omega) \right], \end{aligned}$$

$$\begin{aligned} \tilde{\Sigma}^{(2)}(\omega)_{22} = -\frac{1}{\pi} P \int d\omega' \left[\frac{2}{\omega-\omega'} + \frac{1}{\omega-2E-U+\omega'} \right] \left[\Gamma^{(0)}_a(\omega') + \Gamma^{(0)}_b(\omega') \right] - \\ -2i\left[\Gamma^{(0)}_a(\omega) + \Gamma^{(0)}_b(\omega) - 2\Gamma^{(1)}_a(\omega) - 2\Gamma^{(1)}_b(\omega) \right] - \\ -i\left[\Gamma^{(0)}_a(2E+U-\omega) + \Gamma^{(0)}_b(2E+U-\omega) \right. \\ \left. -2\Gamma^{(1)}_a(2E+U-\omega) - 2\Gamma^{(1)}_b(2E+U-\omega) \right]. \end{aligned}$$

$$\tilde{\Sigma}^{(3)}(\omega)_{11} = \frac{1}{\pi} P \int d\omega' \left[\frac{1}{\omega - \omega'} + \frac{1}{\omega - 2E - U + \omega'} \right] \left[\Gamma_a^{(1)}(\omega') + \Gamma_b^{(1)}(\omega') \right] -$$
$$- i \left[\Gamma_a^{(1)}(\omega) + \Gamma_b^{(1)}(\omega) - 2\Gamma_a^{(2)}(\omega) - 2\Gamma_b^{(2)}(\omega) + \Gamma_a^{(1)}(2E + U - \omega) \right.$$
$$\left. + \Gamma_b^{(1)}(2E + U - \omega) - 2\Gamma_a^{(2)}(2E + U - \omega) - 2\Gamma_b^{(2)}(2E + U - \omega) \right], \tag{6.7}$$

$$\tilde{\Sigma}^{(3)}(\omega)_{12} = 2i \left[\Gamma_a^{(2)}(\omega) + \Gamma_b^{(2)}(\omega) \right] + 2i \left[\Gamma_a^{(2)}(2E + U - \omega) + \Gamma_b^{(2)}(2E + U - \omega) \right],$$

$$\tilde{\Sigma}^{(3)}(\omega)_{21} = -2i \left[\Gamma_a^{(1)}(\omega) + \Gamma_b^{(1)}(\omega) - \Gamma_a^{(2)}(\omega) - \Gamma_b^{(2)}(\omega) + \Gamma_a^{(1)}(2E + U - \omega) \right.$$
$$\left. + \Gamma_b^{(1)}(2E + U - \omega) - \Gamma_a^{(2)}(2E + U - \omega) - \Gamma_b^{(2)}(2E + U - \omega) \right],$$

$$\tilde{\Sigma}^{(3)}(\omega)_{22} = -\frac{1}{\pi} P \int d\omega' \left[\frac{1}{\omega - \omega'} + \frac{1}{\omega - 2E - U + \omega'} \right] \left[\Gamma_a^{(1)}(\omega') + \Gamma_b^{(1)}(\omega') \right] -$$
$$- i \left[\Gamma_a^{(1)}(\omega) + \Gamma_b^{(1)}(\omega) - 2\Gamma_a^{(2)}(\omega) - 2\Gamma_b^{(2)}(\omega) + \Gamma_a^{(1)}(2E + U - \omega) \right.$$
$$\left. + \Gamma_b^{(1)}(2E + U - \omega) - 2\Gamma_a^{(2)}(2E + U - \omega) - 2\Gamma_b^{(2)}(2E + U - \omega) \right].$$

The integrals in the r.h.s. of formulae (6.5)-(6.7) may be divergent at definite values of the frequency ω which will be called the divergence points. Although the functions $\tilde{\Sigma}^{(\alpha)}(\omega)_{ij}$, i, j = 1, 2, contain the small tunnelling coupling constants $V_a(\mathbf{k})$ and $V_b(\mathbf{k})$, near each divergence point some of them may become comparable with the finite terms in $Y(\omega)$. When $Y(\omega)$ vanishes due to the cancellation between the finite terms and those containing divergent integrals, there appear the resonances. Therefore in order to study the resonances it is necessary to investigate the divergence of the integrals in the r.h.s. of the formulae (6.5)-(6.7).

The functions $\tilde{\Sigma}^{(\alpha)}(\omega)_{11}$ and $\tilde{\Sigma}^{(\alpha)}(\omega)_{22}$ contain the dispersion integrals with the spectral functions $\Gamma_{a,b}^{(\alpha)}(\omega)$. Denote μ_a and μ_b the chemical potentials of the systems of conducting electrons in the leads "a" and "b". From the definition (6.4) with

$$E_{a,b}(\mathbf{k}) = E_{a,b}^{(0)}(\mathbf{k}) - \mu_{a,b} \,,$$

where $E_{a,b}^{(0)}(\mathbf{k})$ are the kinetic energies of the conducting electrons in the leads, $E_{a,b}^{(0)}(\mathbf{k}) \geq 0$, it follows that $\Gamma_a^{(\alpha)}(\omega)$ vanishes at $\omega < -\mu_a$ and similarly for $\Gamma_b^{(n)}(\omega)$. Therefore the dispersion integrals in formulae (6.5)-(6.7) have the form

$$K_{a,b}^{(n)} = P \int_{-\mu_{a,b}}^{\Omega_{a,b}} d\omega' \frac{\Gamma_{a,b}^{(n)}(\omega')}{\omega' - \omega} \tag{6.8}$$

and

$$L_{a,b}^{(n)} = P \int_{-\mu_{a,b}}^{\Omega_{a,b}} d\omega' \frac{\Gamma_{a,b}^{(n)}(\omega')}{\omega' - 2E - U + \omega} \,, \tag{6.9}$$

where Ω_a is the top of the energy band of the conducting electrons in the leads "a" and similarly for Ω_b. For the study of the divergence of the integrals we replace approximately the values of $\Gamma_a^{(\alpha)}(\omega)$ in the interval $-\mu < \omega < \Omega_a$ by a constant Γ_a and similarly for $\Gamma_b^{(n)}(\omega)$. Then at zero temperature

$$\begin{aligned} K_{a,b}^{(0)} &= \Gamma_{a,b} I_{a,b} , \\ K_{a,b}^{(1)} &= K_{a,b}^{(2)} = \Gamma_{a,b} I'_{a,b} , \\ L_{a,b}^{(0)} &= \Gamma_{a,b} J_{a,b} , \\ L_{a,b}^{(1)} &= L_{a,b}^{(2)} = \Gamma_{a,b} J'_{a,b} , \end{aligned} \tag{6.10}$$

with

$$I_{a,b} = P \int_{-\mu_{a,b}}^{\Omega_{a,b}} d\omega' \frac{1}{\omega - \omega'} , \tag{6.11}$$

$$I'_{a,b} = P \int_{-\mu_{a,b}}^{0} d\omega' \frac{1}{\omega - \omega'} , \tag{6.12}$$

$$J_{a,b} = P \int_{-\mu_{a,b}}^{\Omega_{a,b}} d\omega' \frac{1}{\omega - 2E - U + \omega'} , \tag{6.13}$$

$$J'_{a,b} = P \int_{-\mu_{a,b}}^{0} d\omega' \frac{1}{\omega - 2E - U + \omega'} . \tag{6.14}$$

Usually $\Omega_a(\Omega_b)$ is very large in comparison with $\mu_a(\mu_b)$ and ω. Therefore we have

$$I_{a,b} \approx \ln\left|\frac{\omega + \mu_{a,b}}{\Omega_{a,b} - \mu_{a,b}}\right| , \tag{6.15}$$

$$I'_{a,b} \approx \ln\left|\frac{\omega + \mu_{a,b}}{\omega}\right| , \tag{6.16}$$

$$J_{a,b} \approx \ln\left|\frac{\Omega_{a,b} - \mu_{a,b}}{\omega - 2E - U - \mu_{a,b}}\right| , \tag{6.17}$$

$$J'_{a,b} \approx \ln\left|\frac{\omega - 2E - U}{\omega - 2E - U - \mu_{a,b}}\right| . \tag{6.18}$$

It is obvious that I_a is divergent at $\omega \to -\mu_a$, I'_a is divergent at $\omega \to -\mu_a$ and $\omega \to 0$, J_a is divergent at $\omega \to 2E + U + \mu_a$ and J'_a is divergent at $\omega \to 2E + U + \mu_a$ and $\omega \to 2E + U$. For I_b, I'_b, J_b and J'_b we have similar results.

If the temperature T of the system is low enough, but does not vanish,

$$\mu_{a,b} >> kT > 0\,,$$

then instead of the divergence of $I'_{a,b}$ at the Fermi surface $\omega \to 0$ we have the limit

$$\lim_{\omega\to 0} I'_{a,b} = \ln\frac{2e\mu_{a,b}}{kT}\,, \tag{6.19}$$

and, similarly, instead of the divergence of $J'_{a,b}$ at $\omega \to 2E+U$ we have the limit

$$\lim_{\omega\to 2E+U} J'_{a,b} = -\ln\frac{2e\mu_{a,b}}{kT}\,. \tag{6.20}$$

For the simplicity we set $\mu_a = \mu_a = \mu$.

From the results of the study of resonances of Green function $\tilde{G}^{c\bar{c}}(\omega)_{11}$ and the explicit expressions (5.21)-(5.28) determining this function we obtain its asymptotic behavious at the divergence points of $\tilde{\Sigma}^{(\alpha)}(\omega)_{ij}$:

a. As $\omega \to -\mu$ and at low temperature $T \approx 0$, the Green function in equation (5.21) has asymptotic form:

$$\begin{aligned}\tilde{G}^{c\bar{c}}(\omega)_{11} \approx &-\frac{\left[1-n+\frac{1}{2}(E+\mu)\frac{\lambda}{\Gamma}\right]U}{E+2U+\mu}\,\frac{1}{\frac{1}{2}(E+\mu)+\frac{2\Gamma}{\pi}\ln\left|\frac{\omega+\mu}{\Omega}\right|+2i\Gamma} \\ &-\frac{E+\mu+\left[1+n-(E+U+\mu)\frac{\lambda}{\Gamma}\right]U}{E+2U+\mu}\,\frac{1}{(E+U+\mu)+\frac{2\Gamma}{\pi}\ln\left|\frac{\omega+\mu}{\Omega}\right|+2i\Gamma}\,.\end{aligned} \tag{6.21}$$

If $E+\mu>0$, then $\tilde{G}^{c\bar{c}}(\omega)_{11}$ has two resonances at two points

$$\omega_1^{(\pm)} = -\mu \pm \Omega e^{-\frac{\pi}{4\Gamma}(E+\mu)} \tag{6.22}$$

and two resonances at two points

$$\omega_2^{(\pm)} = -\mu \pm \Omega e^{-\frac{\pi}{2\Gamma}(E+U+\mu)}\,. \tag{6.23}$$

Between these four resonances there are the dips. If $E+\mu<0$ but $E+U+\mu>0$, then $\tilde{G}^{c\bar{c}}(\omega)_{11}$ has only two resonances at the points $\omega_2^{(\pm)}$. If $E+U+\mu<0$, then in the neighbourhood of the point $\omega=-\mu$, the Green function $\tilde{G}^{c\bar{c}}(\omega)_{11}$ has no resonance. All four points $\omega_1^{(\pm)}$ and $\omega_2^{(\pm)}$ are very close to the point $\omega=-\mu$ and the resonances at $\omega_1^{(\pm)}$ and $\omega_2^{(\pm)}$ look like a resonance at $\omega=-\mu$. The origin of these resonances is the presence of the Fano quasi-bound state at the lower edge of the energy band of the conducting electrons. If they exist, they would be called the Fano resonances.

b. As $\omega \to 0$ and at $T = 0$, the Green functions $\tilde{G}^{c\bar{c}}(\omega)_{11}$ has asymptotic form

$$\tilde{G}^{c\bar{c}}(\omega)_{11} \approx -\frac{E+(1-n)U}{E(E+U)+\frac{2\Gamma U}{\pi}\ln\left|\frac{\mu}{\omega}\right|+2i(3E+2U)\Gamma}. \tag{6.24}$$

If $E(E+U)<0$, then $\tilde{G}^{c\bar{c}}(\omega)_{11}$ has two resonances at the points

$$\omega_3^{(\pm)} = \pm\mu\exp\left\{-\frac{\pi}{2\Gamma}\frac{|E(E+U)|}{U}\right\}, \tag{6.25}$$

which are very close to the point $\omega = 0$. At $\omega = 0$ and $0 < T < T_K$,

$$T_K = \frac{1}{k}\mu\exp\left\{-\frac{\pi}{2}\frac{|E(E+U)|}{\Gamma U}\right\}, \tag{6.26}$$

where k is the Boltzmann constant, instead of formula (6.24) we have

$$\tilde{G}^{c\bar{c}}(0)_{11} \approx \frac{\pi}{2}\frac{E+(1-n)U}{\Gamma U}\frac{1}{\ln|T/T_K|-i\pi(3E+2U)\Gamma}. \tag{6.27}$$

The resonances in the neighbourhood of the point $\omega = 0$ have the same physical origin as the Kondo effect due to the scattering of electrons by a magnetic impurity. They are the Kondo resonances.

c. As $\omega \to 2E+U$ and at $T = 0$, the Green function $\tilde{G}^{c\bar{c}}(\omega)_{11}$ has asymptotic form:

$$\tilde{G}^{c\bar{c}}(\omega)_{11} \approx \frac{E+nU}{E(E+U)-\frac{2\Gamma U}{\pi}\ln\left|\frac{\mu}{\omega-2E-U}\right|-2iE\Gamma}. \tag{6.28}$$

Therefore if $E(E+U)>0$, then $\tilde{G}^{c\bar{c}}(\omega)_{11}$ has also two resonances at the points

$$\omega_4^{(\pm)} = 2E+U \pm \mu e^{-\frac{\pi}{2\Gamma}\frac{E(E+U)}{U}} \tag{6.29}$$

which are very close to the point $\omega = 2E+U$. At $\omega = 2E+U$ and $0 < T < T_K'$,

$$T_K' = \frac{1}{k}\mu\exp\left\{-\frac{\pi}{2}\frac{E(E+U)}{\Gamma U}\right\}, \tag{6.30}$$

instead of formula (6.28) we have

$$\tilde{G}^{c\bar{c}}(2E+U)_{11} \approx \frac{\pi}{2}\frac{E+nU}{\Gamma U}\frac{1}{\ln|T/T_K'|-i\pi E\Gamma}. \tag{6.31}$$

The resonances in the neighbourhood of the point $\omega = 2E+U$ are the Kondo resonances of the crossing terms.

d. As $\omega \to 2E+U+\mu$ and at low temperature $T \approx 0$, the Green functions $\tilde{G}^{c\bar{c}}(\omega)_{11}$ has asymptotic form:

$$\tilde{G}^{c\bar{c}}(\omega)_{11} \approx \frac{1-n+(E+U+\mu)\frac{\lambda}{\Gamma}}{(E+U+\mu)+\frac{2\Gamma}{\pi}\ln\left|\frac{\omega-2E-U-\mu}{\Omega}\right|-2i\Gamma}\frac{U}{E+\mu}. \tag{6.32}$$

If $E+U+\mu>0$, then $\tilde{G}^{c\bar{c}}(\omega)_{11}$ has two resonances at the points

$$\omega_5^{(\pm)} = 2E+U+\mu \pm \Omega e^{-\frac{\pi}{2}\frac{E+U+\mu}{\Gamma}}, \tag{6.33}$$

which are very close to the point $\omega = 2E+U+\mu$. They are the Fano resonances of the crossing terms.

7. Conclusion

The present Chapter is an introductory review of the Keldysh non-equilibrium Green functions of electrons in simplest nanosystems: isolated single-level QD and single-level QD connected with two conducting leads. In the case of an isolated single-level QD the closed system of a finite number of differential equations for a finite number of Green functions was established by using the Heisenberg quantum equations of motion for the electron destruction and creation operators. The exact expressions of the Green functions were derived. In the case of the nanosystem consisting of a single-level QD connected with two conducting leads there does not exist a finite closed system of differential equations for some finite number of Green functions. In the differential equations for n-point Green functions there appear the contributions from (n+2)-point Green functions. Therefore, the exact system of differential equations contains an infinite number of equations for an infinite number of Green functions. In order to truncate this infinite system of differential equations we have applied the mean-field approximation to the products of four electron quantum operators and limited at the terms of the second order with respect to the effective tunnelling coupling constants. As the result we have derived a closed system of Dyson equations for two types of 2-point Green functions. All the crossing terms are included into the equations. The exact solution of the system of Dyson equations may have the resonances of four types in the dependence on the physical parameters of the system: the Kondo resonances at the Fermi surface, whose origin is similar to that of the Kondo effect in the scattering of electrons on magnetic impurities, the Fano resonances due to the presence of the electron quasi-bound state at the lower edge of the energy band of the conducting electrons, the Kondo resonances in the crossing terms and the Fano resonances in the crossing terms. The analytical asymptotic expressions of the single-electron Green function at these resonances were derived. These results agree well with the numerical calculations in references on the electron Green functions in QD (Yeyati et al., 1993; Costi et al., 1994; Izumida et al., 1997, 1998, 2001; Sakai et al., 1999; Torio et al., 2002).

The theoretical study of the non-equilibrium Green functions of electrons in QDs would signify the beginning of the development of the quantum dynamics of physical processes in QD-based nanodevices. The next step would be the elaboration of the theory of non-equilibrium Green functions of phonons in QDs as well as of electrons and phonons of interacting electron-phonon systems in QDs. The quantum dynamical theory of QD-based optoelectronic and photonic nanodevices necessitates also the study of non-equilibrium Green functions of electrons and phonons confined in QDs in the presence of the electron-phonon interactions as well as the interaction of photons with confined electron-phonon systems. The methods and reasonnings presented in this Chapter could be generalized for the application to the study of all above-mentioned non-equilibrium Green functions.

8. Acknowledgment

I would like to express the gratitude to Institute of Materials Science and Vietnam Academy of Science and Technology for the support to my work on the subject of this review during many years. I thank also Academician Nguyen Van Hieu for suggesting the main ideas of the series of publications on this subject.

9. References

Abrikosov, A. A; Gorkov, L. P. & Dzyaloshinski, I. E. (1975). *Methods of Quantum Field Theory in Statistical Physics,* Dover Publications, ISBN 0486632288, New York

Bjoken, J. D & Drell, S. D. (1964). *Relativistic Quantum Mechanics,* Mc Graw-Hill, ISBN 0070054932, New York

Bruus, H. & Flensberg, K. (2004). *Many-Body Quantum Theory in Condensed Matter Physics,* Oxford University Press, ISBN 0198566336, Oxford New York

Choi, M.-S.; Sánchez, D. & López, R. (2004). Kondo effect in a quantum dot coupled to ferromagnetic leads: A numerical renormalization group analysis, *Physical Review Letters,* Vol. 92, No. 5, p. 056601 (4 pages); ISSN 0031-9007 (print), 1079-7114 (online)

Chou, K. C.; Su, Z. B.; Lao, B. L. & Yu, L. (1985). Equilibrium and nonequilibrium formalisms made unified, *Physics Reports,* Vol. 118, Nos. 1&2, pp. 1-131, ISSN 0370-1573

Costi, T. A.; Hewson, A. C. & Zlatic, V. (1994). Transport coefficients of the Anderson model via the numerical renormalization group, *Journal of Physics: Condensed Matter,* Vol. 6, No. 13, p. 2519 (1994), ISSN 0953-8984 (print), 1361-648X (online)

Craco, L & Kang, K. (1999). Perturbation treatment for transport through a quantum dot, *Physical Review B,* Vol. 59, No. 19, pp. 12244–12247; ISSN 1098-0121 (print), 1550-235X (online)

Fujii, T. & Ueda, K. (2003). Perturbative approach to the nonequilibrium Kondo effect in a quantum dot, *Physical Review B,* Vol. 68, No. 15, p.155310 (5 pages); ISSN 1098-0121 (print), 1550-235X (online)

Haken, H. (1976). *Quantum Field Theory of Solids: An Introduction,* North-Holland Pub. Co., ISBN 0720405459, Amsterdam

Hershfield, S.; Davies, J. H. & Wilkins J. W. (1991). Probing the Kondo resonance by resonant tunneling through an Anderson impurity, *Physical Review Letters,* Vol. 67, No. 26 , pp. 3720-3723; ISSN 0031-9007 (print), 1079-7114 (online)

Inoshita, T.; Shimizu, A.; Kuramoto, Y & Sakaki, H. (1993). Correlated electron transport through a quantum dot: The multiple-level effect, *Physical Review B,* Vol. 48, No. 19, pp. 14725–14728; ISSN 1098-0121 (print), 1550-235X (online)

Itzykson, C. & Zuber, J. B. (1987). *Quantum Field Theory*, Mc Graw-Hill, ISBN 0070320713, New York

Izumida,W.; Sakai, O. & Shimizu, Y. (1997). Many body effects on electron tunneling through quantum dots in an Aharonov-Bohm circuit, many body effects on electron tunneling through quantum dots in an Aharonov-Bohm circuit, *Journal of the Physical Society of Japan*, Vol. 66, No. 3, pp. 717-726; ISSN 0031-9015 (print), 1347-4073 (online)

Izumida, W.; Sakai, O. & Shimizu, Y. (1998). Kondo effect in single quantum dot systems - Study with numerical renormalization group method, *Journal of the Physical Society of Japan*, Vol. 67, No. 7, pp. 2444-2454; ISSN 0031-9015 (print), 1347-4073 (online)

Izumida, W.; Sakai, O. & Suzuki, S. (2001). Kondo effect in tunneling through a quantum dot, *Journal of the Physical Society of Japan*, Vol. 70, No. 4, pp. 1045-1053; ISSN 0031-9015 (print), 1347-4073 (online)

Kapusta, J. I. (1989). *Finite Temperature Field Theory*, Cambridge University Press, ISBN 0-521-35155-3, Cambridge (UK)

Keldysh, L. V. (1965). Diagram technique for nonequilibrium processes, *Soviet Physics- JETP, Vol.* 20, No. 4, pp. 1018-1026, ISSN 0038-5646

Konig, J. & Gefen, Y. (2005). Nonmonotonic charge occupation in double dots, *Physical Review B*, Vol. 71, No. 20, p. 201308(R) (4 pages), ISSN 1098-0121 (print), ISSN 1550-235X (online)

Le Bellac, M. (1996). *Thermal Field Theory*, Cambridge University Press, ISBN 0-521-65477-7, Cambridge (UK)

Meir, Y.; Wingreen, N. S. & Lee, P. A. (1991). Transport through a strongly interacting electron system: Theory of periodic conductance oscillations, *Physical Review Letters*, Vol. 66, No. 23, pp. 3048-3051; ISSN 0031-9007 (print), 1079-7114 (online)

Meir, Y.; Wingreen, N. S. & Lee, P. A. (1993). Low-temperature transport through a quantum dot: The Anderson model out of equilibrium, *Physical Review Letters*, Vol. 70, No. 17, pp. 2601-2604; ISSN 0031-9007 (print), 1079-7114 (online)

Ng, T. K. (1993). Nonlinear resonant tunneling through an Anderson impurity at low temperature, *Physical Review Letters*, Vol. 70, No. 23, pp. 3635-3638; ISSN 0031-9007 (print), 1079-7114 (online)

Nguyen Van Hieu & Nguyen Bich Ha (2005). Quantum theory of electron transport through single-level quantum dot, *Advances in Natural Sciences*, Vol. 6, No. 1, pp. 1-18, ISSN 0866-708X

Nguyen Van Hieu & Nguyen Bich Ha (2006). Time-dependent Green functions of quantum dots at finite temperature, *Advances in Natural Sciences*, Vol. 7, No. 2, pp.153-165, ISSN 0866-708X

Nguyen Van Hieu; Nguyen Bich Ha & Nguyen Van Hop (2006a). Dyson equations for Green functions of electrons in open single-level quantum dot, *Advances in Natural Sciences*, Vol. 7, No. 1, pp. 1-12, ISSN 0866-708X

Nguyen Van Hieu; Nguyen Bich Ha; Gerdt, V. P.; Chuluunbaatar, O.; Gusev, A. A.; Pali, Yu. G. & Nguyen Van Hop (2006b). Analytical asymptotic expressions for the Green's function of the electron in a single-level quantum dot at the Kondo and the Fano resonances, *Journal of the Korean Physical Society*, Vol. 53, No. 96, pp. 3645-3649, ISSN 0374-4884 (print), 1976-8524 (online)

Peskin, M. E. & Schroeder, D. V. (1995). *An Introduction to Quantum Field Theory*, Addison-Wesley, ISBN 0201503972, New York

Pustilnik, M. & Glasman, L. (2004). Kondo effect in quantum dots, *Journal of Physics: Condensed Matter*, Vol. 16, No. 16, p. R513; ISSN 0953-8984 (print), 1361-648X (online)

Sakai, O.; Suzuki, S.; Izumida, W. & Oguri, A. (1999). Kondo effect in electron tunneling through quantum dots - Study with quantum Monte Carlo method, *Journal of the Physical Society of Japan*, Vol. 68, No. 5, pp. 1640-1650; ISSN 0031-9015 (print), 1347-4073 (online)

Swirkowicz, R.; Barnas, J. & Wilczynski, M. (2003). Nonequilibrium Kondo effect in quantum dots, *Physical Review B*, Vol. 68, No. 19, p. 195318 (10 pages); ISSN 1098-0121 (print), 1550-235X (online)

Swirkowicz, R.; Wilczynski, M. & Barnas, J. (2006). Spin-polarized transport through a single-level quantum dot in the Kondo regime, *Journal of Physics: Condensed Matter*, Vol. 18, No. 7, p. 2291; ISSN 0953-8984 (print), 1361-648X (online)

Takagi, O. & Saso, T. (1999a). Magnetic field effects on transport properties of a quantum dot studied by modified perturbation theory, *Journal of the Physical Society of Japan*, Vol. 68, No. 6, pp.1997-2005; ISSN 0031-9015 (print), 1347-4073 (online)

Takagi, O. & Saso, T. (1999b). Modified perturbation theory applied to the Kondo-type transport through a quantum dot under magnetic field, *Journal of the Physical Society of Japan*, 68, No. 9, pp. 2894-2897; ISSN 0031-9015 (print), 1347-4073 (online)

Torio, M. E.; Hallberg, K.; Ceccatto, A. H. & Proetto C. R. (2002). Kondo resonances and Fano antiresonances in transport through quantum dots, *Physical Review B*, Vol. 65, No. 8, p. 085302 (5 pages); ISSN 1098-0121 (print), 1550-235X (online)

Wingreen, N. S. & Meir, Y. (1994). Anderson model out of equilibrium: Noncrossing-approximation approach to transport through a quantum dot, *Physical Review B*, Vol. 49, No. 16, pp. 11040–11052; ISSN 1098-0121 (print), 1550-235X (online)

Yeyati, A. L.; Martin-Rodero, A. & Flores, F. (1993). Electron correlation resonances in the transport through a single quantum level, *Physical Review Letters*, Vol. 71, No. 18 , pp. 2991–2994; ISSN 0031-9007 (print), 1079-7114 (online)

8

The Thermopower of a Quantum Dot Coupled to Luttinger Liquid System

Kai-Hua Yang[1], Yang Chen[1], Huai-Yu Wang[2] and Yan-Ju Wu[3]
[1]*College of Applied Sciences, Beijing University of Technology*
[2]*Department of Physics, Tsinghua University*
[3]*College of Applied Sciences, Beijing University of Technology*
China

1. Introduction

With the progress in nanofabrication technique and nanometer scale materials, research on electron transport properties of mesoscopic systems has become a very active field in condensed matter physics. Considerable researches have been mainly focused on charge transport in nanostructures and nanodevices. Besides the charge transport, a detailed understanding of heat transport through mesoscopic systems is of equally importance (Afonin, 1995; Small, 2003) because they can provide additional information on the kinetics of carriers not available in the measurement of current voltage characteristics (Heremans, 2004). For instance, thermoelectric properties are very sensitive to dimensionality, the electronic spectrum near the Fermi level, scattering processes (Koch, 2004), electron-phonon coupling strength (Yang1, 2010) and electron-hole symmetry (Small, 2003). There have been several theoretical studies on the thermopower S, which mainly focused on quantum dot (QD) coupled to the normal Fermi liquid (FL) leads (Boese, 2001; Dong, 2002; Kim, 2003; Krawiec, 2007; Yang1, 2010), denoted hereafter as FL-QD-FL. As for systems containing a quantum dot coupled to one-dimensional (1D) interacting electron leads, although their charge transport phenomena have been investigated (Yang2, 2001; Yang3, 2010), yet there have been much less efforts devoted to the thermoelectric properties of them (Kane, 1996; Krive, 2001; Romanovsky, 2002). It is well known that the 1D interacting electron systems can be described by the Luttinger liquied (LL) theory (Luttinger, 1963), which holds some unique features such as spin-charge separation, suppression of the electron tunneling density of states, power-law dependence of the electrical conductance on temperature and bias voltage, etc.. The LL behaviour has been experimentally reported in single- and multi-wall carbon nanotubes (Bockrath, 1999; Kanda, 2004; Yao, 1999) and fractional quantum Hall edge states (Chang, 1996). Recently, the use of carbon nanotubes as a thermoelectric material has gained great interest due to their 1D structure. The thermopower of single-walled carbon nanotubes have been measured in experiments (Bradley, 2000; Choi, 1999; Collins, 2000; Hone, 1998; Kong, 2005; Small, 2003). For example, Kong *et al.* have shown a linear temperature dependence of the thermopower at low temperature (Kong, 2005). Small *et al.* have observed strong modulations of thermopower as the function of gate voltage Vg in individual Carbon nanotubes (Small, 2003). Dresselhaus *et al.* have found that the low-dimensional thermoelectric materials performed better than bulk ones (Dresselhaus, 1999). Several theoretical works have been developed to predict the enhancement of thermopower in nanoscaled materials by the intralead electron interaction (Kane, 1996; Krive, 2001; Krive2, 2001; Romanovsky, 2002). Krive *et al.* (Kane, 1996; Krive, 2001) used a phenomenological approach to investigate the thermopower of a LL wire containing an impurity. In spite of the above work, an explicit thermopower formula in the LL leads was not given out. In the following, we use the notation S to denote the thermopower S of systems comprosing LL and S_0 to those comprising noninteracting FL. Theoretically, the thermopower S of a LL with an impurity can be represented by the thermopower S_0 multiplied by an interaction-dependent renormalization factor. Alternatively, one may intentionally introduce a QD into the LL, denoted as LL-QD-LL. Thus, he may attach a QD, instead of an impurity atom, to the end of a carbon nanotube. A quantum dot is experimentally more controllable than an impurity. For instance, the energy of a quantum dot can be tuned by the gate voltage V_g.

In this chapter, we will first give the stationary thermopower formula of a QD coupled to LL leads through tunneling junctions, a system of LL-QD-LL (see Fig.1) by applying the nonequilibrium Green function technique (Haug, 1996) instead of phenomenal theories. And then we later turn our attention to the time-dependent phenomena. The generalized thermopower formula is obtained under time-dependent gate voltage. Although there are many studies on time-dependent nonequilibrium transport, the research on the time-dependent thermopower and the formula under the ac field are still lack. Here we will fill the blanks for the low dimension system.

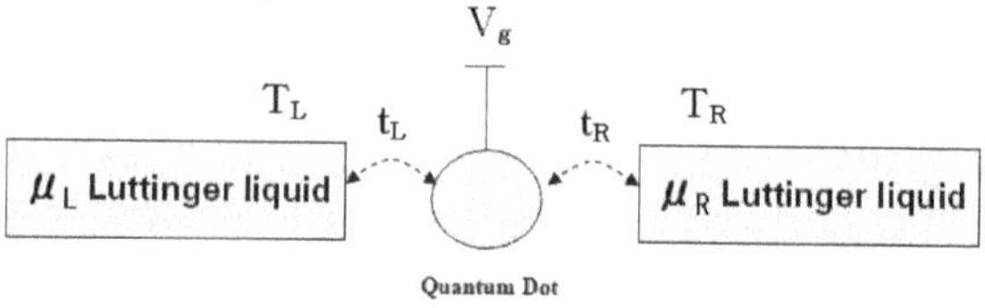

Fig. 1. The two-terminal electron transport through a single-level quantum dot weakly coupled to the Luttinger liquid leads with the chemical potentials μ_L and μ_R. Electrons tunnel from one lead to another by hopping on and off the dot level with the energy ε. The position of the dot levels with respect to the Fermi energy can be uniformly shifted by applying a voltage V_g to the gate electrode.

2. The Model

In the considered LL-QD-LL system, the QD is weakly connected with semi-infinite LL electrodes. The Hamiltonian of this system includes three parts:

$$H = H_{leads} + H_{dot} + H_T, \tag{1}$$

which represents the Hamiltonians of the left and right LL leads ($H_{leads} = H_L + H_R$), the central dot and the tunneling interactions between them H_T, respectively. Firstly, we present a detailed discussion of bosonization for a continuum model of length L with open boundary conditions (Eggert0, 1992; Fabrizio, 1995; Furusaki0, 1994) and electron-electron interaction.

The specify Hamiltonian of the LL leads can be easily written

$$H_{L/R} = H_{0\mathrm{L/R}} + H_{intL/R}, \tag{2}$$

where the first term represents the kinetic energy,

$$H_0 = \sum_{\sigma=\uparrow,\downarrow} \int_0^L dx \psi_s^\dagger(x)\varepsilon(-i\partial_x)\psi_s(x), \tag{3}$$

and the second one describes the electron-electron interaction,

$$H_{int} = \frac{1}{2}\sum_{\sigma\sigma'} \int dxdy \psi_\sigma^\dagger(x)\psi_{\sigma'}^\dagger(y)U_{\sigma\sigma'}(x-y)\psi_{\sigma'}(y)\psi_\sigma(x), \tag{4}$$

ε_k is the dispersion law of the 1D band, and $\psi_\sigma(x)$ is the spin σ electron annihilation operator subject to the open boundary conditions:

$$\psi_\sigma(0) = \psi_\sigma(L) = 0. \tag{5}$$

We apply the boundary conditions Eq.(5) to expand electron annihilation operator ψ which takes the form

$$\psi_\sigma(x) = \sqrt{\frac{2}{L}} \sum_k \sin(kx) c_{\sigma k}, \tag{6}$$

with $k = \pi n/L$, n being a positive integer. Usually a 1D system with periodic boundary conditions has two Fermi points $\pm k_F$. Here we only have single Fermi point given by $k = k_F$. The 1D fermion field ψ_σ can be expanded about the Fermi point k_F in terms of the left moving and right moving fields as

$$\psi_\sigma(x) = e^{ik_F x}\psi_{\sigma R}(x) + e^{-ik_F x}\psi_{\sigma L}(x). \tag{7}$$

In the case of periodic boundary conditions (Haldane, 1981), these left moving and right moving fields are not independent and satisfy

$$\psi_{\sigma L}(x) = -\psi_{\sigma R}(-x). \tag{8}$$

Then the fermion fields automatically satisfy the boundary conditions

$$\psi_\sigma(0) = 0, \tag{9}$$

whereas the condition

$$\psi_\sigma(L) = 0 \tag{10}$$

implies that the operator $\psi_{\sigma R}(x)$ should obey

$$\psi_{\sigma R}(-L) = \psi_{\sigma R}(L). \tag{11}$$

Therefore, we can actually work with the right moving field only, the left moving one is then defined by the above relation. Thus the field $\psi_{\sigma R}(x)$ can be defined for all x, and obeying the periodicity condition with the period $2L$:

$$\psi_{\sigma R}(x+2L) = \psi_{\sigma R}(x). \tag{12}$$

In terms of the right moving field, the kinetic energy terms in the Hamiltonian Eq.(1) takes the form

$$H_0 = v_F \sum_{\sigma} \int_{-L}^{L} dx \psi^{\dagger}_{\sigma R}(x)(-i\partial_x)\psi_{\sigma R}(x) \tag{13}$$

where we have linearized the electron spectrum. The single fermion operators for right-moving electrons with spin σ on lead α can be bosonized in the position representation by applying the periodic boundary condition Eq.(12) as

$$\psi_{\sigma R}(x) \approx \frac{\eta_\sigma}{\sqrt{2L}} e^{ik_F x} e^{i\phi_\sigma(x)}. \tag{14}$$

The operator η_σ is real fermion and satisfies the anti-commutation relations

$$\{\eta_\sigma, \eta'_\sigma\} = \delta_{\sigma\sigma'}, \tag{15}$$

with δ is Delta function. Eq.(15) assure the correct anti-commutation rules for electron operators with different σ. In order to calculate the correlation function, a method of dealing with this was suggested by Luther and Perchel (Luther, 1974). It used a limiting process, where the wave function contained s parameter α', and the limit $\alpha' \to 0$ is taken at the end of the calculation of the correlation function. Using the parameter α', we can represent the electron operator $\psi_{\sigma R}(x)$ as

$$\psi_{\sigma R}(x) \approx \lim_{\alpha' \to 0} \frac{1}{\sqrt{2\pi\alpha'}} e^{ik_F x} e^{i\phi_\sigma(x)}. \tag{16}$$

Where α' is a short-distance cutoff of the order of the reciprocal of the Fermi wave number k_F. The phase field $\phi_\sigma(x)$ satisfies periodic boundary condition:

$$\phi_\sigma(x+2L) = \phi_\sigma(x) \tag{17}$$

and can be expressed as follows

$$\phi_\sigma(x) = \sum_{q>0} \sqrt{\frac{\pi}{qL}} e^{iqx - \alpha' q/2} a_q + H.c, \tag{18}$$

here, $a^{\dagger}_q$ and a_q are the creation and annihilation operators of bosons. These operators satisfy the canonical bosonic commutation relations $[a_q, a^{\dagger}_{q'}] = \delta(q-q')$. $q = \pi n/L$, n is an integer. The density of right moving electrons is given by

$$\rho_{\sigma R}(x) \approx \frac{\partial_x \phi_\sigma(x)}{2\pi} \tag{19}$$

Applying the boundary conditions Eq.(8), we have

$$\rho_{\sigma L}(-x) = \rho_{\sigma R}(x). \tag{20}$$

The bosonized form of the kinetic energy is

$$H_0 = \pi v_F \sum_\sigma \int_{-L}^{L} dx : \rho_{\sigma R}(x)\rho_{\sigma R}(x) := v_F \sum_{\sigma q>0} q b^\dagger_{\sigma q} b_{\sigma q}, \tag{21}$$

where colon represents the normal order form of the operators.

In order to deal with electron-electron interacting terms in Hamiltonian Eq.(1), we continue to make use of the above bosonization procedure expressing the electron interaction Hamiltonian in terms of the right moving Fermi field $\psi_{\sigma R}$ only.

Before we turn to the interaction effects, we introduce the bosonic variables corresponding to charge and spin excitations:

$$b_{\rho(\sigma)q} = \frac{1}{\sqrt{2}}(b_{\uparrow q} \pm b_{\downarrow q}) \tag{22}$$

and

$$\rho_{\rho(\sigma)} = \frac{1}{\sqrt{2}}(\rho_{\uparrow q} \pm \rho_{\downarrow q}). \tag{23}$$

The interaction part of the Hamiltonian contains several terms classified as: the diagonal terms in the electron densities and the mixing left and right densities term. Consequently, the Hamiltonian becomes

$$H = \sum_{\nu=\rho(\sigma)} \{\sum_{q>0} v^0_\nu q[b^\dagger_{\nu q} b_{\nu q} - \frac{g_{2\nu}}{4\pi}(b_{\nu q} b_{\nu q} + b^\dagger_{\nu q} b^\dagger_{\nu q})] \tag{24}$$

where

$$v_{\nu q} = v_F + \frac{g_{4\nu} + g_{2\nu}}{2\pi}, \tag{25}$$

where v_F is the Fermi velocity, g_4 and g_2 represent forward scatterings; in our work, we will not consider the backscattering interaction. In the absence of backscattering, the Luttinger Hamiltonian, H_{LL}, is exactly soluble using the technique of bosonization. In order to express the Hamiltonian in diagonal form, we introduce the canonically conjugate Boson operators, in a standard way by the Bogolubov rotation,

$$b_{\nu q} = \cosh(\varphi_\nu)\widetilde{b}_{\nu q} - \sinh(\varphi_\nu)\widetilde{b}^\dagger_{\nu q} \tag{26}$$

where

$$\tanh(2\varphi_\nu) = -\frac{\widetilde{g}_\nu}{2\pi v^0_\nu} \tag{27}$$

The Hamiltonian can be achieved by the canonical transformation in terms of $\widetilde{b}_{\nu q}$ and $\widetilde{b}^\dagger_{\nu q}$

$$\widetilde{H} = UHU^\dagger = \sum_{\nu q>0} v_\nu q b^\dagger_{\nu q} b_{\nu q}, \tag{28}$$

where

$$v_\nu = \frac{v^0_\nu}{\cosh(2\varphi_\nu)}. \tag{29}$$

The unitary operator U is defined by

$$U = exp\{\frac{1}{2}\sum_{\nu,q>0}\varphi_\nu(b^\dagger_{\nu q}b^\dagger_{\nu q} - b_{\nu q}b_{\nu q})\} \tag{30}$$

In the next step we find how the Fermi operators transform by applying U. Employing the method of Mattis and Lieb (Mattis, 1964), after lengthy but straightforward calculations, we arrive at the expression for the electron annihilation operator in terms of free bosons for the case of the interacting Fermi system with open boundaries:

$$U\psi_{\sigma R}(x,t)U^\dagger \approx \frac{\eta_\sigma}{\sqrt{2\pi\alpha'}}\exp\{i\sum_\nu \varepsilon_{\nu\sigma}[\frac{c_\nu}{\sqrt{2}}\phi_\nu(x - v_\nu t) - \frac{s_\nu}{\sqrt{2}}\phi_\nu(-x - v_\nu t)]\} \tag{31}$$

where $\varepsilon_\nu\sigma$ is $+1$ unless $\sigma =\downarrow$ and $\nu = \sigma$ when its value is -1. We have defined

$$c_\nu = \cosh(\varphi_\nu), s_\nu = \sinh(\varphi_\nu). \tag{32}$$

In the continuum limit, the Hamiltonian can be expressed

$$H_{L/R} = \hbar v_c \int_0^\infty k a_k^\dagger a_k \mathrm{d}k. \tag{33}$$

This Hamiltonian describes the propagation of the charge density fluctuations in the leads with renormalized velocity v_c. From Krönig's relation (Krönig, 1935), the kinetic term has been written in a quadratic form of the density operators, because the bosons are defined as excitations above an N particle ground state, Hamiltonian must include terms that include the energy of the different bosonic ground states. These terms are not required for the calculations in this chapter, and are hence omitted.

The Hamiltonian of the single-level QD takes the form of

$$H_{dot} = \varepsilon d^\dagger d, \tag{34}$$

where ε is the energy of the electron on the dot, and $d^\dagger$ and d are fermionic creation and annihilation operators satisfying canonical commutation relation $\{d, d^\dagger\} = 1$.

The tunneling Hamiltonian is given by the standard expression:

$$H_T = \sum_\alpha (t_\alpha d^\dagger \psi_\alpha + h.c.), \tag{35}$$

where t_α is the electron tunneling constant and $\psi^\dagger_\alpha, \psi_\alpha$ $(\alpha = L/R)$ are the Fermi field operators at the end points of the left/right lead. The operator ψ_α could be written in a "bosonized" form (Furusaki, 1998)

$$\psi_\alpha = \sqrt{\frac{2}{\pi\alpha'}}\exp\left[\int_0^\infty \mathrm{d}q \frac{e^{(-\alpha' q/2)}}{\sqrt{2K_{\rho\alpha}q}}(a_{q\alpha} - a^\dagger_{q\alpha}) + \sigma\int_0^\infty dq\frac{e^{(-\alpha' q/2)}}{\sqrt{2q}}(b_{q\alpha} - b^\dagger_{q\alpha})\right], \tag{36}$$

where and $K_{\rho\alpha} = e^{2\varphi_v}$ is the interaction parameter in the "fermionic" form of the LL Hamiltonian (33), which restricts the LL parameter g to vary between 0 and 1. The noninteracting case corresponds to $v_c = v_F$ and $K_{\rho\alpha} = 1$. For repulsive interactions, $K_{\rho\alpha} < 1$. Because of the SU(2) spin symmetry under no magnetic filed, $K_\sigma = 1$. Thus the correlation functions the end point of the left LL lead without the coupling to the quantum dot $\langle\psi_\sigma^\dagger(0,t)\psi_\sigma(0,0)\rangle$ can be obtained after long calculation

$$\langle\psi_\sigma^\dagger(0,t)\psi_\sigma(0,0)\rangle_L = \frac{c_A}{\alpha'}\{\frac{i\Lambda}{\pi T}\sinh[\frac{\pi T(t-i\delta)}{\hbar}]\}^{-1/g_L}. \tag{37}$$

Where c_A is a dimensionless constant of order 1, Λ is a high-energy cutoff or a band width, δ is positive infinitesimal, and $g_L^{-1} = \frac{1}{2}(1/K_{\rho L}+1)$. $\psi_\sigma(0,t) = e^{iH_L t/\hbar}\psi_\sigma(0,0)e^{-iH_L t/\hbar}$. Similarly, the correlation function at the end point of the the right lead is obtained as the above method. The electron-electron interaction parameters of the left and right LL leads are assumed equal $g_L = g_R = g$ for convenience.

3. The thermopower formula under no ac field

The charge current J_L flowing from the left lead L into the quantum dot can be evaluated as follows:

$$J_L(t) = -\frac{e}{\hbar}\left\langle\frac{d}{dt}N_L\right\rangle = \frac{ie}{\hbar}\left\langle t_L d^\dagger(t)\psi_L(t) - h.c.\right\rangle. \tag{38}$$

We introduce the time-diagonal parts of the correlation functions: $G_{dL}^<(t,t') = i\langle\psi_L^\dagger(t')d(t)\rangle$ and $G_{Ld}^<(t,t') = i\langle d^\dagger(t')\psi_L(t)\rangle$. With the help of the Langreth analytic continuation rules (Haug, 1996). By means of them, it is easy to express the current as $J_L = 2eRe(t_L^* G_{Ld}^<(t,t'))$. After applying Langreth theorem of analytic continuation, the average current can then be expressed as

$$J_L = \frac{e}{2\pi}|t_L|^2\int d\omega Re[G_d^r(\omega)g_L^<(\omega) + G_d^<(\omega)g_L^a(\omega)]. \tag{39}$$

In terms of a long derivation, we can easily establish an expression for the expectation value of the electric current

$$J_L = \frac{e}{2\pi}|t_L|^2|t_R|^2\int d\omega G_d^r G_d^a\left[g_L^<(\omega)g_R^>(\omega) - g_L^>(\omega)g_R^<(\omega)\right], \tag{40}$$

where $G_d^{r(a)}$ is retarded (advanced) Green function of the quantum dot and $\Gamma_{L/R}$, proportional to $|t_{L/R}|^2$, describes the effective level broadening of the dot. $g_\alpha^{<(>)}(\omega)$ is the Fourier transform of the lesser (greater) Green function at the end point of the left (right) LL lead without the coupling to the QD, which has been obtained by (Furusaki, 1998):

$$g_\alpha^{<,>}(\omega) = \pm i\frac{T_\alpha}{|t_\alpha|^2}\exp[\mp(\omega-\mu_\alpha)/2T_\alpha]\gamma_\alpha(\omega-\mu_\alpha), \tag{41}$$

Now we define the Luttinger liquid distribution functions $F_{L/R}^{<,>}$ as

$$F_\alpha^{<,>}(\omega) = \frac{1}{2\pi}e^{\frac{\mp(\omega-\mu_\alpha)}{2T_\alpha}}\left(\frac{\pi T_\alpha}{\Lambda}\right)^{1/g-1}\frac{|\Gamma[\frac{1}{2g}+i\frac{\omega-\mu_\alpha}{2\pi T_\alpha}]|^2}{\Gamma(1/g)} \quad (\alpha = L/R). \tag{42}$$

The function $F^{<}(\omega)$ is the electron occupation number for interacting electrons which is analogous to the Fermi distribution function $f(\omega)$ of noninteracting electrons and $F^{>}(\omega)$ is analogous to $1 - f(\omega)$ of FL leads. $T_{L/R}$ is temperature and $\mu_{L/R}$ the chemical potential of the left or right lead where $\mu_L = \mu + \eta V$ and $\mu_R = \mu + (\eta - 1)$. $\Gamma(z)$ is the Gamma function. Following the derivation in Ref. (Yang3, 2010), the current can be obtained as

$$J_L = \frac{e}{2\pi} \int d\omega T(\omega) \left[F_L^{<}(\omega) F_R^{>}(\omega) - F_L^{>}(\omega) F_R^{<}(\omega) \right], \tag{43}$$

with $T(\omega) = \Gamma_L \Gamma_R G_d^r(\omega) G_d^a(\omega)$ is the transmission probability. If $g = 1$, Eq. (43) will degrade to the usual well-known current expression for a FL-QD-FL system.

Our goal is to find the general thermopower formula of the model described by the Hamiltonian Eq. (62). The thermopower S is defined in terms of the voltage V generated across the quantum dot when temperature gradient $\Delta T = T_L - T_R$ is much less than T_L and T_R and when current J is zero (Cutler, 1969):

$$S \equiv - \lim_{\Delta T \to 0} \frac{V}{\Delta T} |_{J=0} = - \frac{1}{eT} \frac{L_{12}}{L_{11}}. \tag{44}$$

where L_{12} and L_{11} are linear response coefficients when the current J_L is presented by small bias voltages and small temperature gradients ΔT:

$$J_L = L_{11} \frac{\delta \mu}{T} + L_{12} \frac{\delta T}{T^2} = \frac{e}{2\pi} \int d\omega T(\omega) \left\{ \left[\frac{\partial F(\omega)}{\partial \mu} \right]_T \delta V + \left[\frac{\partial F(\omega)}{\partial T} \right]_\mu \delta(\Delta T) \right\}, \tag{45}$$

where $F(\omega) = F_L^{<}(\omega) F_R^{>}(\omega) - F_L^{>}(\omega) F_R^{<}(\omega)$. Comparing both sides of the Eqs.(78) (let $e = 1, \hbar = 1$), we obtain

$$L_{11} = \frac{T}{2\pi} \int d\omega T(\omega) \left[\frac{\partial F(\omega)}{\partial \mu} \right]_T, \tag{46}$$

$$L_{12} = \frac{T^2}{2\pi} \int d\omega T(\omega) \left[\frac{\partial F(\omega)}{\partial T} \right]_\mu. \tag{47}$$

The formulas Eqs.(46) and Eqs.(47) are independent of the approximation adopted in deriving the retarded (advanced) Green function. However, the partial derivatives $\frac{\partial F}{\partial \mu}$ and $\frac{\partial F}{\partial T}$ are not yet expressed evidently. In the following we will show the explicit expression for L_{11} and L_{12}. The linear expansion of the Luttinger liqiud distribution function becomes

$$F_\alpha(\omega) = F(\omega) + \frac{\partial F_\alpha(\omega)}{\partial \mu_\alpha} |_{\mu_\alpha = \mu, T_\alpha = T} (\mu_\alpha - \mu) + \frac{\partial F_\alpha(\omega)}{\partial T_\alpha} |_{\mu_\alpha = \mu, T_\alpha = T} (T_\alpha - T). \tag{48}$$

In order to achieve a compact expression, we define $F_1 = F_L^{<} F_R^{>}$ and $F_2 = F_R^{<} F_L^{>}$, and expand them to the first order derivatives:

$$F_1 = F_1(\mu, T) + \frac{\partial F_1}{\partial V} |_T \delta V + \frac{\partial F_1}{\partial (\Delta T)} |_\mu \delta(\Delta T) \tag{49}$$

and

$$F_2 = F_2(\mu, T) + \frac{\partial F_2}{\partial V} |_T \delta V + \frac{\partial F_2}{\partial (\Delta T)} |_\mu \delta(\Delta T), \tag{50}$$

where $F_1(\mu, T)$ and $F_2(\mu, T)$ are the equilibrium LL distribution functions, and $F_1(\mu, T) = F_2(\mu, T)$. Then

$$F_1 - F_2 = \frac{\partial(F_1 - F_2)}{\partial V}|_T \delta V + \frac{\partial(F_1 - F_2)}{\partial(\Delta T)}|_\mu \delta(\Delta T)$$
$$= \frac{\partial F}{\partial V}|_T \delta V + \frac{\partial F}{\partial(\Delta T)}|_\mu \delta(\Delta T). \tag{51}$$

Substituting of Eq.(51) into Eq.(43)enables one to obtain the expressions of $\frac{\partial F}{\partial V}$ and $\frac{\partial F}{\partial(\Delta T)}$ required in Eqs. (46) and (47). We arrive at that

$$\frac{\partial F_L^{<,>}}{\partial V} = \eta \left\{ \pm \frac{1}{2T} F^{<,>} - \frac{i}{2\pi T} \Psi(\frac{1}{2g} + i\frac{\omega - \mu}{2\pi T}) F^{<,>} + \frac{i}{2\pi T} \Psi(\frac{1}{2g} - i\frac{\omega - \mu}{2\pi T}) F^{<,>} \right\}, \tag{52}$$

and

$$\frac{\partial F_R^{<,>}}{\partial V} = (\eta - 1) \left\{ \pm \frac{1}{2T} F^{<,>} - \frac{i}{2\pi T} \Psi(\frac{1}{2g} + i\frac{\omega - \mu}{2\pi T}) F^{<,>} + \frac{i}{2\pi T} \Psi(\frac{1}{2g} - i\frac{\omega - \mu}{2\pi T}) F^{<,>} \right\}, \tag{53}$$

In derivation we have used the relation $|\Gamma(x + iy)|^2 = \Gamma(x + iy)\Gamma(x - iy)$ and $\Gamma'(z) = \psi(z)\Gamma(z)$ with $\psi(z)$ is the Digamma function. Then substituting the Eqs.(52) and Eqs.(53) into the Eq. (51), we obtain

$$\frac{\partial F}{\partial V} = \frac{1}{T} F^> F^<. \tag{54}$$

With the same deriving process, we obtain the partial derivation with respect to temperature as

$$\frac{\partial F}{\partial(\Delta T)} = \frac{\omega - \mu}{T^2} F^> F^<. \tag{55}$$

It follows from Eqs. (54), (55), (46) and (47) that

$$L_{11} = \frac{T}{h} \int d\omega T(\omega) \frac{1}{T} F^> F^< \tag{56}$$

and

$$L_{12} = \frac{T^2}{h} \int d\omega T(\omega) \frac{\omega - \mu}{T^2} F^> F^<, \tag{57}$$

with $T(\omega)|_{\delta V = 0, \delta(\Delta T) = 0}$. And they become functions related to the QD density of states and LL distribution function. We stress that Eqs. (56) and (57) are the linear response coeffcients in a LL-QD-LL system. These equations will naturally degrade to those of a FL-QD-FL system if $g = 1$. The thermopower can be obtained by the equation Eq. (44) in which the current equals to zero. Substituting Eqs. (56) and (57) into Eq. (44), we have

$$S = \frac{1}{T} \frac{\int d\omega(\omega - \mu) T(\omega) F^> F^<}{\int d\omega T(\omega) F^> F^<}. \tag{58}$$

As shown in Eq. (78), when temperature difference between the leads is zero, conductance is then given by $G = e^2 L_{11}/T$. Comparison of the explicit expressions of the conductance G and thermopower S exhibits that the latter contains information different from the former.

In calculation, the Green functions of the QD are required as shown in Eq. (40). The retarded Green function is defined by $G^r(t) = -i\theta(t)\langle\{d(t), d^\dagger(0)\}\rangle$ and can be derived by means of the equation of motion method. Its analytical expression is

$$G_d^r(\omega) = \frac{1}{\omega - \varepsilon - \Sigma^r(\omega)}, \tag{59}$$

where the retarded self-energy is originated from the tunneling into the leads and is given by:

$$\Sigma^r(\omega) = -\frac{i}{2} \sum_{\alpha=L,R} \Gamma_\alpha [F_\alpha^<(\omega) + F_\alpha^>(\omega)]. \tag{60}$$

In the next section we will give our numerical results and discuss the thermoelectric properties.

4. Numerical results

The expressions (56) and (57)enable us to calculate numerically the conductance and thermopower as functions of the applied voltage and temperature. It is assumed that the system is of structural symmetric: $\Gamma_L = \Gamma_R = \Gamma$. In calculation we take the coupling strength Γ as the energy unit and set the Fermi level of the lead to be zero. Then the energy level ε of the QD represents the gate voltage Vg. No other bias is applied, *i.e.*, we always consider the zero bias case.

Figures 1(a) and (b) show the gate voltage dependence of the conductance and thermopower, respectively. The conductance varies smoothly, which is in agreement with the previous scanning gate microscopy experiments (Small2, 2003; Woodside, 2002). The thermopower S varies rapidly with the variation of the gate voltage and can reach a large absolute value at low temperature. Obviously, the gate voltage violates electron-hole asymmetry and its value tunes the thermopower. Experiments did show the features (Small, 2003; Small3, 2004).

From Fig. 1 it is seen that the conductance is an even function of ε , while the thermopower is an odd function: $S(\varepsilon) = -S(-\varepsilon)$, which is coincide to experiments (Staring, 1993). It is easily understood that the Hamiltonian in this paper possesses electron-hole symmetry: as V_g is changed to $-V_g$, the form of the Hamiltonian remains unchanged if the electrons are simultaneously converted to holes. This is the foundation of discussing the symmetry relations for the dependence of the G and S on V_g. When both bias voltage V and the current J_L change their signs, the sign of the conductance $G = dJ_L/dV$ remains unchanged. Subsequently, $G(-\varepsilon) = G(\varepsilon)$. On the other hand, the temperature difference is irrelevant to the change of the current carriers, *i.e.*, the kinetic coefficient $L_{12} = dJ_L/d(\delta T)$, changes sign: $L_{12}(\varepsilon) = -L_{12}(-\varepsilon)$ when V_g changes sign. Thus we conclude that $G(-\varepsilon) = G(\varepsilon)$. Note that G is proportional to L_{11}. Therefore, the thermopower is also an odd function $S = -\frac{1}{e^2T}\frac{L_{12}}{L_{11}}$ of the gate voltage: $S(\varepsilon) = -S(-\varepsilon)$. The result is qualitative agreement with the experiments (Dzurak, 1997; Egger, 1997; Moller, 1998) where there was no evidence of other significant contributions of transport mechanisms, such as phonon drag, to the observed thermopower.

Comparing the curves G in Fig. 1(a) and S in Fig. 1(b), we find that the sign of S coincides with dG/dV_g. The reason of the lowering of the conductance with the energy level increasing of the quantum dot is ascribed to the change of the electron tunneling from the resonant tunneling to sequential behavior.

Now we turn to the effect of temperature on the conductance and thermopower. Figure 2 plots their curves. Figure 2(a) shows that at low temperature the conductance scales as power laws with respect to temperature, $G(T) \propto T^{\alpha}$ where $\alpha = 2/g - 2$ ($g < 1$). This functional form and the power index are in good agreement with experimental results (Bockrath, 1999; Kong, 2005; Yao, 1999). Some theoretical works with respect to impurity-contained systems gained the same conclusion (Dresselhaus, 1999; Krive, 2001; Krive2, 2001; Romanovsky, 2002). The temperature-dependent power-law scalings of conductance is associated with the suppression of tunneling to a LL in which the density of states vanishes as a power law in the energy of the tunneling electron, and the suppression becomes stronger with the decrease of g, which manifests a signature for electron-electron correlations (Harman, 2002; Kane, 1996). With increasing temperature, the mechanism of electron transport gradually turns from a resonant tunneling-like process to a sequential process. At higher temperature, the conductance is inverse to the temperature. This reflects that the effect of the electron-electron interaction on transport mechanism decays. In the temperature range between, there appears a conductance peak.

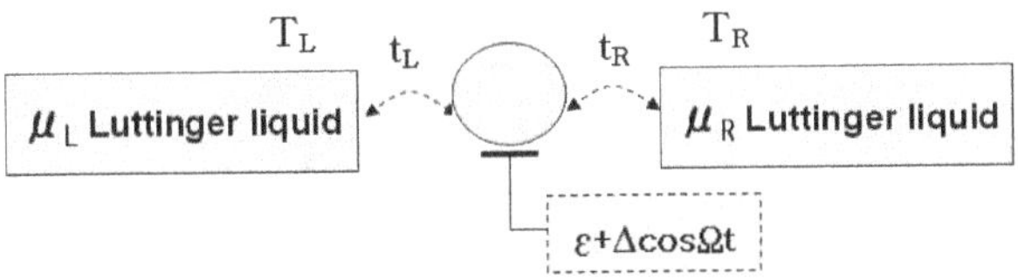

Fig. 2. The schematic picture of the two-terminal electron transport through a single-level quantum dot weakly coupled to the Luttinger liquid leads with the chemical potentials μ_L and μ_R . The position of the dot levels with respect to the Fermi energy can be uniformly shifted by applying a voltage V_g to the gate electrode.

In Fig. 2(b), the thermopower S shows linear behavior as temperature rises from zero. This is because in the low temperature regime electron tunneling transport mechanism is dominant. This behavior is the same as that of a LL containing an impurity (Krive2, 2001) and in agreement with experiments (Hone, 1998; Kong, 2005). With the electron-electron interaction enhancement, the thermopower is also increased, which has been proved in experiment (Lerner, 2008). We fit numerically the thermopower relation between the thermopower S of the LL and S_0 of Fermi liquid at low temperature with $S = (\frac{3}{2g} - \frac{1}{2})S_0(T)$ where $S_0(T) \propto T$. The electron-electron interaction in LL systems enhances and renormalizes the thermopower. In the limit of strong interaction $g << 1$, this thermopower S can be expressed as $S(T) \propto S_0(T)/g$. In this case, the thermopower of the LL is enhanced by a factor of order of magnitude of $3/(2g)$. Figure 2(b) reveals that at a fix temperature, a smaller g results in a lager S value. Hence, a larger slope of the $S - T$ curve at low temperature means a stronger interaction in LL leads. It is worth to note that the thermopower S of LL is much greater than the value S_0 of FL ($g = 1$), which reflects that the intralead electron interactions in the LL enhance the electron-hole asymmetry. With further increasing temperature a peak-like structure emerges. This is due to the mechanism at low temperature of electron transport switching from a tunnelling process to a diffusive process at high temperature. It is worth mentioning that the result is qualitatively agreement with the works (Romanovsky, 2002) with respect to an impurity in the LL lead connected to noninteracting electrons or a FL. At low temperature, a small potential barrier can strongly influence the transport properties of a LL system, so that the thermopower induced by electron backscattering dominates. This behavior is similar the case of an impurity (Romanovsky, 2002), where the impurity backscattering is considered to be a main origin of the thermopower. The impurity can also be modeled as tunnelling junctions between two decoupled semi-infinite LLs (Collins, 2000), and the tunneling junction between impurity and the LL is described by the tunneling Hamiltonian (Barnabe, 2005; Goldstein, 2010)[45-47].

At high temperature, the thermomotion of electrons become predominant and the interaction between them will be less important. Thus discrepancy between the LL and FL systems will disappear. As a result, in the high temperature limit, S becomes identical to S, as shown in Fig. 2(b). We recall that in a weak interaction system, the thermopower S_M is related to conductance G by Mott's formula (Kane, 1996):

$$S_M = -\frac{\pi^2}{3}\frac{k_B^2 T}{e}\frac{\partial lnG}{\partial \mu}, \tag{61}$$

which was originally derived for bulk systems. Note that this approximation is independent of the specific form of the transmission probability $T(\omega)$. The quantity S_M is different from S_0 of a noninteraction FL. Dependence of the zero bias conductance G on the chemical potential can be in practice measured under the variation of the gate voltage V_g. Since the gate voltage shifts the energy levels of the QD, one may assume that $\frac{\partial lnG}{\partial \mu}, \frac{\partial lnG}{\partial V_g}$. Then Eq. (26) becomes $S_M = -\frac{\pi^2}{3}\frac{k_B^2 T}{e}\frac{dlnG}{dV_g}|_{E_F}$. Figure 3 shows the variation of the thermopower S obtained from Eq. (23) and S_M from Mott relation Eq. (26) at $T = 1.0$ for four electron-electron interactions. It is seen that the relation between and G holds qualitatively for weak electron-electron interaction (Appleyard, 2000; Kane, 1996; Krive, 2001). However, even in the noninteraction case $g = 1$, there is some quantitative difference between S and S_M. The difference is obviously enhanced by the strong electron-electron interaction. Experiments (Bockrath2, 2001) evidenced the deviation from the Mott formula Eq. (26). We interpret it as a manifestation of many-body effects in the 1D electron gas. The intralead electron interactions affect the thermopower through the dependence of the transmission probability on electron-electron interaction.

From the above numerical results, we can observe that both the thermopower and conductance manifest linear temperature-dependent power-law scaling, a behavior the same as that of an impurity-contained LL system (Kane, 1996) at low temperature. The electron-electron interaction in the leads brings a significant improvement of the thermopower, a conclusion similar to that of a LL with an impurity (Krive, 2001; Krive2, 2001). As is well known, in a perfectly electron-hole symmetric system, the thermopower $S = 0$. The strong suppression of thermopower arises from the exact counteraction of the currents of electrons and holes induced by temperature gradient, which results in a zero net electric current. Only when the electron-hole symmetry is broken, the nonzero thermopower emerges. Our numerical fittings show that at low temperature the thermopower S can be expressed by the thermopower S_0 of noninteracting electrons multiplied by an

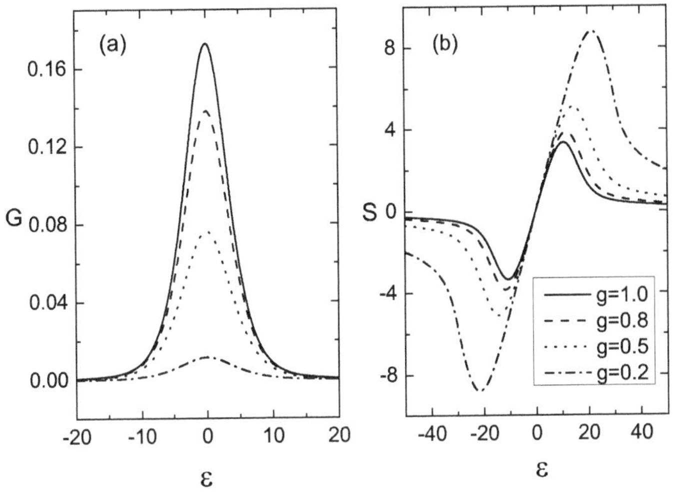

Fig. 3. The dependence of the conductance G and thermopower S on the gate voltage with $T = 1.0$ for g=0.2, 0.5, 0.8 and 1.0. The thermopower is strongly modulated by the gate voltage.

interaction factor as: $S = (3/2g - 1/2)S_0(T)$. In the limit of the strong intralead electron interaction, we have $S(T) \approx 3S_0(T)/2g$ which has an additional $3/2$ factor compared to $S(T) \approx S_0(T)/g$ of the impurity-induced thermopower in 1D systems (Romanovsky, 2002). A slight deviation from the electron-hole symmetry will cause a considerable thermopower. In low-dimensional materials, the electron-hole asymmetry is usually strong and can be modulated experimentally by tuning external parameters, such as gate voltage and magnetic field. Our results reveal that how the thermopower of a LL system containing a QD is modulated by tuning the gate voltage.

5. The thermopower formula with a time-dependent gate voltage

The thermopower formula has been derived at the stationary system. Below we will derive an expression for the time-dependent thermopower for the Luttinger liquid leads connected to the central region. It is well known that more rich physics could be exploited if the QD device is subject to a microwave irradiation field. The perturbations of ac fields can give rise to some very interesting phenomena, such as photon-electron pumping effect, the turnstile effect, the sideband effect, and photon-assisted tunneling (Blick, 1995; Kouwenhoven et al., 1997; Tien, 1963). It has been reported that the microwave spectroscopy is a possible tool to probe the energy spectrum of small quantum systems (Wiel, 2002). So the photon-assisted tunneling could provide a new way of understanding the electron-electron influence on the transport properties of the dot. Indeed, the influence of the ac field on the current-voltage characteristics in the strongly correlated interaction model was discussed by some authors. The essential effect of photon-assisted tunneling on transport properties is that the electrons tunneling through the system can exchange energy with microwave fields, opening new inelastic tunneling channels and introducing many effects. The measurement of the thermopower at ac field frequency in the order of GHz regime may offer more value information on the electron interaction. However, the explicit thermopower formula under the time-dependent gate has been still lacking. Here we will fill the blanks. We start out by introducing a model for a QD coupled to the LL leads under a time-dependent gate voltage. The Hamiltonian of the system (see Fig.2) can be described as follows

$$H = H_{leads} + H_{\mathrm{D}} + H_{\mathrm{T}}. \tag{62}$$

where $H_{leads} = H_L + H_R$ represents the Hamiltonians of the left and right LL leads and its standard form is given as above, $H_{\mathrm{D}} = \varepsilon(t)d^\dagger d$ is the Hamiltonian of the QD, with $\{d^\dagger, d\}$ the creation/annihilation operators of the electron in the QD, $\varepsilon(t) = \varepsilon + \Delta cos\Omega t$, ε is the time-independent single electron energy of the QD without microwave fields, Δ and Ω are the amplitude and frequency of the ac gate voltage, respectively. It causes an alternating current through the dot. H_T is the tunneling Hamiltonian and can be written as

$$H_{\mathrm{T}} = \sum_\alpha (t_\alpha d^\dagger \psi_\alpha + h.c.)). \tag{63}$$

by applying a time-dependent canonical transformation (Bruder, 1994)(hereafter $\hbar = 1$) to Hamiltonian H_D

$$U_1(t) = \exp\left[i\int_{-\infty}^{t} dt' \Delta\cos(\Omega t')d^\dagger d\right]. \tag{64}$$

Under this transformation, we obtain $H'_{\mathrm{D}}(t) = U_1(t)H(t)U_1^{-1}(t) - iU_1(t)\partial_t U_1^{-1}(t) = \varepsilon d^\dagger d$, the time-dependence of the gate voltage ε is removed. Instead, the electron tunnel coupling

$$t_\alpha(t) = t_\alpha \exp\left[-i\int_{-\infty}^{t} dt' \Delta\cos(\Omega t')\right]. \tag{65}$$

is now time-dependent. The current operator which describes tunneling from the L lead into the QD at time t is found to be: (in units of $\hbar = 1$))

$$J_L(t) = ie\left[t_L(t)\psi_L^\dagger d - t_L^*(t)d^\dagger \psi_L\right]. \tag{66}$$

Using nonequilibrium-Green-function technique and Langreth theorem of analytic continuation, the current can then be expressed as:

$$J_L(t) = -2eRe\int dt_1 t_L(t)[G^r(t,t_1)g_L^<(t_1,t) + G^<(t,t_1)g_L^a(t_1,t)]t_L^*(t_1). \tag{67}$$

where $G^r(t,t_1)$ and $G^<(t,t_1)$ are the Green's function of the QD. The retarded Green function $G^r(t,t_1)$ and lesser Green function $G^<(t,t_1)$ can be calculated from the following Dyson equation:

$$G^r(t,t_1) = g^r(t,t_1) + \int\int d\tau d\tau' g^r(t,\tau)\Sigma^r(\tau,\tau')G^r(\tau',t_1). \tag{68}$$

and the Keldysh equation

$$G^<(t,t_1) = \int\int d\tau d\tau' G^r(t,\tau)\Sigma^<(\tau,\tau')G^a(\tau',t_1). \tag{69}$$

where $g^r(t,t_1)$ is the free retarded Green function of isolated dot which depends only on the time difference $t - t_1$. $\Sigma^{r/a,<}(\tau,\tau') = \sum_{\alpha=L,R} t_\alpha^*(\tau)g_\alpha^{r/a,<}(\tau,\tau')t_\alpha(\tau')$ is the self-energy.

We now make Fourier transformation over the two times t and t' which switches from the time-domain into energy representation through a double-time Fourier-transform defined as (Wang, 1999; Xing, 2004)

$$F(\omega,\omega_1) = \int dtdt_1 F(t,t_1)e^{i\omega t}e^{-i\omega_1 t_1} \tag{70}$$

and

$$F(t,t_1) = \int \frac{d\omega}{2\pi}\frac{d\omega_1}{2\pi} F(\omega,\omega_1)e^{-i\omega t}e^{i\omega_1 t_1}. \tag{71}$$

And with the help of the above equation with respect to $\tau = (t+t_1)/2$ and let $t' = t - t_1$.

$$\langle F(t,t_1)\rangle = \lim_{T\to\infty}\frac{1}{2T}\int_{-T}^{T} F(\tau + t'/2, \tau - t'/2)d\tau, \tag{72}$$

we finally obtain following expression for Dyson equation

$$G^r(\omega) = g^r(\omega) + +g^r(\omega)\Sigma^r(\omega,\omega_1)G^r(\omega_1,\omega). \tag{73}$$

and the following expression for the lesser Green function from Eq. (69)

$$G^<(\omega) = G^r(\omega)\Sigma^<(\omega)G^a(\omega). \tag{74}$$

Where the time-average greater (lesser)self-energy $\Sigma^>(\omega)$ $(\Sigma^<(\omega))$ is which can be obtained by the time-average double-time self energy

$$\Sigma^{>,<}(\omega) = \sum_{\alpha=L/R,n} J_n^2(\frac{\Delta}{\Omega})|t_\alpha|^2 g_\alpha^{>,<}(\omega + n\Omega). \tag{75}$$

After using Langreth theorem of analytic continuation, and taking Fourier transformation over the current equation Eq. (67) and the time-averaged tunneling current can be expressed as,

$$I_L = \frac{e}{2\pi}\sum_{m,n} J_m^2 J_n^2 |t_L|^2|t_R|^2 \int d\omega G_d^r G_d^a \left[g_{Ln}^<(\omega)g_{Rm}^>(\omega) - g_{Ln}^>(\omega)g_{Rm}^<(\omega)\right]. \tag{76}$$

with $g_{Lm/Rm}(\omega) = g_{L/R}(\omega + m\Omega)$.

As the above step, we also introduce the electron occupation number for interacting electrons $F(\omega)$, then we finally obtains the photon-assisted tunneling current

$$J = e\sum_{m,n} J_m^2 J_n^2 \int \frac{d\omega}{2\pi}\Gamma_L\Gamma_R G^r G^a (F_{Rm}^>(\omega)F_{Ln}^<(\omega) - F_{Rm}^<(\omega)F_{Ln}^>(\omega)), \tag{77}$$

where $J_m(z)$ is the mth-order Bessel function and $F_{Lm/Rm}(\omega) = F_{L/R}(\omega + m\Omega)$. The more detail derivation process of the time-dependent current formula Eq. (77) can be found in the work (Yang3, 2010).

In the next we only give the time-dependent thermopower formula using the above procedure. Under the small bias voltages and small temperature gradients and with the help of the linear expansion, we have

$$J_L = L_{11}\frac{\delta\mu}{T} + L_{12}\frac{\delta T}{T^2} = \frac{e}{2\pi}\sum_{m,n} J_m^2 J_n^2 \int d\omega T(\omega)\left\{\left[\frac{\partial F(\omega)}{\partial\mu}\right]_T \delta V + \left[\frac{\partial F(\omega)}{\partial T}\right]_\mu \delta(\Delta T)\right\}, \quad (78)$$

where $F_{mn}(\omega) = F_{Ln}^{<}(\omega)F_{Rm}^{>}(\omega) - F_{Ln}^{>}(\omega)F_{Rm}^{<}(\omega)$. omparing both sides of the Eqs.(78) (let $e = 1, \hbar = 1$), we obtain

$$L_{11} = \frac{T}{2\pi}\sum_{m,n} J_m^2 J_n^2 \int d\omega T(\omega)\left[\frac{\partial F_{mn}(\omega)}{\partial\mu}\right]_T, \quad (79)$$

$$L_{12} = \frac{T^2}{2\pi}\sum_{m,n} J_m^2 J_n^2 \int d\omega T(\omega)\left[\frac{\partial F_{mn}(\omega)}{\partial T}\right]_\mu, \quad (80)$$

where the partial derivatives $\frac{\partial F_{mn}}{\partial\mu}$ and $\frac{\partial F_{mn}}{\partial T}$ are not yet expressed evidently. In the following we will show the explicit expression for L_{11} and L_{12}. In order to obtain L_{11} and L_{12}, we must arrive at $\frac{\partial F_{Lm}^{<,>}}{\partial V}$ and $\frac{\partial F_{Lm}^{<,>}}{\partial T}$. With the help of the linear expansion of the Luttinger liqiud distribution function and expand them to the first order derivatives, the Luttinger liqiud distribution function becomes

$$\frac{\partial F_{Lm}^{<,>}}{\partial V} = \eta\left\{\pm\frac{1}{2T} - \frac{i}{2\pi T}\Psi_{+m} + \Psi_{-m}\right\} F_{Lm}^{<,>}, \quad (81)$$

and

$$\frac{\partial F_{Rm}^{<,>}}{\partial V} = (\eta - 1)\left\{\pm\frac{1}{2T} - \frac{i}{2\pi T}\Psi_{+m} + \Psi_{-m}\right\} F_{Rm}^{<,>}, \quad (82)$$

where $\Psi_{\pm m} \equiv \Psi(\frac{1}{2g} \pm i\frac{\omega + m\Omega - \mu}{2\pi T})$. Substituting of Eq.(81) and Eq.(82) into Eq.(77) enables one to obtain the expressions of $\frac{\partial F_{mn}}{\partial V}$ and $\frac{\partial F_{mn}}{\partial(\Delta T)}$ required in Eqs. (79) and (80). After a long calculation using the same steps above we finally obtain

$$\begin{aligned}\frac{\partial F_{mn}}{\partial V} &= \frac{\partial[F_{Lm}^{<}F_{Rn}^{>} - F_{Lm}^{>}F_{Rn}^{<}]}{\partial V} \\ &= F_m^{<}F_n^{>}(\frac{1}{T} + \frac{i}{2\pi T}(\Psi_{-m} - \Psi_{+m} - \Psi_{-n} + \Psi_{+n})),\end{aligned} \quad (83)$$

With the same deriving process, we obtain the partial derivation with respect to temperature as

$$\begin{aligned}\frac{\partial F_{mn}}{\partial V} &= \frac{\partial[F_{Lm}^{<}F_{Rn}^{>} - F_{Lm}^{>}F_{Rn}^{<}]}{\partial T} \\ &= F_m^{<}F_n^{>}[\frac{(\omega_m + \omega_n)}{2T^2} - \frac{i\omega_m}{2\pi T^2}(\Psi_{+m} - \Psi_{-m}) + \frac{i\omega_n}{2\pi T^2}(\Psi_{+n} - \Psi_{-n})],\end{aligned} \quad (84)$$

with $\omega_m = \omega + m\Omega$ and $F_m = F(\omega + m\Omega)$. In terms of the linear expansion of the time-dependent current formula, The coefficients $L_{11/12}$ of the linear response theory can be determined from the corresponding correlation functions.

$$L_{11} = \frac{T}{h}\int d\omega T(\omega) \sum_{m,n=-\infty}^{\infty} J_m^2(\frac{\Delta}{\Omega}) J_n^2(\frac{\Delta}{\Omega}) F_m^< F_n^> \Big[\frac{1}{T} + \frac{i}{2\pi T}(\Psi_{-m} - \Psi_{+m} - \Psi_{-n} + \Psi_{+n})\Big] \quad (85)$$

and

$$L_{12} = \frac{T^2}{h}\int d\omega T(\omega) \sum_{m,n=-\infty}^{\infty} J_m^2(\frac{\Delta}{\Omega}) J_n^2(\frac{\Delta}{\Omega}) F_m^< F_n^> \Big[\frac{(\omega_m + \omega_n)}{2T^2} - \frac{i\omega_m}{2\pi T^2}(\Psi_{+m} - \Psi_{-m}) + \frac{i\omega_n}{2\pi T^2}(\Psi_{+n} - \Psi_{-n})\Big]. \quad (86)$$

From the expression of the coefficients $L_{11/12}$, we can see the coefficients $L_{11/12}$ containing a additional term caused by the time-dependent gate voltage.

The time-dependent zero bias conductance is then given by $G(0) = \frac{e^2}{T} L_{11}$, and the time-dependent thermopower can be obtained from the ratio between voltage gradient ΔV and and temperature gradient ΔT between the two reservoirs, when both left and right time-dependent electric currents cancel

$$S = -\frac{\Delta V}{\Delta T}|_{<I(t)=0>}. \quad (87)$$

Thus the conductance and thermopower take the form

$$G = \frac{1}{h}\int d\omega T(\omega) \sum_{m,n=-\infty}^{\infty} J_m^2(\frac{\Delta}{\Omega}) J_n^2(\frac{\Delta}{\Omega}) F_m^< F_n^> \Big[\frac{1}{T} + \frac{i}{2\pi T}(\Psi_{-m} - \Psi_{+m} - \Psi_{-n} + \Psi_{+n})\Big], \quad (88)$$

and

$$S = \frac{\int d\omega T(\omega) \sum_{m,n=-\infty}^{\infty} J_m^2(\frac{\Delta}{\Omega}) J_n^2(\frac{\Delta}{\Omega}) F_m^< F_n^> (\frac{1}{T} + \frac{i}{2\pi T}\psi_{mn})}{T^2 \int d\omega T(\omega) \sum_{m,n=-\infty}^{\infty} J_m^2(\frac{\Delta}{\Omega}) J_n^2(\frac{\Delta}{\Omega}) F_m^< F_n^> [\frac{(\omega_m+\omega_n)}{2T^2} - \frac{i\omega_m}{2\pi T^2}\psi_{\pm m} + \frac{i\omega_n}{2\pi T^2}\psi_{\pm n}]}, \quad (89)$$

where $\psi_{mn} = \Psi_{-m} - \Psi_{+m} - \Psi_{-n} + \Psi_{+n}$, $\psi_{\pm m} = \Psi_{+m} - \Psi_{-m}$ and $\psi_{\pm n} = \Psi_{+n} - \Psi_{-n}$. When no ac filed, the above formula return to the equation (58). This formula describes the time-averaged thermopower through the LL-QD-LL system in the presence of ac fields which contains more information than the equation (58). The numerical results of the time-dependent thermopower will be published in the future.

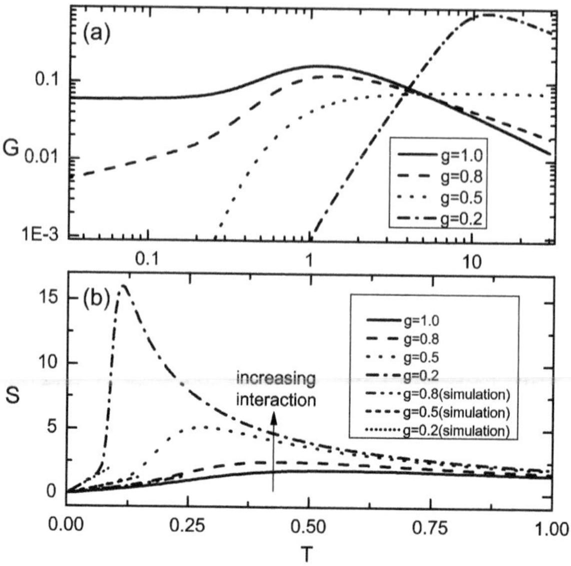

Fig. 4. The conductance (a) and thermopower S (b) as a function of temperature with $\varepsilon = 2.0$ for g=0.2, 0.5, 0.8 and 1.0. At low temperature, the conductance exhibits a power-law dependence of the temperature and the thermopower manifests the linear and positive temperature dependence, respectively. The interaction factor g can be inferred from the slopes of thermopower. With the enhancement of the electron-electron, the thermopower is increased.

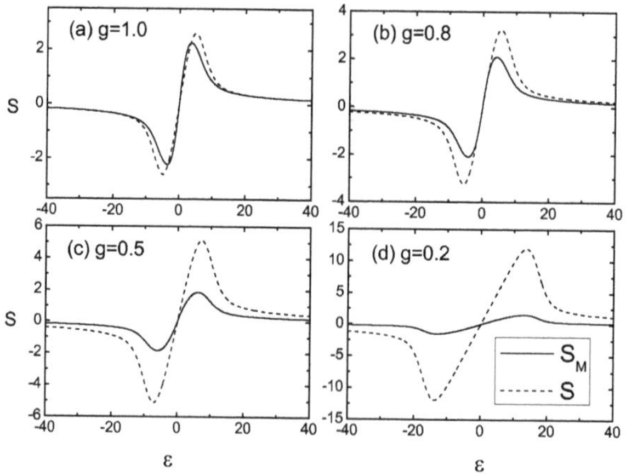

Fig. 5. Thermopowers S calculated by Eq. (23) (dash lines) and SM calculated by Mott relation Eq. (26) (solid lines) at T=1.0 for (a) g=1.0, (b) g=0.8, (c) g=0.5 and (d) g=0.2.

6. Acknowledgment

This work is supported by the Beijing Novel Program 2005B11, by Beijing Natural Science Foundation (1112003) and Fund of Beijing Municipal Education Commission.

7. References

Afonin V.V. & Rudin A.M. ,(1995).Thermoelectric resonant transport through the Anderson impurities. *Phys. Rev. B*, Vol. 51, No.24, June 1995, 18025–18028.

Appleyard N. J., Nicholls J. T., Pepper M., Tribe W. R., Simmons M. Y.& Ritchie D. A., (2000).Direction-resolved transport and possible many-body effects in one-dimensional thermopower. *Phys. Rev. B*, vol. 62, December 2000, 16275(R)–16278(R).

Barnabé-Théiault X. , Sedeki A., Meden V. & Schóhammer K.,(2005).Junctions of one-dimensional quantum wires: Correlation effects in transport.*Phys. Rev. B*, vol. 71,May 2005, 205327-1–205327-13.

Blick R.H., et al.,(1995).Photon-assisted tunneling through a quantum dot at high microwave frequencies. *Appl. Phys. Lett.* vol. 67,October 1995,3924-3926.

Bockrath M.,Cobden D. H.,Lu J.,Rinzler A. G.,Smalley R. E.,Balents L. & McEuen P. L., (1999).Luttinger-liquid behaviour in carbon nanotubes.*Nature* vol. 397, 598-601.

Bockrath M.,Liang W.,Bozovic D., Hafner J. H., Lieber C.M., Tinkham M.& Park H., (2001). Resonant Electron Scattering by Defects in Single-Walled Carbon Nanotubes.*Science* vol. 291, 283-285.

Boese D. & Fazio R.,(2001). Thermoelectric effects in Kondo-correlated quantum dots. *Europhys. Lett.* vol. 56, November 2001, 576–582.

Bradley K., Jhi S.-H., Collins P.G., Hone J., Cohen M.L., Louie S.G.& Zettl A.,(2000).Is the Intrinsic Thermoelectric Power of Carbon Nanotubes Positive? *Phys. Rev. Lett.* vol. 85, November 2000, 4361-4364.

Bruder C. & Schoeller H.,(1994).Charging effects in ultrasmall quantum dots in the presence of time-varing fields. *Phys. Rev. Lett.* vol. 72,February 1994,1076-1079.

Chang A. M., Pfeiffer L. N. & West K. W. (1996).Observation of Chiral Luttinger Behavior in Electron Tunneling into Fractional Quantum Hall Edges.*Phys. Rev. Lett.* vol. 77, September 1996, 2538–2541.

Choi E.S., Suh D.S., Kim G.T., Kim D.C., Park Y.W., Liu K., Duesberg G.& Roth S.,(1999)Magnetothermopower of singlewall carbon nanotubenewtwork . *Synth. Met.* vol. 103, 2504-2505.

Collins P.G., Bradley K., Ishigami M. & Zettl A., (2000).Extreme Oxygen Sensitivity of Electronic Properties of Carbon Nanotubes *Science* vol. 287, 1801-1804.

Cutler M. & Mott N. F.,(1969). *Phys. Rev.*, vol. 181, 1336.

Dong B. & Lei X.L. ,(2002). Effect of the Kondo correlation on the thermopower in a quantum dot. *J. Phys.: Condens. Matter* vol. 14,November 2002, 11747-11756.

Dresselhaus MS, Dresselhaus G, Sun X, Zhang Z, Cronin SB& Koga T., (1999).Low-dimensional thermoelectric materials. *Phys Solid State* vol. 41,May 1999,679-682.

Dzurak A. S., Smith C. G., Barnes C. H. W., Pepper M., Martin-Moreno L., Liang C. T., Ritchie D. A. & Jones G. A. C., (1997).Thermoelectric signature of the excitation spectrum of a quantum dot.*Phys. Rev. B* vol. 55, April 1997, 10197(R)-10200(R).

Egger R., Gogolin A.,(1997). Effective Low-Energy Theory for Correlated Carbon Nanotubes. *Phys. Rev. Lett.* vol. 79,December 1997, 5082-5085.

Eggert S. & Affleck I.,(1992).Magnetic impurities in half-integer-spin Heisenberg antiferromagnetic chains. *Phys. Rev. B*,vol. 46,May 1992, 10866-10883; Eggert S., Johannesson H., & Mattsson A.,(1996).Boundary Effects on Spectral Properties of Interacting Electrons in One Dimension. *Phys. Rev. Lett.*, vol. 76,February 1996, 1505-1508.

Fabrizio M. & Gogolin A. O.,(1995).Interacting one-dimensional electron gas with open boundaries. *Phys. Rev. B* vol. 51,June 1995, 17827-17841.

Furusaki A., (1998). Resonant tunneling through a quantum dot weakly coupled to quantum wires or quantum Hall edge states.*Phys. Rev. B* vol. 57,March 1998, 7141-7148.

Furusaki A. & Nagaosa N.,(1994),Kondo effect in a Tomonaga-Luttinger liquid. *Phys. Rev. Lett.* 72,February 1994, 892-895.

Goldstein M. & Berkovits, (2010).Density of states of a dissipative quantum dot coupled to a quantum wire. *Phys. Rev. B* vol. 82, December 2010,235315-1–235315-9; Goldstein M., Gefen Y., & Berkovits R.,(2011). Entanglement entropy and quantum phase transitions in quantum dots coupled to Luttinger liquid wires. *Phys. Rev. B*, vol. 83, June 2011, 245112-1–245112-11.

Haldane F.D.,(1981).'Luttinger liquid theory' of one-dimensional quantum fluids. I. Properties of the Luttinger model and their extension to the general 1D interacting spinless Fermi gas. *J. Phys. C* vol. 14,July 1981, 2585-2610.

Harman T.C., Taylor P. J., Walsh M. P.& LaForge B. E.,(2002).Quantum Dot Superlattice Thermoelectric Materials and Devices. *Science* vol. 297, 2229–2232 .

Haug H., Jauho A.-P., (1996)*Quantum Kinetics in Transport and Optics of Semiconductors*, (Springer-Verlag).

Heremans J.P. ,Thrush C.M. & D.T. Morelli,(2004).Thermopower enhancement in lead telluride nanostructures . *Phys. Rev. B* ,Vol. 70, September 2004, 115334-1–115334-5, and references cited therein.

Hone J., Ellwood I.,Muno M., AriMizel,Cohen M.L., Zettl A., Rinzler A.G. & Smalley R.E. , Thermoelectric Power of Single-Walled Carbon Nanotubes.(1998). *Phys. Rev. Lett.* vol. 80, February 1998, 1042-1045.

Kanda A, Tsukagoshi K, Aoyagi Y & Ootuka Y (2004). Gate-Voltage Dependence of Zero-Bias Anomalies in Multiwall Carbon Nanotubes. *Phys. Rev. Lett.* vol. 92,January 2004. 036801-1–036801-4.

Kane C.L. & Fisher M.P.A.,(1996).Thermal Transport in a Luttinger Liquid . *Phys. Rev. Lett.* vol. 76, September 1995, 3192–3195.

Kim T.-S. & Hershfield S., (2003). Thermoelectric effects of an Aharonov-Bohm interferometer with an embedded quantum dot in the Kondo regime . *Phys. Rev. B* vol. 67,April 2003, 165313-1–165313-15.

Koch J., Oppen F. von ,Oreg Y., & Eran Sela,(2004). Thermopower of single-molecule devices.*Phys. Rev. B* ,Vol.70, November 2004, 195107-1–195107-12.

Kong W.J., Lu L., Zhu H.W., Wei B.Q.& Wu D.H.,(2005). *J. Phys.: Condens. Matter* vol. 17,March 2005, 1923-1928.

Kouwenhoven L.P., Schön G. & Soun L.L.,(1997). Mesoscopic Electron Transport. Kluwer Academic Publisher,ISBN 0-7923-4737-4, The Netherlands.

Krawiec M. & Wysokinski K.I.,(2007). Thermoelectric phenomena in a quantum dot asymmetrically coupled to external leads. *Phys. Rev. B* vol.75, April 2007, 155330-1–155330-6;Krawiec M. & Wysokinski K.I.,(2006) Thermoelectric effects in strongly interacting quantum dot coupled to ferromagnetic leads. *Phys. Rev. B* vol. 73, Fermber 2006, 075307-1–075307-7; *Physica B* vol. 378, 2006, 933–934.

Krive I.V. ,Bogachek E.N.,Scherbakov A.G. & Landman Uzi,(2001). Interaction enhanced thermopower in a Luttinger liquid. *Phys. Rev. B* vol. 63, February 2001, 113101-1–113101-4čż Romanovsky I.A. ,Krive I.V. ,Bogachek E.N. & Landman U.,(2002).Thermopower of an infinite Luttinger liquid.*Phys. Rev. B* vol. 65, Fermber 2002, 075115-1–075115-9.

Krive I.V. , Romanovsky I.A., Bogachek E.N., Scherbakov A.G. & Landman U.,(2001). Thermoelectric effects in a Luttinger liquid. *Low Temp. Phys.* vol. 27, Octember 2001, 821-830.

Krönig R. de L.,(1935).Zur neutrinotheorie des lichtes III. *Physica* 2,August 1935, 968-980.

Lerner I. V., Yudson V. I.,& Yurkevich I. V.,(2008).Quantum Wire Hybridized With a Single-Level Impurity. *Phys. Rev.Lett.* vol. 100,June 2008, 256805-1–256805-4 .

Luther A.& Peschel I.,(1974).Single-particle states, Kohn anomaly, and pairing fluctuations in one dimension. *Phys. Rev. B*, vol. 9, April 1974, 2911-2919.

Luttinger J. M.,(1963).An Exactly Soluble Model of a Many-Fermion System. *J. Math. Phys.*, vol. 4, September 1963, 1154-1162.

Mattis D.C. & Lieb E.H.,(1965).Exact Solution of a Many-Fermion System and Its Associated Boson Field. *J. Math. Phys.* vol. 6,September 1964,February,304-312.

Möller S., Buhmann H., Godijn S. F.& Molenkamp L. W.,(1998). Charging Energy of a Chaotic Quantum Dot. *Phys. Rev. Lett.* vol. 81,June 1998, 5197-5200.

Romanovsky I.A. ,Krive I.V.,Bogachek E.N. & Landman U. ,(2002).Thermopower of an infinite Luttinger liquid.*Phys. Rev. B* vol. 65, Fermber 2002, 075115-1–075115-9.

Small J. P. , Perez K. M. & Kim P.,(2003).Modulation of Thermoelectric Power of Individual Carbon Nanotubes. *Phys. Rev. Lett.*,Vol. 91,No.25,December 2003, 256801-1–256801-4.

Small J.P., Shi L.& Kim P., (2003).Mesoscopic thermal and thermoelectric measurements of individual carbon nanotubes. *Solid State Commun* vol. 127,March 2003, 181-186.

Small J. P.& Kim P.,(2004). THERMOPOWER MEASUREMENT OF INDIVIDUAL SINGLE WALLED CARBON NANOTUBES. *Microscale Thermophysical Engineering* .vol.8,July 2003, 1–5.

Staring A.A.M., Molenkamp L.W., Alphenhaar B.W., Houten H. van, Buyk O.J.A., Mabesoone M.A.A., Beenakker C.W.J. & Foxon C.T.,(1993).Coulomb-Blockade Oscillations in the Thermopower of a Quantum Dot. *Europhys. Lett.* vol. 22,April 1993, 57-62.

Tien P.K. & Gordon J.P.,(1963).Multiphoton Process Observed in the Interaction of Microwave Fields with the Tunneling between Superconductor Films. *Phys. Rev.* vol. 129, January 1963, 647-677.

Wang B., Wang J.& Guo H.,(1999).Current Partition: A Nonequilibrium Green's Function Approach. *Phys. Rev. Lett.* vol. 82, January 1999, 398-401.

Wiel W.G. van der, Franceschi S. De, Elzerman J.M., Fujisawa T., Tarucha S. & Kouwenhoven L.P.,(2003).Electron transport through double quantum dots. *Rev. Mod. Phys.*, vol. 75,December 2002,1-22.

Woodside M.T.& McEuen P. L., (2002).Scanned Probe Imaging of Single-Electron Charge States in Nanotube Quantum Dots. *Science* vol. 296, 1098–1101.

Xing Y.X., Wang B., Wei Y.D., Wang B.G. & Wang J.,(2004).Spin pump in the presence of a superconducting lead. *Phys. Rev. B* vol. 70,December 2004, 245324-1–245324-8.

Yang K.-H. ,Wu Y.-P. & Zhao Y.L. ,(2010). The shot noise in a vibrating molecular dot in the Kondo regime. *Europhys. Lett.*, vol. 89, February 2010, 37008-1–37008-6.

Yang Yi-feng & Lin Tsung-han ,(2001)Nonequilibrium transport through a quantum dot weakly coupled to Luttinger liquids. *Phys. Rev. B* vol. 64, Nomber 2001,233314-1–233314-4.

Yang K.-H.,Wu Y.-P. & Zhao Y.-L. ,(2010).The shot noise of the quantum dot weakly coupled to Luttinger liquid. *Phys. Lett. A* vol. 374, December,2009, 917–922; Yang K.-H.,Wu Y.-J.,Wu Y.-P.& Zhao Y.-L.,(2011).Photon-assisted shot noise through a quantum dot coupled to Luttinger liquid . *Phys. Lett. A* vol.375,December 2010, 747–755; Yang K.-H. & Zhao Y.-L., (2010).Phonon-assisted shot noise through a quantum dot weakly coupled to Luttinger liquid. *Physica E* vol. 42,May 2010, 2324–2330.

Yao Z.,Postma H. W. Ch. ,Balents L., & Dekker C. ,(1999). Carbon nanotube intramolecular junctions.*Nature* vol. 402, 273-276.

Permissions

The contributors of this book come from diverse backgrounds, making this book a truly international effort. This book will bring forth new frontiers with its revolutionizing research information and detailed analysis of the nascent developments around the world.

We would like to thank Ameenah N. Al-Ahmadi, PhD, for lending her expertise to make the book truly unique. She has played a crucial role in the development of this book. Without her invaluable contribution this book wouldn't have been possible. She has made vital efforts to compile up to date information on the varied aspects of this subject to make this book a valuable addition to the collection of many professionals and students.

This book was conceptualized with the vision of imparting up-to-date information and advanced data in this field. To ensure the same, a matchless editorial board was set up. Every individual on the board went through rigorous rounds of assessment to prove their worth. After which they invested a large part of their time researching and compiling the most relevant data for our readers. Conferences and sessions were held from time to time between the editorial board and the contributing authors to present the data in the most comprehensible form. The editorial team has worked tirelessly to provide valuable and valid information to help people across the globe.

Every chapter published in this book has been scrutinized by our experts. Their significance has been extensively debated. The topics covered herein carry significant findings which will fuel the growth of the discipline. They may even be implemented as practical applications or may be referred to as a beginning point for another development. Chapters in this book were first published by InTech; hereby published with permission under the Creative Commons Attribution License or equivalent.

The editorial board has been involved in producing this book since its inception. They have spent rigorous hours researching and exploring the diverse topics which have resulted in the successful publishing of this book. They have passed on their knowledge of decades through this book. To expedite this challenging task, the publisher supported the team at every step. A small team of assistant editors was also appointed to further simplify the editing procedure and attain best results for the readers.

Our editorial team has been hand-picked from every corner of the world. Their multi-ethnicity adds dynamic inputs to the discussions which result in innovative outcomes. These outcomes are then further discussed with the researchers and contributors who give their valuable feedback and opinion regarding the same. The feedback is then collaborated with the researches and they are edited in a comprehensive manner to aid the understanding of the subject.

Apart from the editorial board, the designing team has also invested a significant amount of their time in understanding the subject and creating the most relevant covers. They scrutinized every image to scout for the most suitable representation of the subject and create an appropriate cover for the book.

The publishing team has been involved in this book since its early stages. They were actively engaged in every process, be it collecting the data, connecting with the contributors or procuring relevant information. The team has been an ardent support to the editorial, designing and production team. Their endless efforts to recruit the best for this project, has resulted in the accomplishment of this book. They are a veteran in the field of academics and their pool of knowledge is as vast as their experience in printing. Their expertise and guidance has proved useful at every step. Their uncompromising quality standards have made this book an exceptional effort. Their encouragement from time to time has been an inspiration for everyone.

The publisher and the editorial board hope that this book will prove to be a valuable piece of knowledge for researchers, students, practitioners and scholars across the globe.

List of Contributors

Eliade Stefanescu
Center of Advanced Studies in Physics of the Romanian Academy, Romania

Dong Ho Wu and Bernard R. Matis
Naval Research Laboratory, Washington, DC, USA

Dmitry Filatov and Vladimir Shengurov
Technical Physics Research Institute, N.I. Lobachevskii University of Nizhny Novgorod, Russia

E Niyaz Nurgazizov, Pavel Borodin and Anastas Bukharaev
K. Zavoisky Kazan' Physical-Technical Institute, Kazan' Scientific Centre, Russian Academy of Sciences, Russia

I. Filikhin, S.G. Matinyan and B. Vlahovic
North Carolina Central University, USA

Leonardo Kleber Castelano
Departamento de Física, Universidade Federal de São Carlos, Brazil

Guo-Qiang Hai
Instituto de Física de São Carlos, Universidade de São Paulo, Brazil

Mu-Tao Lee
Departamento de Química, Universidade Federal de São Carlos, Brazil

Minjie Ma
Computational Nanoelectronics and Nano-Device Laboratory, Electrical and Computer Engineering Department, National University of Singapore, Singapore

Mansoor Bin Abdul Jalil
Computational Nanoelectronics and Nano-Device Laboratory, Electrical and Computer Engineering Department, National University of Singapore, Singapore
Information Storage Materials Laboratory, Department of Electrical and Computer Engineering, National University of Singapore, Singapore

Seng Ghee Tan
Computational Nanoelectronics and Nano-Device Laboratory, Electrical and Computer Engineering Department, National University of Singapore, Singapore
Data Storage Institute, A *STAR (Agency of Science, Technology and Research), Singapore

Nguyen Bich Ha
Institute of Materials Science, Vietnam Academy of Science and Technology, Cau Giay, Hanoi, Vietnam

Kai-Hua Yang and Yang Chen
College of Applied Sciences, Beijing University of Technology, China

Huai-Yu Wang
Department of Physics, Tsinghua University, China

Yan-Ju Wu
College of Applied Sciences, Beijing University of Technology, China

Printed in the USA
CPSIA information can be obtained
at www.ICGtesting.com
JSHW011422221024
72173JS00004B/633

9 781632 383815